Lecture Notes in
Mathematics

Edited by A. Dold and B. Eckmann

pproximation
lications
atics and Physics

1985

Edited by J. Gilewicz, M. Pindor, W. Siemaszko

Springer-Verlag
Berlin Heidelberg New York London Paris Tokyo

Editors

Jacek Gilewicz
Centre de Physique Théorique, CNRS
Luminy, Case 907, 13288 Marseille Cedex 9, France

Maciej Pindor
Instytut Fizyki Teoretycznej, Uniwersytet Warszawski
ul. Hoża 69, 00-681 Warszawa, Poland

Wojciech Siemaszko
Instytut Matematyki i Fizyki, Politechnika Rzeszowska
ul. Poznańska 2, P.O. Box 85, 35-084 Rzeszów, Poland

Mathematics Subject Classification (1980): Primary: 30B70, 30E10, 41A20
Secondary: 12D10, 41A21, 81G05

ISBN 3-540-17212-2 Springer-Verlag Berlin Heidelberg New York
ISBN 0-387-17212-2 Springer-Verlag New York Berlin Heidelberg

Library of Congress Cataloging-in-Publication Data. Rational approximation and its applications in mathematics and physics. (Lecture notes in mathematics; 1237) 1. Approximation theory–Congresses. 2. Fractions, Continued–Congresses. I. Gilewicz, Jacek, 1937-. II. Pindor, M. (Maciej), 1914-. III. Siemaszko, W. (Wojciech), 1945-. IV. Series: Lecture notes in mathematics (Springer-Verlag); 1237. QA3.L28 no. 1237 510 s 87-4289 [QA221] [513'.24]
ISBN 0-387-17212-2 (U.S.)

Printed in Germany

Printing and binding: Druckhaus Beltz, Hemsbach/Bergstr.
2146/3140-543210

FOREWORD

One of the main purposes of the Łancut Conference was the direct exchange of experiences and results between specialists in rational approximation who have not had any occasion to meet until now. At the first European meetings in Marseille -Toulon (1975), Lille (1978), Antwerp (1979), Amsterdam (1980), Bad Honnef (1983), Bar le Duc (1984), Segovia (1985), Marseille(1985) there were only a few participants from Poland. The first French -Polish meeting on rational approximation at Warsaw, took place in June 1981. The proof of the need for such meetings as that of Łancut, for exemple, is the constatation made by Ukrainian mathematicians that some of their results obtained more than ten years before, though published in a journal little known in the West, were rediscovered in 1985.
We would like to explain the reasons for the choice of the topics of the Conference. As is well known, continued fractions and rational approximations constitute the same domain expressed in different languages. Considering these problems in a wider aspect of approximation theory is necessary today for their further development. It should not be forgotten that interest in these problems comes from their spectacular applications to numerical and physical problems. Thus, all these subjects have found their place in the program of the Łancut Conference.
A very serious disease did not allow our friend, Prof. Dr. Helmut Werner, to participate in the Conference. He did his best in sending us his article ... a few days before he died. A little In Memoriam is devoted to him. We would like to thank his wife, Mrs. Ingrid Werner, for writing a few words about her husband's life. We also thank his devoted secretary, Mrs. Elisabeth Becker, and his colleague, Dr. Paul Janssen, for their cooperation. All Helmut's friends' thoughts are expressed by Dr. Annie Cuyt.
The Conference was sponsored by the "Komitet Nauk Fizycznych PAN" of Warsaw, the "Towarzystwo Naukowe" of Rzeszów, the "Instytut Fizyki Teoretycznej" of Warsaw University and by the "Politechnika Rzeszowska im. I. Łukasiewicza" of Rzeszów, to all of whom our grateful thanks.
The Organizing Committee of the Łancut Conference expresses its gratitude to Springer-Verlag for kindly publishing the Proceedings in the series Lecture Notes in Mathematics.

IN MEMORIAM

Prof. Dr. Helmut Werner
1931-1985

November 22, 1985

Dear friend and colleague,

"Nur wer den Gipfel des Berges erstiegen, vermag in die weiteste Ferne zu sehn". However true this proverb may be, today it announced the sad event of your decease. At the conference in Łańcut all of us were still hoping that you would get better again. Although your farewell was not completely unexpected, it came far too soon.

Many have known you and worked with you and I'm sure that as many have loved and appreciated you. You were always such a busy man. Even when your health was not what it used to be anymore, you only felt really happy when you could be very active and were trying to do several things simultaneously. You were often short of time but never short of appointments! You were also a very precise man. When mathematical formulas had to be checked, when a paper had to be written down, you took all the time to make sure that everything was correct. If necessary you went over the same thing several times until you found it satisfactory.

We have all learnt from you and so have many students at the Universities of Münster and Bonn. What's more, you also cared about people. The large number of reports on the use of mathematics and computer science in medicine, especially to improve the situation of the blind, can testify to this

I am sure to speak for all the participants when I say: "May you rest in peace !"

Annie Cuyt

My husband was born on March 22, 1931 in Zwenkau near Leipzig. His father was a teacher at the Gymnasium. He went to school in his home town and in Leipzig, and after his Abitur in 1949 he was allowed to start studying mathematics and physics at the University of Leipzig which was a great privilege at that time.

In 1951 he moved with his parents to the German Federal Republic and continued his studies at the University of Göttingen. In between the terms of the academic year he worked in the oil fields near Lingen/Ems to earn his living. Later on he earned some money being a teaching assistant. His teachers were the Professors Beckert, Hölder and Kähler in Leipzig and the Professors Deuring, Heinz, Kaluza, Rellich and Siegel in Göttingen. He specialized in partial differential equations and prepared his thesis with Prof. Rellich who died of cancer before the thesis was finished. Prof. Siegel and Prof. Heinz, at that time assistant of Prof. Rellich, accepted his work and he got his doctor's degree in 1956.

A few months later we got married.

While preparing for his doctor's degree he could work at the Max-Planck Institute of Physics using (mainly at night) one of the first computers available in Germany. He was fascinated by that kind of work and therefore preferred a position in industry, at the AEG Research Center, rather than a position as assistant in pure mathematics at the university.

But soon he accepted an invitation to teach as assistant professor at the University of Southern California in Los Angeles. We stayed in California for two years and came back to Germany because my husband had met Prof. Collatz in Los Angeles who offered him the possibility to get his Habilitation at the University of Hamburg. Since some research had already been done in Los Angeles he obtained the Habilitation early 1962. Then he started teaching at the University of Hamburg. This period was interrupted by half a year of teaching and researching at Stanford University in California. There he was offered a full professorship to erect the Institute for Numerical and Instrumental Mathematics and the Computation Center of the University of Münster. He accepted the offer and we moved to Münster in 1964.

At the beginning my husband spent a lot of time running over the building plans for the Computation Center, deciding what kind of computer had to be bought, trying to get money for it and training students to work with it. Ten years later, when the IBM 360-50 became too small for the university he had to go through this procedure again.

He always tried to help a lot of people from other faculties of the university who wanted to use the computer for their own research projects. Over the years he developed very intense contacts with many colleagues not only from Science but also from Medicine, Theology, the Humanities and others. He became a member of the Sonderforschungsbereich Mittelalterforschung, helping historians to handle huge amounts of data and laborious publication procedures. This stimulated his interest in providing a text editing system.

As long as I have known him, in every private or professional situation, he had an open mind for other people's problems trying to help them with what he knew about mathematics and computer science. When we were newly wedded and visiting my girl friend who had married a blind teacher, we learnt a lot about braille. This inspired him to develop an automatic braille program for computers. During the next 25 years he developed this project to the extent that it is now used in Germany, Switzerland and Austria for the production of braille printing of all kind. Recently he was awarded the Louis Braille price (in 1984) and the Carl-Strehl medal (in

1985) for this work. When he met a professor in ophtalmology who tried to help patients having problems with their three-dimensional sight due to the removal of a lense in one of the eyes, he developed formulas which not only made use of spectacles, as was common practice, but also of contact lenses. He arranged those formulas in such a way that any ophtalmologist all over the country could easily use them.

In the late seventies more than 50 people were working at the Computation Center in Münster, including an academically trained staff of 23 and another 6 researchers at the Institute of Numerical Mathematics. He was Fachgutachter for mathematics for the Deutsche Forschungsgemeinschaft between 1972 and 1980. Though this responsibility took a lot of time and energy he found it very stimulating. Besides this he was a member of the senate of the Sonderforschungsbereiche of the Deutsche Forschungsgemeinschaft from 1974 till 1982. In this way he was closely connected to the most recent research projects.

He never wanted to return to pure research again (like at AEG) or accept a position as manager. He enjoyed giving lectures and advising students and doctorands. In total 26 students got their doctor's degree inspired by him. In 1980 he became director of the Institute of Applied Mathematics and the Department of Functional Analysis and Numerical Mathematics at the University of Bonn. He continued his research and teaching there, also being a member of the Sonderforschungsbereich 72 (applied mathematics).

He had to stop lecturing in the middle of a term, one week before Whit Sunday in 1985. He entered the hospital the next day because he was very much in pain but he hoped to be able to continue his lecturing after Whit Sunday. During all the following months, up to the last two weeks, he had some of his older students come to the hospital for discussions or examinations. He published a great number of technical notes and scientific papers. He wrote 11 books, of which many were reprinted, and was editor of another 10.

In 1978 he became a member of the Akademie der Naturforscher-Leopoldina in Halle and was very happy about it because it enabled him to make friends with many colleagues from his home region.

He loved the professional and social contact with colleagues all over the world. On his last main lecturing tour in September 1984, already ill with cancer, we visited several universities in China and he lectured almost every day. He was very sorry not to be able to come to Łańcut in Poland anymore as we had planned and hoped till the last moment. However, his last scientific work will appear in these proceedings together with the work of those mathematicians he felt so close to.

When travelling he took every opportunity to enrich himself culturally, using night hours to attend concerts or visit musea – he really got excited about modern paintings. What's more, he always tried to plan these things so that I could share his opportunities.

Working at home, he always listened to music, mainly Bach, Mozart, Brahms, Mahler and Prokoviev. In his spare time he enjoyed reading books on modern history and art or do some handicraft, especially with wood. He collected music on tapes, books and maps, lending the latter out to who ever needed them. On Sundays the family used to make excursions by bicycle or car, most of the times to a point from where the whole landscape could be overlooked. Several times he biked from Münster to Texel, together with our twins, while our youngest daughter and I went by car with the luggage for our holidays. During our last family holiday in 1983 we toured the western part of the USA and Helmut showed us all the places he had got to know at various earlier occasions.

So when he was seriously ill he had a profound reservoir of mathematical problems to be solved, of favourite art to enjoy, of fine experiences to remember and a lot of friends to care for and who cared for him !

He did hope to get his strength back again, supporting his doctors in every physical and mental way, but on the other hand he was prepared to accept his fate, if necessary, having "set his house in order".

He died on November 22, 1985.

Ingrid Werner

LIST OF PARTICIPANTS

ANTOLIN Juan
APTEKAREV Alexandre I.
BIAŁKOWSKI Grzegorz
CUYT Annie
DE BRUIN Marcel G.
DRAUX André
DUNHAM B. Charles
GILEWICZ Jacek
GRAGG William B.
GRAVES-MORRIS Peter
GUZIŃSKI Wojciech
JACOBSEN Lisa
JANIK Adam
JONES B. William
KOVACHEVA Ralitza
KUCHMINSKAYA Khristina I.
LAMBERT Franklin J.
LEWANOWICZ Stanisław
LEWICKI Grzegorz
ŁOSIAK Janina
MOUSSA Pierre
OLEJNICZAK Andrzej
PASZKOWSKI Stefan
PINDOR Maciej
PLEŚNIAK Wiesław
RUSHEWEYH Stephan
SIEMASZKO Wojciech
SKOROBOGAT'KO Vitalij Ya.
SMARZEWSKI Ryszard
SMOLUK Antoni
STAHL Herbert
STANKIEWICZ Jan
SZUSTALEWICZ Adam
VERDONK Brigitte
WAADELAND Haakon
WRONICZ Zygmunt
ZIĘTAK Krystyna

CONTENTS

Survey papers

Polynomial and rational approximation

Continued fractions

Problems related to physics

Miscellanea

A survey of bounds for the zeros of analytic functions obtained by continued fraction methods.

M.G. De Bruin
Department of Mathematics
University of Amsterdam
Roetersstraat 15
1018 WB Amsterdam
Nederland

J. Gilewicz
CNRS - Luminy
Case 907
Centre de Physique Théorique
13288 Marseille Cedex 9
France

H.-J. Runckel
Abteilung Mathematik IV
Universität Ulm
Oberer Eselsberg
D-7900 Ulm
D.B.R.

1. Introduction

We mainly consider sequences of polynomials $q_n(z)$, which satisfy a three term recurrence relation

(1) $q_n(z)=b_n(z)q_{n-1}(z)-a_n(z)q_{n-2}(z)$, $n \in \mathbb{N}$

where $q_{-1}=0$, $q_0=1$, and where a_n, b_n are complex polynomials $\not\equiv 0$. All results below are concerned with the construction of various subsets of $\mathbb{C}$ from the coefficients of a_n, b_n, such that these subsets contain all zeros of $q_n(z)$ for $n \geq 2$ or $2 \leq n \leq N$. Most of these results easily can be extended to power series.

2. The first continued fraction method

The method which is underlying all relevant proofs consists in associating the sequence q_n from (1) to a sequence of Moebius transforms and utilizing their various mapping properties.
For example, (1) yields

$$q_{n-1}/q_n = \frac{1}{b_n-a_n(q_{n-2}/q_{n-1})} \quad , n \geq 1 ,$$

or, using $\mu_n:=q_{n-1}/q_n$, $n \geq 2$, $\mu_1:=1/b_1$, $\mu_0:=0$,

$$\mu_n = \frac{1}{b_n-a_n\mu_{n-1}} \quad , n \geq 1.$$

Hence, we obtain for each $N \geq 2$ the finite continued fraction representation

(2) $\mu_N = \frac{1}{b_N} - \frac{a_N}{b_{N-1}} - \cdots - \frac{a_2}{b_1}$.

(For definitions and notations see [19], [25], [34].)
Therefore, μ_N can be written as a composition of Moebius transforms

$$\begin{aligned}\mu_N &= T_N \circ T_{N-1} \circ \dots \circ T_2(\mu_1)\\ &= T_N \circ T_{N-1} \circ \dots \circ T_2 \circ T_1(0) \quad , \text{ where}\end{aligned}$$

$$T_n(u) := \frac{1}{b_n-a_n u} \; , n \geq 2, \; T_1(u) := \frac{1}{b_1-u} \quad .$$

Applying equivalence transformations to the continued fraction (2) yields other representations of μ_N as composition of suitable Moebius transforms.
Observing that $\mu_n = T_n \circ T_{n-1} \circ \ldots \circ T_1(0)$ holds for $n \geq 1$, conditions on T_n of the following type are formulated. Choose $V_n \subset \hat{\mathbb{C}} := \mathbb{C} \cup \{\infty\}$, such that the boundary of each V_n is a circle or straight line, $n \geq 0$. Assume, furthermore, that $0 \in V_0$ and $T_n(V_{n-1}) \subset V_n$ holds for $n \geq 1$. Since $\mu_0 = 0$ and $\mu_n = T_n(\mu_{n-1}) \in V_n$ for $n \geq 1$, the condition $\infty \notin V_N$ yields $\mu_N \neq \infty$. In most of the applications the V_n are closed half-planes and $T_n(V_{n-1})$ are closed disks. Therefore, $\mu_2, \ldots, \mu_N \neq \infty$ holds in this case. In many special cases this then implies $q_2, \ldots, q_N \neq 0$.

The above formulated conditions on T_n are conditions on the polynomials $a_n(z)$, $b_n(z)$ leading to the required subsets of $\mathbb{C}$ which contain the zeros of $q_2(z), \ldots, q_N(z)$. In this connection see also chapter 10 of [34].

3. The Parabola theorem of E.B. Saff and R.S. Varga

As an application of the first continued fraction method we now consider polynomials $q_n(z)$ satisfying

(3) $q_n(z) = (z+\beta_n)q_{n-1}(z) - \alpha_n z\, q_{n-2}(z)$, $n \in \mathbb{N}$

where $q_{-1} = 0$, $q_0 = 1$, α_{n+1}, $\beta_n \in \mathbb{C}$, $\alpha_{n+1} \neq 0$, $n \geq 1$.

Let, for example, $f(z) = \sum_{\nu=0}^{\infty} c_\nu z^\nu$ be a formal power series and put

$c_{-\nu} := 0$ for $\nu \geq 1$, $A_m^{(0)} := 1$ and

$$A_m^{(n)} := \begin{vmatrix} c_m & c_{m-1} & \cdots & c_{m-n+1} \\ c_{m+1} & c_m & \cdots & c_{m-n+2} \\ & & & \\ c_{m+n-1} & \cdots\cdots & & c_m \end{vmatrix} , \quad m \geq 0,\ n \geq 1.$$

If $A_m^{(n)} \neq 0$ for all $m, n \geq 0$, and if $U_{m,n}(z)$ and $V_{m,n}(z)$ denote the Padé-numerator and Padé-denominator ([10], [25], [34]) of the (m,n)-Padé approximant to f(z), then one obtains (see [9], [27], [32])

<u>Proposition 1.</u> For <u>fixed</u> $n \geq 0$ $q_m(z) := U_{m,n}(z) A_m^{(n)} / A_m^{(n+1)}$

satisfies (3) with $\alpha_{m+1} = A_{m+1}^{(n)} A_{m-1}^{(n+1)} / A_m^{(n)} A_m^{(n+1)}$,

$\beta_m = A_{m-1}^{(n+1)} A_m^{(n)} / A_{m-1}^{(n)} A_m^{(n+1)}$, and

$\beta_m - \alpha_m = A_m^{(n)} A_{m-1}^{(n+2)} / A_m^{(n+1)} A_{m-1}^{(n+1)}$, $m \geq 1$ $(\alpha_1 = 0)$

Especially, $U_{m,0}(z)$ is the m-th partial sum of $f(z)$ and in this case

(4) $\beta_m = \alpha_{m+1} = c_{m-1}/c_m$, $m \geq 1$.

Proposition 2. For fixed $m \geq 0$ $q_n(z) := V_{m,n}(-z) A_m^{(n)} / A_{m+1}^{(n)}$

satisfies (3) with $\alpha_{n+1} = A_m^{(n+1)} A_{m+1}^{(n-1)} / A_m^{(n)} A_{m+1}^{(n)}$,

$\beta_n = A_m^{(n)} A_{m+1}^{(n-1)} / A_m^{(n-1)} A_{m+1}^{(n)}$, and

$\beta_n - \alpha_n = A_m^{(n)} A_{m+2}^{(n-1)} / A_{m+1}^{(n)} A_{m+1}^{(n-1)}$, $n \geq 1$ $(\alpha_1 = 0)$.

If $f(z) = e^z$, i.e. $c_\nu = 1/\nu!$, $\nu \geq 0$, then

(5) $$A_m^{(n)} = \prod_{j=1}^{n} \frac{(j-1)!}{(m+j-1)!}, \quad m, n \geq 1 \quad \text{(see [32])}$$

Theorem 1 (Parabola theorem of E.B. Saff and R.S. Varga).

Assume that the polynomials $q_n(z)$, $n \geq 1$, satisfy (3) such that $\beta_n > 0$, $1 \leq n \leq N$, $\alpha_n > 0$, $2 \leq n \leq N$ and $D_N := \min_{1 \leq n \leq N} (\beta_n - \alpha_n) > 0$ (with $\alpha_1 = 0$). Then $q_n(z) \neq 0$ for $2 \leq n \leq N$ and all $z \in P$, where

$$P := \{\zeta \in \mathbb{C} : |\zeta| \leq \operatorname{Re}\zeta + 2 D_N\} .$$

For the proof of Theorem 1 and all examples below, see [32]. Using (4), this result immediately can be applied to partial sums of power series $f(z) = \sum_{\nu=0}^{\infty} c_\nu z^\nu$, if $c_\nu > 0$, $\nu \geq 0$ and $\dfrac{c_{\nu-1}}{c_\nu} - \dfrac{c_{\nu-2}}{c_{\nu-1}} > 0$, $\nu \geq 2$.

In particular, $\sum_{\nu=0}^{m} \frac{1}{\nu!} z^\nu$ is $\neq 0$ for $m \geq 2$ and all $z \in \mathbb{C}$ satisfying $|z| \leq \operatorname{Re} z + 2$, since $D_N = 1$ for each $N \geq 2$ in this case. See also [23], [24], and [33].

Besides many other examples also generalized Bessel polynomials are treated successfully in [32] by means of Theorem 1. The n-th generalized Besselpolynomial $Y_n^{(\delta)}$ is defined by (see [16], [32])

(6) $$Y_n^{(\delta)} := 1 + \sum_{j=1}^{n} \binom{n}{j} (n+\delta+1)(n+\delta+2)\ldots(n+\delta+j)(-z/2)^j$$

for $n \in \mathbb{N}$, $\delta \in \mathbb{C}$.

For each <u>fixed</u> $m \geq 2$ $q_n^{(m+\delta)}(z) := z^n Y_n^{(m+\delta-n)}(-2/z)$ satisfies (3) with $\beta_n = n+m+\delta$, $\alpha_n = n-1$, $n \geq 1$.

Observe that the substitution $z \to -2/z$ maps the exterior of P (in Theorem 1) onto the interior of a cardioid region. Then Theorem 1 yields for n=m [32].

<u>Theorem 2</u>. If $\delta \in \mathbb{R}$ and $m+\delta+1=c>0$, then all zeros of $Y_m^{(\delta)}(z)$, $m \geq 2$, are contained in the open cardioid region

$$C := \{\zeta = re^{i\theta} \in \mathbb{C} : 0<r<(1+\cos\theta)/c, \ |\theta|<\pi\}.$$

For further results concerning the zeros of generalized Bessel polynomials see [20], [22], [26] and section 8 of this paper. In [6] K. Dočev showed that all zeros of $Y_m^{(\delta)}(z)$, $m+\mathrm{Re}\,\delta+1=c>0$ are contained in $\{z \in \mathbb{C} : |z| \leq 2/c\}$. Especially, M.G. de Bruin, E.B. Saff and R.S. Varga have proved in [1], [31], [33], that the result of Theorem 2 is sharp in the sense that each boundary point of $\{z=re^{i\theta} \in \mathbb{C} : 0<r<(1+\cos\theta), |\theta|<\pi\}$ is an accumulation point of zeros of the set of normalized Bessel polynomials $Y_m^{(\delta)}(z/(m+\delta+1))$, $m \in \mathbb{N}$, $m+\delta+1>0$.

4. Results of M.G. de Bruin

By applying a method of proof which is similar to the first continued fraction method (introduced by E.B. Saff and R.S. Varga in [32]) M.G. de Bruin obtained the following results.

<u>Theorem 3</u>. Let the sequence of polynomials $\{P_n(z)\}_{n=0}^{\infty}$ be generated by

$$P_n(z) = (1+a_n z)P_{n-1}(z) + b_n z\, P_{n-2}(z) + c_n z^2\, P_{n-3}(z), \ n \geq 3, \quad \text{where}$$

$$P_0(z)=1,\ P_1(z)=1+a_1 z,\ P_2(z)=(1+a_1 z)(1+a_2 z)+b_2 z, \text{ with } a_n \in \mathbb{R}\setminus\{0\},$$

$n \geq 1$, $b_n \in \mathbb{R}$, $n \geq 2$, $c_n \in \mathbb{R}\setminus\{0\}$, $n \geq 3$.

Let $\{A_n^{(k)}\}_{n=1}^{\infty}$ $(k=1,2,\ldots,5)$ be sequences of positive real numbers and define the sets V_k $(k=1,2,\ldots 5)$ of complex numbers z as follows

$$V_1 := \left\{ \begin{array}{l} 1 + a_1 \operatorname{Re} z \geqq A_1^{(1)} , \\[2ex] 1 + a_2 \operatorname{Re} z + b_2 \dfrac{\operatorname{Re} z + a_1 |z|^2}{|1+a_1 z|^2} \geqq A_2^{(1)} , \\[3ex] 1 + a_n \operatorname{Re} z + \dfrac{2b_n A_{n-2}^{(1)} \operatorname{Re} z + c_n \operatorname{Re}(z^2)}{4\, A_{n-2}^{(1)}\; A_{n-1}^{(1)}} - \\[3ex] - \dfrac{2|b_n z| A_{n-2}^{(1)} + 3|c_n z^2|}{4\, A_{n-2}^{(1)}\; A_{n-1}^{(1)}} \geqq A_n^{(1)} , \; n \geqq 3 \end{array} \right.$$

$$V_2 := \left\{ \begin{array}{l} 1 + a_1 \operatorname{Re} z \leqq -A_1^{(2)} , \\[2ex] 1 + a_2 \operatorname{Re} z + b_2 \dfrac{\operatorname{Re} z + a_1 |z|^2}{|1+a_1 z|^2} \leqq -A_2^{(2)} , \\[3ex] 1 + a_n \operatorname{Re} z - \dfrac{2b_n A_{n-2}^{(2)} \operatorname{Re} z - c_n \operatorname{Re}(z^2)}{4\, A_{n-2}^{(2)}\; A_{n-1}^{(2)}} + \\[3ex] + \dfrac{2|b_n z| A_{n-2}^{(2)} + 3|c_n z^2|}{4\, A_{n-2}^{(2)}\; A_{n-1}^{(1)}} \leqq -A_n^{(2)} , \; n \geqq 3 \end{array} \right.$$

$$V_3 := \left\{ \begin{array}{l} \operatorname{Re} z + a_1 |z|^2 \geqq A_1^{(3)} |z|^2 , \\[2ex] \operatorname{Re} z + a_2 |z|^2 + b_2 \dfrac{1 + a_1 \operatorname{Re} z}{|1+a_1 z|^2} |z|^2 \geqq -A_2^{(3)} |z|^2 \\[3ex] \operatorname{Re} z + a_n |z|^2 + \dfrac{2b_n A_{n-1}^{(3)} + c_n}{4\, A_{n-2}^{(3)}\, A_{n-1}^{(3)}} \operatorname{Re} z - \\[3ex] - \dfrac{|2b_n\, A_{n-2}^{(3)} + c_n| + 2|c_n|}{4\, A_{n-2}^{(3)}\, A_{n-1}^{(3)}} |z| \geqq A_n^{(3)} |z|^2 , \; n \geqq 3 . \end{array} \right.$$

$$V_4 := \begin{cases} \operatorname{Re} z + a_1|z|^2 \leqq -A_1^{(4)}|z|^2 , \\ \operatorname{Re} z + a_2|z|^2 + b_2 \dfrac{1+a_1 \operatorname{Re} z}{|1+a_1 z|^2} |z|^2 \leqq -A_2^{(4)}|z|^2 , \\ \operatorname{Re} z + a_n|z|^2 - \dfrac{2b_n A_{n-2}^{(4)} - c_n}{4\, A_{n-2}^{(4)} A_{n-1}^{(4)}} \operatorname{Re} z - \\ - \dfrac{|2b_n A_{n-2}^{(4)} - c_n| + 2|c_n|}{4\, A_{n-2}^{(4)} A_{n-1}^{(4)}} |z| \leqq -A_n^{(4)}|z|^2 , \quad n \geqq 3 . \end{cases}$$

$$V_5 := \begin{cases} |1+a_1 z| \geqq A_1^{(5)} , \\ |1+a_2 z| \geqq A_2^{(5)} + |b_2 z|/A_1^{(5)} \\ |1+a_n z| \geqq A_n^{(5)} + \dfrac{|b_n z|}{A_{n-1}^{(5)}} + \dfrac{|c_n z^2|}{A_{n-2}^{(5)} A_{n-1}^{(5)}} , \quad n \geqq 3 . \end{cases}$$

Then all zeros of the whole sequence $\{P_n(z)\}_{n=1}^{\infty}$ are contained in $\mathbb{C} \setminus \bigcup_{k=1}^{5} V_k$.

Theorem 4. Let the sequence of polynomials $\{Q_n(z)\}_{n=0}^{\infty}$ be generated by

$$Q_n(z) = (1+a_n z) Q_{n-1}(z) + b_n z^2 Q_{n-2}(z) + c_n z^3 Q_{n-3}(z) , \quad n \geqq 3 ,$$

where $Q_0(z)=1$, $Q_1(z)=1+a_1 z$, $Q_2(z)=(1+a_1 z)(1+a_2 z)+b_2 z^2$ and $a_n>0, n\geqq 1$; $b_n>0$, $n\geqq 2$; $c_n>0$, $n\geqq 3$. Let $\{A_n^{(k)}\}_{n=1}^{\infty}$ $(k=1,\ldots,4)$ be sequences of positive real numbers and define the sets W_k, $k=1,\ldots,4$ of complex numbers z as follows

$$W_1 := \begin{cases} \operatorname{Re} z + a_1|z|^2 \geq A_1^{(1)}|z|^2 , \\ \operatorname{Re} z + a_2|z|^2 + b_2 \dfrac{\operatorname{Re} z + a_1|z|^2}{|1+a_1 z|^2}|z|^2 \geq A_2^{(1)}|z|^2 , \\ \operatorname{Re} z + \left(a_n - \dfrac{c_n^2}{8A_{n-1}^{(1)}\, A_{n-2}^{(1)}(2A_{n-2}^{(1)}\, b_n + c_n)}\right)|z|^2 \geq \\ \geq A_n^{(1)}|z|^2 , \quad n \geq 3. \end{cases}$$

$$W_2 := \begin{cases} \operatorname{Re} z + a_1|z|^2 \leq -A_1^{(2)}|z|^2 \\ \operatorname{Re} z + a_2|z|^2 + b_2 \dfrac{\operatorname{Re} z + a_1|z|^2}{|1+a_1 z|^2}|z|^2 \leq -A_2^{(2)}|z|^2, \\ \operatorname{Re} z + \left(a_n + \dfrac{P_n}{4\, A_{n-1}^{(2)}\, A_{n-2}^{(2)}}\right)|z|^2 \leq -A_n^{(2)}|z|^2, \ n \geq 3 , \\ \text{where } P_n := 4(c_n - A_{n-2}^{(2)}\, b_n) \text{ if } A_{n-2}^{(2)}\, b_n \leq 3c_n/4, \\ \text{and } P_n := \dfrac{c_n^2}{2(2A_{n-2}^{(2)}\, b_n - c_n)} \text{ if } A_{n-2}^{(2)}\, b_n \geq 3c_n/4, \ n \geq 3 . \end{cases}$$

$$W_3 := \begin{cases} |1+a_1 z| \geq A_1^{(3)}|z|^2 , \\ |1+a_2 z| \geq \left(A_2^{(3)} + \dfrac{b_1}{A_1^{(3)}}\right)|z|^2 , \\ |1+a_n z| \geq \left(A_n^{(3)} + \dfrac{b_n}{A_{n-1}^{(3)}} + \dfrac{c_n}{A_{n-1}^{(3)}\, A_{n-2}^{(3)}}\right)|z|^2 , \ n \geq 3 . \end{cases}$$

$$W_4 := \begin{cases} |\operatorname{Im} z| \geq A_1^{(4)} |z|^2 , \\ \left|\operatorname{Im} z \left(1 - \dfrac{b_2|z|^2}{|1+a_1 z|^2}\right)\right| \geq A_2^{(4)} |z|^2 , \\ |\operatorname{Im} z| \geq \\ \geq \left(A_n^{(4)} + \dfrac{2(b_n A_{n-2}^{(4)} + c_n) + ((2b_n A_{n-2}^{(4)})^2 + c_n^2)^{1/2}}{4 A_{n-1}^{(4)} A_{n-2}^{(4)}}\right) |z|^2, \ n \geq 3 . \end{cases}$$

Then all zeros of the whole sequence $\{Q_n(z)\}_{n=1}^{\infty}$ are contained in $\mathbb{C} \setminus \bigcup_{k=1}^{4} W_k$.

For the proofs of Theorems 3 and 4 and many special examples see [3].

Theorem 5. Let $n \in \mathbb{N}$ be fixed and assume that the sequence of polynomials $\{P_k(z)\}_{k=1}^{\infty}$ satisfies

$$P_k(z) = (z+\beta_k)P_{k-1}(z) + \alpha_{n,k}\, z\, P_{k-2}(z) + \ldots + \alpha_{1,k} z\, P_{k-n-1}(z), k \geq 1,$$

where $P_0(z)=1$, $P_{-k}(z)=0$ $(k=1,\ldots,n)$ and where β_k, $\alpha_{j,k} \in \mathbb{C}$, $j=1,\ldots,n$, $k \geq 1$, such that $\beta_k\, \alpha_{1,k} \neq 0$ for $k \geq 1$.

Let z be a complex number for which there exists a sequence $\{A_k\}_{k=1}^{\infty}$ of positive real numbers with

$$|z+\beta_1| \geq A_1 ,$$

$$|z+\beta_j| \geq A_j + |\alpha_{n,j} z|/A_{j-1} + |\alpha_{n-1,j} z|/(A_{j-1}A_{j-2}) + \ldots$$
$$\ldots + |\alpha_{n+2-j,j} z|/(A_{j-1}A_{j-2}\ldots A_1), \ 2 \leq j \leq n,$$

$$|z+\beta_j| \geq A_j + |\alpha_{n,j}\, z|/A_{j-1} + |\alpha_{n-1,j}\, z|/(A_{j-1}A_{j-2}) + \ldots$$
$$\ldots + |\alpha_{1,j}\, z|/(A_{j-1}A_{j-2}\ldots A_{j-n}) , \ j \geq n+1 .$$

Then $P_k(z) \neq 0$ for all $k \geq 1$.

As a corollary this yields

Theorem 6. With the same notations as in Theorem 5 put

$b_m := \inf\{|\beta_j| : j \geq 1\}$, $b_M := \sup\{|\beta_j| : j \geq 1\}$,

$a := \sup\{|\alpha_{n+2-j,k}| : 2 \leq j \leq n+1, k \geq j\}$ and assume that $b_M < \infty$ and $a < \infty$.

For an arbitrary $A > 0$ define

$D_A := \{z \in \mathbb{C} : |z| \geq b_M + A + a|z|(A^{-1} + A^{-2} + \ldots + A^{1-n})\}$,
and

$E_A := \{z \in \mathbb{C} : b_m \geq |z| + A + a|z|(A^{-1} + A^{-2} + \ldots + A^{1-n})\}$.

Then $P_k(z) \neq 0$ for all $k \geq 1$ and all $z \in \bigcup_{A>0} (D_A \cup E_A)$.

For the proofs of Theorems 5 and 6 and examples see [5]. In [2],[4] generalized Padé-approximation is considered which leads to recursively defined polynomials satisfying or being related to recurrence relations considered in Theorems 3-6.

5. The results of J. Gilewicz and E. Leopold

By applying suitable special cases of the first continued fraction method J. Gilewicz and E. Leopold obtained the following general results.

Theorem 7, Let $P_1, \ldots, P_N$, $N>1$, be complex polynomials which satisfy

$$P_{n+1}(z) = (b_n + b'_n z) P_n(z) - (a_n + a'_n z + a''_n z^2) P_{n-1}(z), \text{ where } b'_n \neq 0,\ 0 \leq n < N,$$

and where $P_{-1} \equiv 0$, $P_0 \neq 0$.

Put $A^*_N := \{z \in \mathbb{C} : a_n + a'_n z + a''_n z \neq 0,\ 0 \leq n < N\}$.

(i) Assume that $b_0 \neq 0$ and $4|a_n b'_n| < |b'_{n-1} b^2_n|$ holds for $1 \leq n < N$ and put

$I_0 := (0, |b_0/b'_0|)$, and for $1 \leq n < N$

$$I_n := \left(\left|\frac{b_n}{2b'_n}\right| - \left|\left|\frac{b_n}{2b'_n}\right|^2 - \left|\frac{a_n}{b'_{n-1} b'_n}\right|\right|^{1/2},\ \left|\frac{b_n}{2b'_n}\right| + \left|\left|\frac{b_n}{2b'_n}\right|^2 - \left|\frac{a_n}{b'_{n-1} b'_n}\right|\right|^{1/2} \right).$$

If $I^*_N := \bigcap_{0 \leq n < N} I_n \neq \emptyset$, then

$P^{*\prime}_N := \{z \in A^*_N : |z| \leq \max_{d \in I^*_N} (\min_{1 \leq n < N} x_n(d))\}$ contains no zero of $P_1, \ldots, P_N$.

Here $x_n(d)$, $1 \leq n < N$, is defined for $d \in I^*_N$ as follows.

If $a''_n \neq 0$, then

$$x_n(d) := \frac{1}{2|a''_n|} (|b'_{n-1} b'_n| d - |a'_n| - |b'_{n-1} b_n| + \Delta_n^{1/2}), \text{ where}$$

$$\Delta_n := |b'_{n-1} b'_n| (|b'_{n-1} b'_n| - 4|a''_n|) d^2 + 2|b'_{n-1}| (2|a''_n b_n| - |b'_{n-1} b'_n b_n| - |a'_n b'_n|) d$$
$$+ (|a'_n| + |b'_{n-1} b_n|)^2 - 4|a_n a''_n|, \text{ and}$$

if $a''_n = 0$, then

$$x_n(d) := d - \frac{|a_n| + |a'_n| d}{|b'_{n-1}|(|b_n| - |b'_n| d) + |a'_n|} .$$

(ii) Assume that $4|a''_n| < |b'_{n-1} b'_n|$ holds for $1 \leq n < N$ and put $d_0 := |b_0/b'_0|$ and for $1 \leq n < N$

$$d_n := \frac{|b'_{n-1}|(|b'_{n-1} b'_n b_n| - 2|a''_n b_n| + |a'_n b'_n|) + 2\Delta'^{1/2}_n}{|b'_{n-1} b_n|(|b'_{n-1} b'_n| - 4|a''_n|)} , \quad \text{where}$$

$$\Delta'_n := |a''_n b'_{n-1}|(|a''_n b'_{n-1} b^2_n| + |a'_n b'_n|(|a'_n| + |b'_{n-1} b_n|) + |a_n b'_n|(|b'_{n-1} b'_n| - 4|a''_n|)).$$

Next, define $J^*_N := [\max_{0 \leq n < N} d_n, \infty)$ and

$$P^{*\,''}_N := \bigcup_{d \in J^*_N} \{z \in A^*_N : \max_{1 \leq n < N} x'_n(d) \leq |z| \leq \min_{1 \leq n < N} x''_n(d)\},$$ where $x'_n(d)$ and $x''_n(d)$ are defined for $d \in J^*_N$ as follows.

If $a''_n \neq 0$, then

$$x'_n(d) := \frac{1}{2|a''_n|}(|b'_{n-1} b'_n| d - |a'_n| - |b'_{n-1} b_n| - \Delta^{1/2}_n) \text{ and}$$

$x''_n(d) := x_n(d)$.

If $a''_n = 0$, then $x'_n(d) := x_n(d)$ and $x''_n := +\infty$. Here $\Delta_n, x_n(d)$ are defined by the same formulas as in (i).

Under these assumptions $P^{*\,''}_N$ does not contain any zero of $P_1, \ldots, P_N$.

If all conditions in (i) and (ii) are satisfied, then $P^*_N := P^{*\,'}_N \cup P^{*\,''}_N$ does not contain any zero of $P_1, \ldots, P_N$.

The following corollary is concerned with recurrence relations which easily can be transformed into one of type (3).

<u>Corollary 1.</u> Let $P_1, \ldots, P_N$, $N > 1$, be complex polynomials which satisfy

$$P_{n+1}(z) = (1 + t_n z) P_n(z) - u_n z\, P_{n-1}(z), \text{ where } t_n u_n \in \mathbb{C},\ t_n u_n \neq 0 \text{ for } 0 \leq n < N$$

and where $P_{-1} \equiv 0$, $P_0 \neq 0$.

Then all zeros of $P_1, \ldots, P_N$ are located in

$$\Gamma_N := \{z \in C : \max_{d \in I^*_N} (\min_{0 < n < N} x_n(d)) < |z| < \min_{d \in J^*_N} (\max_{0 < n < N} x_n(d)\}, \text{ where}$$

$$I^*_N := (0, \min_{0 < n < N} |t_n|^{-1}) \cap (0, |t_0|^{-1}] ,$$

$$J_N^*:=\Big[\max_{0\le n<N}\frac{|t_{n-1}|+|u_n|}{|t_{n-1}t_n|},\infty\Big),\ (t_{-1}=1,u_0=0)\ \text{and}\ x_n(d):=\frac{|t_{n-1}|(1-|t_n|d)d}{|t_{n-1}|(1-|t_n|d)+|u_n|},\quad 0<n<N.$$

The next corollary is concerned with a type of recurrence relation which also is satisfied by orthogonal polynomials

Corollary 2. Let $P_1,\ldots,P_N$, $N>1$ be complex polynomials which satisfy

$P_{n+1}(z) = (b_n+b_n'z)P_n(z)-a_nP_{n-1}(z)$, where $b_n'a_n \in \mathbb{C}$, $b_n'a_n\neq 0$, $0\le n<N$, and where $P_{-1}\equiv 0$, $P_0\neq 0$.

Then all zeros of $P_1,\ldots,P_N$ are contained in

$$D_N:=\{z\in\mathbb{C}:|z|<\min_{d\in J_N}\max_{0<n<N}x_n(d)\},\quad\text{where}$$

$$J_N:=\Big[\max_{0\le n<N}|b_n/b_n'|,\infty\Big)\ \text{and}\ x_n(d):=d-\frac{|a_n|}{|b_{n-1}'|(|b_n|-|b_n'|d)}.$$

Moreover, if $b_0\neq 0$ and $4|a_nb_n'|<|b_{n-1}'b_n^2|$ is satisfied for $0<n<N$ and if I_N^* (in Theorem 7) is $\neq\emptyset$, then all zeros of $P_1,\ldots,P_N$ are contained in

$$\Gamma_N:=\{z\in\mathbb{C}:\max_{d\in I_N^*}\min_{0<n<N}x_n(d)<|z|<\min_{d\in J_N}\max_{0<n<N}x_n(d)\}.$$

For the proof of Theorem 7 and Corollaries 1,2 see [11].

Theorem 8. Let $P_1,\ldots,P_N$, $N>1$, be complex polynomials satisfying

$$P_{n+1}(z)=(b_n+b_n'z)\,P_n(z)-A_n(z)\,P_{n-1}(z),$$

where $b_n'>0$, $b_n\in\mathbb{C}$ and $A_n(z)\in\mathbb{C}[z]$ is of dégree ≤ 2, $0\le n<N$ and where $P_{-1}\equiv 0$, $P_0\neq 0$.

Put $A_N^*:=\{z\in\mathbb{C}:A_n(z)\neq 0,\ 0\le n<N\}$,

$$f_n(z,d):=\frac{|A_{n+1}(z)|+\operatorname{Re}A_{n+1}(z)}{2\,b_n'\,(\operatorname{Re}z-d)},$$

$$\tilde f_n(z,d):=\frac{|A_{n+1}(z)|-\operatorname{Re}A_{n+1}(z)}{2\,b_n'(\operatorname{Im}z-d)},\quad 0\le n<N-1,$$

$$g_n(d):=\operatorname{Re}b_n+b_n'd,\ \tilde g_n(d):=\operatorname{Im}b_n+b_n'd,\ 0\le n<N,$$

and

$$J_{1,N}:=\Big(\max_{0<n<N}(-\operatorname{Re}b_n/b_n'),+\infty\Big)\cap\Big[-\operatorname{Re}b_0/b_0',+\infty\Big),$$

$$J_{2,N}:=\Big(-\infty,\min_{0<n<N}(-\operatorname{Re}b_n/b_n')\Big)\cap\Big(-\infty,-\operatorname{Re}b_0/b_0'\Big],$$

$$J_{3,N}:=\Big(\max_{0<n<N}(-\operatorname{Im}b_n/b_n'),+\infty\Big)\cap\Big[-\operatorname{Im}b_0/b_0',+\infty\Big),$$

$J_{4,N} := (-\infty, \min_{0<n<N} (-\mathrm{Im}\, b_n / b_n')) \cap (-\infty, -\mathrm{Im}\, b_0 / b_0']$

Furthermore, define

$$P_{1,N} := \bigcup_{d \in J_{1,N}} \left(\bigcap_{0 \leq n < N-1} \{z \in A_N^* : f_n(z,d) \leq g_{n+1}(d),\ \mathrm{Re}\, z > d\} \right),$$

$$P_{2,N} := \bigcup_{d \in J_{2,N}} \left(\bigcap_{0 \leq n < N-1} \{z \in A_N^* : f_n(z,d) \geq g_{n+1}(d),\ \mathrm{Re}\, z < d\} \right),$$

$$P_{3,N} := \bigcup_{d \in J_{3,N}} \left(\bigcap_{0 \leq n < N-1} \{z \in A_N^* : \tilde{f}_n(z,d) \leq \tilde{g}_{n+1}(d),\ \mathrm{Im}\, z > d\} \right),$$

$$P_{4,N} := \bigcup_{d \in J_{4,N}} \left(\bigcap_{0 \leq n < N-1} \{z \in A_N^* : \tilde{f}_n(z,d) \geq \tilde{g}_{n+1}(d),\ \mathrm{Im}\, z < d\} \right).$$

Then $P_1, \ldots, P_N$ have no zeros in $\bigcup_{1 \leq i \leq 4} P_{i,N}$.

The next theorem is again concerned with a recurrence relation of type (3).

__Theorem 9 (Conic Theorem).__ Let $P_1, \ldots, P_N$ $(N > 1)$ be real polynomials satisfying

$P_{n+1}(z) = (b_n + b_n' z) P_n(z) - a_n z\, P_{n-1}(z)$, where

$a_n > 0$, $b_n' > 0$, $b_n \in \mathbb{R}$, $0 \leq n < N$ and where $P_{-1} \equiv 0$, $P_0 \not\equiv 0$.

Put $\phi(z,d) := \dfrac{|z| + \mathrm{Re}\, z}{2(\mathrm{Re}\, z - d)}$, $\tilde{\phi}(z,d) := \dfrac{|z| - \mathrm{Re}\, z}{2(\mathrm{Im}\, z - d)}$,

$\psi_n(d) := b_{n-1}'(b_n + b_n' d)/a_n$, $\tilde{\psi}_n(d) := b_{n-1}' b_n' d / a_n$, $0 < n < N$.

Furthermore, define

$$\Gamma_{1,N} := \bigcup_{\substack{d \geq -b_0/b_0', \\ d > \max_{0<n<N} (-b_n/b_n')}} \{z \in \mathbb{C} : \phi(z,d) \leq \min_{0<n<N} \psi_n(d),\ \mathrm{Re}\, z > d\},$$

$$\Gamma_{2,N} := \bigcup_{\substack{d \leq -b_0/b_0', \\ d < \min_{0<n<N}(-b_n/b_n')}} \{z \in \mathbb{C} : \phi(z,d) \geq \max_{0<n<N} \psi_n(d),\ \mathrm{Re}\, z < d\},$$

$$\Gamma_{3,N} := \bigcup_{d>0} \{z \in \mathbb{C} : \tilde{\phi}(z,d) \leq \min_{0<n<N} \tilde{\psi}_n(d),\ \mathrm{Im}\, z > d\},$$

$$\Gamma_{4,N} := \bigcup_{d<0} \{z \in \mathbb{C} : \tilde{\phi}(z,d) \geq \max_{0<n<N} \tilde{\psi}_n(d),\ \mathrm{Im}\, z < d\}.$$

Then $P_1, \ldots, P_N$ have no zeros in $\bigcup_{1 \leq i \leq 4} \Gamma_{i,N}$.

Next, again a recurrence relation for orthogonal polynomials is considered.

Theorem 10. Let $P_1,\dots,P_N$, $N>1$, be real polynomials satisfying

$$P_{n+1}(z) = (b_n+b_n'z)P_N(z) - a_nP_{n-1}(z), \quad \text{where}$$

$$a_n>0,\ b_n'>0,\ b_n \in \mathbb{R},\ 0\leq n<N, \text{ and where } P_{-1}\equiv 0,\ P_0\not\equiv 0.$$

Put

$$m(x):= \min_{0<n<N} \frac{a_n}{b'_{n-1}}(b_n+b_n'x)^{-1} \quad \text{for } x \leq x_1:= \min_{0\leq n<N}(-b_n/b_n')$$

and

$$M(x):= \max_{0<n<N} \frac{a_n}{b'_{n-1}}(b_n+b_n'x)^{-1} \quad \text{for } x \geq x_2:= \max_{0\leq n<N}(-b_n/b_n') .$$

Then all zeros of $P_1,\dots,P_N$ are contained in $(\max_{x\leq x_1}(x+m(x)), \min_{x\geq x_2}(x+M(x)))$.

For the proofs of Theorems 8-10 see [21].
Further refinements of the preceding results are treated in [13], [14], [15]. See also [12] and [22].

6. The second continued fraction method

This method differs only slightly from the first continued fraction method. Again, a sequence of complex polynomials is given which satisfies (1). We now associate each $q_n(z)$ to the continued fraction

$$w_n(z) = \frac{p_n(z)}{q_n(z)} = \frac{a_1}{b_1} - \frac{a_2}{b_2} - \cdots - \frac{a_n}{b_n}, \quad n \in \mathbb{N}.$$

Then the p_n, $n\in\mathbb{N}$, also satisfy (1) with $p_0=0_1$, $p_{-1}=1$. Because of $p_n q_{n-1} - q_n p_{n-1} = a_1a_2\dots a_n$, $n\in\mathbb{N}$, p_n and q_n do not vanish simultaneously, if $a_1a_2\dots a_n \neq 0$. In this case $q_n=0$ holds iff $w_n=\infty$.

Again we want to find conditions on a_n,b_n, $n\in\mathbb{N}$ (i.e. determine subsets of the z-plane), such that $w_n(z)\neq\infty$ holds for $2\leq n\leq N$ or all $n\in\mathbb{N}$. Therefore, we put $w_n(z)=s_1\circ s_2\circ\dots\circ s_n(0)$ with suitable (not uniquely determined) Moebius transformations $s_n(u)$, $n\in\mathbb{N}, u\in\overline{\mathbb{C}}$. To be more explicit, we next choose halfplanes

$$H_n:=\{\zeta\in\mathbb{C}:\operatorname{Re} e^{i\phi_n}(\zeta+d_n)\geq 0\}\cup\{\infty\}\subset\overline{\mathbb{C}} ,$$

where $\phi_n \in \mathbb{R}$, $d_n \in \mathbb{C}$, $n\in\mathbb{N}$, and we assume that the following conditions are satisfied.

(a) $D_n: = s_n(H_n)\subset\mathbb{C}$ is a disk for $2 \leq n \leq N$, i.e.

$\operatorname{Re} e^{i\phi_n}(b_n-d_n) > 0, \quad 2 \leq n \leq N$.

(b) $D_n \subset H_{n-1}$ for $2 \leq n \leq N$, i.e.

$|a_n| + \operatorname{Re} e^{i(\phi_{n-1}+\phi_n)} a_n \leq 2(\operatorname{Re} e^{i\phi_{n-1}} d_{n-1})(\operatorname{Re} e^{i\phi_n}(b_n-d_n)), \ 2 \leq n \leq N$

(c) $o \in \mathring{H}_n$, the interior of H_n, $2 \leq n \leq N$, i.e.

$\operatorname{Re} e^{i\phi_n} d_n > 0, \ 2 \leq n \leq N$,

(d) $-b_1 \notin \mathring{D}_2$, the interior of D_2, or the stronger condition $-b_1 \notin \mathring{H}_1$, i.e.

$\operatorname{Re} e^{i\phi_1}(b_1-d_1) \geq 0$.

If these conditions are satisfied, then $s_2 \circ s_3 \circ \ldots \circ s_n(0) \in \mathring{D}_2 \subset \mathring{H}_1$, and, hence, $w_n(z) \neq \infty$ holds for $2 \leq n \leq N$.
If instead of (c) only $o \in H_n$ is required, then condition (d) has to be modified appropriately.
By suitably choosing the parameters $\phi_n \in \mathbb{R}$ and $d_n \in \mathbb{C}$ for $n \geq 1$, various results are obtained. Instead of stating general results obtainable by this method, we want to demonstrate this method by discussing several special cases which can be described more easily.

7. A generalization of the parabola theorem

Assume again that (1) has the following special form

(3) $q_n(z) = (z+\beta_n)q_{n-1}(z) - \alpha_n z q_{n-2}(z), \ n \geq 1$, where $\alpha_n, \beta_n \in \mathbb{C}, \ \alpha_n \neq 0$.

The following theorem is a special case of a more general theorem derived in [26].

Theorem 11. Assume that $a_n = |a_n| e^{i\psi_o}$, $n \geq 2$, $\beta_n - \alpha_n = |\beta_n - \alpha_n| e^{i\psi_1}$, $n \geq 1$ $(\alpha_1 = 0)$, with fixed $\psi_0, \psi_1 \in \mathbb{R}$. Next, assume that $\rho := \inf_{n \geq 2} |\beta_n - \alpha_n| > 0$.
Finally, put $d := \rho e^{i\psi_1}$ and let $\phi \in \mathbb{R}$ be chosen such that $\operatorname{Re} e^{i\phi}(\beta_1 - d) \geq 0$, $\cos(\phi+\psi_o) > 0$, and $\cos(\phi+\psi_1) > 0$.
Then $q_n(z) \neq 0$ for $n \geq 2$ and all $z \in P'$, where
$P' = \{\zeta \in \mathbb{C} : |\zeta| \leq \operatorname{Re} e^{-i\psi_o} \zeta + 2\rho \cos(\phi+\psi_o)\cos(\phi+\psi_1)\}$.
If $|\psi_1 - \psi_o| < \pi$ and $\phi := -(\psi_o + \psi_1)/2$, then $q_n(z) \neq 0$ for $n \geq 2$ and all $z \in P \supset P'$, where

$P = \{\zeta\in\mathbb{C}: |\zeta| \leq \text{Re } e^{-i\psi_o}\zeta + 2\rho\cos^2((\psi_1-\psi_o)/2)\}$.

If the assumptions are satisfied for $2\leq n\leq N$ only and if $\rho := \min_{2\leq n\leq N} |\beta_n-\alpha_n| > 0$, then $q_n(z)\neq 0$ for $2\leq n\leq N$ and all $z\in P'$ or $z\in P$ as above.

Theorem 11 contains as special case ($\psi_0=\psi_1=0$) Theorem 1.

In [17] P. Henrici generalized Theorem 1 and obtained

<u>Theorem 12.</u> Let $\{\beta_n\}_{n=1}^{\infty}$, $\{\varepsilon_n\}_{n=1}^{\infty}$ be sequences of positive real numbers such that $\alpha := \inf_{n>0}\{\beta_n-\varepsilon_n\} > 0$ and put
$P_\alpha := \{\zeta\in\mathbb{C}: |\zeta| < \text{Re }\zeta + 2\alpha\}$.
Next, let $\{z_n\}_{n=1}^{\infty}$ be a sequence of complex numbers, such that $z_n\in P_\alpha$ for $n\in\mathbb{N}$.

If the sequence $\{q_n\}_{n=1}^{\infty}$ satisfies

(7) $q_n = (\beta_n+z_{n+1})q_{n-1} - \varepsilon_n z_n q_{n-2}$, $n\geq 1$,

where $q_0=1$, $q_{-1}=0$, then $q_n\neq 0$ holds for $n\geq 0$.

As an application of Theorem 12 let $\{x_n\}_{n=0}^{\infty}$ be a sequence of real numbers and let Φ be a real function defined at least at the points x_n. Define $\alpha_k := \Phi[x_0, x_1, \ldots, x_k]$ to be the k-th divided difference of Φ and assume that $\alpha_k\neq 0$ for all k. If $f_n(z) := \alpha_0 + \sum_{j=1}^{n} \alpha_j (z-x_0)\ldots(z-x_{j-1})$ denotes the Newton interpolation polynomial, then $q_n(z) := \alpha_n^{-1} f_n(z)$ satisfies $q_n = (\frac{\alpha_{n-1}}{\alpha_n} + z - x_{n-1})q_{n-1} - (z-x_{n-1})\frac{\alpha_{n-2}}{\alpha_{n-1}} q_{n-2}$, $n\geq 1$, $q_0=1$, $q_{-1}=0$, which is of type (7), if $z_n := z-x_{n-1}$, $\beta_n := \frac{\alpha_{n-1}}{\alpha_n} + x_n - x_{n-1}$, $\varepsilon_n := \frac{\alpha_{n-2}}{\alpha_{n-1}}$.

Then $\alpha := \inf_{n>0}\left\{\frac{\alpha_{n-1}}{\alpha_n} - \frac{\alpha_{n-2}}{\alpha_{n-1}} + x_n - x_{n-1}\right\}$.

If $\alpha > 0$ and $\{x_n\}$ is nondecreasing, then $P_n(z)\neq 0$ for $z-x_{n-1}\in P_\alpha$. If $\alpha>0$ and $\{x_n\}$ is nonincreasing, then $P_n(z)\neq 0$ for $z\in P_\alpha$.
Especially, if $\Phi(x)=e^x$ and $x_n=-nh$ with $h>0$, then $\alpha_k = \frac{1}{k!}(\frac{1-e^{-h}}{h})^k$.

Therefore, $f_n(z) := \sum_{k=0}^{n} \frac{1}{k!}(1-e^{-h})^k (\frac{z}{h})^k$ is $\neq 0$ for all n and all $z\in P_\alpha$, where $\alpha := \frac{h}{e^h-1}$.

8. Complex generalized Bessel polynomials

Similarly as Theorem 2 was derived from Theorem 1, now Theorem 11 yields (see [26] and (6))

Theorem 13. If $m+\delta+1$ lies on the parabola $|z|+\operatorname{Re} z = 2c>0$, then all zeros of $Y_m^{(\delta)}(z)$, $m \geq 2$, are contained in the open cardioid region

$$C=\{\zeta = re^{i\phi} \in \mathbb{C} : 0<r<(1+\cos\phi)/c,\ |\phi|<\pi\}.$$

For $m+\delta+1=c>0$ this again is Theorem 2.

9. Bessel functions and Lommel polynomials.

The ν-th Bessel function is defined by $J_\nu(z)=(z/2)^\nu \Phi_\nu(z)$, where

$$\Phi_\nu(z) := \sum_{n=o}^{\infty} (-1)^n (z/2)^{2n}/n!\Gamma(\nu+n+1),\ \nu, z\in\mathbb{C},\ \nu \neq -1, -2, \ldots .$$

Then (see [29])

$$\frac{\Phi_{\nu+1}(z)}{\Phi_\nu(z)} = \frac{1}{\nu+1} - \frac{(z/2)^2}{\nu+2} - \frac{(z/2)^2}{\nu+3} - \ldots \text{ or}$$

(8) $$(\nu+1)\frac{\Phi_{\nu+1}(z)}{\Phi_\nu(z)} = \frac{1}{1} - \frac{c_1c_2z^2}{1} - \frac{c_2c_3z^2}{1} - \ldots \text{ holds}$$

for $z\in\mathbb{C}$, where $c_n := 1/2(\nu+n)$, $n\geq 1$.

If the results of [29] are applied to the denominator polynomials of the continued fraction (8), then one obtains

Theorem 14. If $\nu=\nu_1+i\nu_2$, $\nu_1,\nu_2 \in \mathbb{R}$, satisfies $0 \leq \arg(\nu+1) < \pi$, then $\Phi_\nu(z) \neq 0$ holds for

(a) all $z=|z|e^{i\phi}$ satisfying $\arg(\nu+1) < \phi < \pi$ or $\arg(\nu+1) + \pi < \phi < 2\pi$,

(b) all $z \in \mathbb{C}$ satisfying $|\operatorname{Im} z| < \nu_2$ (if $\nu_2 > 0$) .

Furthermore,

(c) if $0 \leq \arg(\nu+1) < \pi/2$, then in addition to (a) and (b) $\Phi_\nu(z) \neq 0$ also holds for all $z\in\mathbb{C}$ satisfying $|\operatorname{Re} z| < ((\nu_1+1)(\nu_1+2))^{1/2}$

Remark The "large" zeros ζ of $\Phi_\nu(z)$ satisfy $\operatorname{Im}\zeta \approx \nu_2 \frac{\pi}{2}$ for $\operatorname{Re}\zeta > 0$ and $\operatorname{Im}\zeta \approx -\nu_2 \frac{\pi}{2}$ for $\operatorname{Re}\zeta < 0$.

Similar methods can be applied to the Lommel polynomials

$$g_{m,\nu}(z) := \sum_{0\le n\le m/2} (-1)^n \binom{m-n}{n} z^n \frac{\Gamma(\nu+m-n+1)}{\Gamma(\nu+n+1)}, \quad z,\nu\in\mathbb{C},\ \nu \neq -1,-2,\ldots,$$

which satisfy

$$\frac{g_{m-1,\nu}(z)}{g_{m,\nu}(z)} = \frac{1}{\nu+m} - \frac{z}{\nu+m-1} - \cdots - \frac{z}{\nu+1} \quad \text{, or}$$

$$(\nu+m)\frac{g_{m-1,\nu}(z)}{g_{m,\nu}(z)} = \frac{1}{1} - \frac{c_1c_2z}{1} - \cdots - \frac{c_{m-1}c_mz}{1} \quad ,$$

where $c_n := 1/(\nu+m-n+1)$, $1\le n\le m$.

After replacing z by $z^{1/2}$, a similar reasoning as the one which led to Theorem 14 yields (see [29])

Theorem 15. If $\nu=\nu_1+i\nu_2$, $\nu_1,\nu_2 \in \mathbb{R}$ satisfies $0\le\arg(\nu+1)<\pi$, and if $m\in\mathbb{N}$ is fixed, then $g_{m,\nu} \neq 0$ holds for

(a) all $z = |z|e^{i\phi}$ satisfying $2\arg(\nu+1)<\phi<2\pi+2\arg(\nu+m)$,

(b) all $z \in \mathbb{C}$ such that $|z| \le \operatorname{Re} z + \nu_2^2/2$.

Furtermore,

(c) if $0\le\arg(\nu+1) < \pi/2$, then in addition to (a) and (b) $g_{m,\nu}(z)\neq 0$ holds for all $z \in \mathbb{C}$ such that $|z| \le -\operatorname{Re} z + (\nu_1+1)(\nu_1+2)/2$.

10. Zero-free angular regions

We again assume that the polynomials $q_n(z)$ satisfy (3) with $\alpha_{n+1},\beta_n > 0$ for $n\ge 1$.
Then the following results are derived in [27].

Theorem 16. For fixed $N\ge 2$ put $B_N := \max_{1\le n\le N} \beta_n$, $Q_N := \max_{2\le n\le N} (\alpha_n/\beta_n)$.

Then $q_n(z) \neq 0$ for $1\le n\le N$ and all $z=re^{i\phi}$ satisfying $r=r(\phi)>0$ and

$$r(\phi) \ge 2\, B_N \frac{Q_N-\cos\phi}{1-\cos\phi} \quad (r(0) = +\infty \text{ if } Q_N > 1) .$$

Theorem 17. If $D_N := \min_{1\le n\le N} (\beta_n-\alpha_n)>0$ $(\alpha_1=0)$ and, hence, $Q_N\le 1-D_N/B_N<1$, then $q_n(z)\neq 0$ for $1\le n\le N$ and all $z=re^{i\phi}$ satisfying

$$\cos\phi \ge Q_N - \frac{D_N}{B_N} \text{ , i.e. } |\phi|\le\phi_N := \arccos\left(Q_N - \frac{D_N}{B_N}\right) .$$

These theorems immediately can be applied to the numerator - and denominator polynomials of Padé-approximants to the power series $\sum_{\nu=0}^{\infty} c_\nu z^\nu$, where all $A_m^{(n)}$ are > 0 which occur in Propositions 1 and 2.

Especially Theorems 16 and 17 can be applied to the partial sums $f_n(z) = \sum_{\nu=0}^{n} c_\nu z^\nu$, $1 \leq n \leq N$, provided $c_\nu > 0$, $0 \leq \nu \leq N$ and $\begin{vmatrix} c_{\nu-1} & c_{\nu-2} \\ c_\nu & c_{\nu-1} \end{vmatrix} > 0$ for $1 \leq \nu \leq N$ $(c_{-1}=0)$. Then $B_N = \max_{1 \leq \nu \leq N} c_{\nu-1}/c_\nu$,

$Q_N = \max_{2 \leq \nu \leq N} c_{\nu-2}\, c_\nu / c_{\nu-1}^2$ and $D_N = \min_{1 \leq \nu \leq N} \left[\frac{c_{\nu-1}}{c_\nu} - \frac{c_{\nu-2}}{c_{\nu-1}} \right] > 0$.

For $f(z) = e^z$ Theorem 17 yields (see [27]).

Theorem 18. Let $U_{m,n}(z)$ and $V_{m,n}(z)$ be the Padé-numerator and the Padé-denominator of $f(z)=e^z$. Then

(i) For fixed $n \geq 0$ and $M \geq 2$ $U_{m,n}(z) \neq 0$ holds for $1 \leq m \leq M$ and all $z = re^{i\phi}$ satisfying $\cos\phi \geq \frac{M-n-2}{M+n} = 1 - 2\,\frac{n+1}{n+M}$.

(ii) For fixed $m \geq 0$ and $N \geq 2$ $V_{m,n}(-z) \neq 0$ holds for $1 \leq n \leq N$ and all $z = re^{i\phi}$ satisfying $\cos\phi \geq \frac{N-m-2}{N+m} = 1 - 2\,\frac{m+1}{m+N}$.

This result was first proved by E.B. Saff and R.S. Varga in [31].

11. Complex orthogonal polynomials

We now consider polynomials $q_n(z)$ satisfying a recurrence relation (1) of the following special type

$$q_n(z) = (\alpha_n z + \beta_n) q_{n-1}(z) - \gamma_n q_{n-2}(z), \tag{9}$$

where $\alpha_n, \beta_n, \gamma_n \in \mathbb{C}$ and $\alpha_n \gamma_{n+1} \neq 0$ for $n \geq 1$. All classical orthogonal polynomials satisfy (9) with $\gamma_{n+1} > 0$, $\alpha_n, \beta_n \in \mathbb{R}$, $\alpha_n \neq 0$, $n \geq 1$.

The following result is proved in [28].

Theorem 19. Put $T_1 := 0$, $T_n := \gamma_n / \alpha_n \alpha_{n-1}$, $n \geq 2$,

$$F_n^+ := (-\beta_n/\alpha_n) + (T_{n+1})^{1/2} + (T_n)^{1/2},$$

$$F_n^- := (-\beta_n/\alpha_n) - (T_{n+1})^{1/2} - (T_n)^{1/2}, \quad n \geq 1,$$

where $(T_n)^{1/2}$ is chosen such that $0 \leq \arg(T_n)^{1/2} < \pi$ holds.

For each fixed $N \geq 2$ put $\sigma_N := \max_{1 \leq n \leq N} \arg(T_{n+1})^{1/2}$, where T_{N+1} can be replaced by an arbitrary number $\varepsilon_{N+1} \neq 0$, for example $\varepsilon_{N+1} > 0$ and arbitrarily small. Next, we define

$$S_n^+ := \{w \in \mathbb{C} : w = F_n^+ + z, \text{ where } z \neq 0 \text{ and } \pi \leq \arg z \leq \pi + \sigma_N\},$$

$$S_n^- := \{w \in \mathbb{C} : w = F_n^- + z, \text{ where } z \neq 0 \text{ and } 0 \leq \arg z \leq \sigma_N\},$$

and

$$\Delta_n := \{w \in \mathbb{C} : |w + \beta_n/\alpha_n| < |T_{n+1}|^{1/2} + |T_n|^{1/2}\}, \quad 1 \leq n \leq N.$$

If K_N^+ is the closed convex hull of $\bigcup_{n=1}^{N} S_n^+$ and K_N^- is the closed convex hull of $\bigcup_{n=1}^{N} S_n^-$, then all zeros of $q_1(z), \ldots, q_N(z)$ are contained in $K_N^+ \cap K_N^- \cap \left(\bigcup_{n=1}^{N} \Delta_n\right)$.

This theorem can be applied to the Laguerre polynomials $L_n^{(\alpha)}(z)$, which satisfy (9) with $\alpha_n = -1/n$, $\beta_n = (2n+\alpha-1)/n$, $\gamma_{n+1} = (n+\alpha)/(n+1)$, $n \geq 1$, and $\alpha \in \mathbb{C} \setminus \{-1,-2,\ldots\}$. Then

$$F_n^+ = 2n+\alpha-1+((n-1)(n+\alpha-1))^{1/2} + (n(n+\alpha))^{1/2},$$

$$F_n^- = 2n+\alpha-1-((n-1)(n+\alpha-1))^{1/2} - (n(n+\alpha))^{1/2}, \quad n \geq 1,$$

where $n^{1/2} > 0$ and $(n+\alpha)^{1/2}$ is chosen such that $0 \leq \arg(n+\alpha)^{1/2} < \pi$ holds. See also the recent results of E.A. van Doorn in [7], [8].

12. Bounds for the eigenvalues of complex tridiagonal matrices

Let $c_n, d_n, e_n \in \mathbb{C}$, $n \in \mathbb{N}$, be given and put

$$(10) \qquad A_n := \begin{pmatrix} c_1 & d_1 & & & 0 \\ e_1 & c_2 & \ddots & & \\ & \ddots & \ddots & \ddots & \\ & & \ddots & c_{n-1} & d_{n-1} \\ 0 & & & e_{n-1} & c_n \end{pmatrix}, \quad n \geq 2.$$

If I_n denotes the $n \times n$-identity matrix, then $q_n(z) := \det(z I_n - A_n)$, $n \geq 2$, satisfies (9) with $\alpha_n = 1$, $\beta_n = -c_n$, $\gamma_{n+1} = d_n e_n$, $n \geq 1$,

$q_1(z)=z-c_1,\ q_0=1,\ q_{-1}=0,\ d_0e_0=0$.

Remark. If $B\in\mathbb{C}^{n\times n}$, then a non-singular matrix $T\in\mathbb{C}^{n\times n}$ can be computed in finitely many steps, such that $TBT^{-1}=A$ is of Type (10), (see [18, ch.1]).

Now Theorem 19 yields (see [30])

Theorem 20. Assume that $c_n, d_n, e_n \in \mathbb{C}$ and $d_ne_n \neq 0$, $n\in\mathbb{N}$. For $n\in\mathbb{N}$ put (with $d_0e_0=0$)

$$F_n^+ := c_n+(d_ne_n)^{1/2}+(d_{n-1}e_{n-1})^{1/2} ,$$

$$F_n^- := c_n-(d_ne_n)^{1/2}-(d_{n-1}e_{n-1})^{1/2} ,$$

where for each $n\in\mathbb{N}$ $(d_ne_n)^{1/2}$ is chosen such that $0\leq\arg(d_ne_n)^{1/2}<\pi$ holds. For each fixed $N\geq 2$ put $\sigma_N := \max_{1\leq n\leq N} \arg(d_ne_n)^{1/2}$, where d_Ne_N can be replaced by an arbitrary number $\varepsilon_N \neq 0$, for example $\varepsilon_N>0$ and arbitrarily small. Next, define for $1\leq n\leq N$

$$S_n^+ := \{w\in\mathbb{C}: w=F_n^+ + z \text{ , where } z\neq 0 \text{ and } \pi\leq\arg z\leq\pi+\sigma_N\} ,$$

$$S_n^- := \{w\in\mathbb{C}: w=F_n^- + z \text{ , where } z\neq 0 \text{ and } 0\leq\arg z\leq\sigma_N\} ,$$

and

$$\Delta_n := \{w\in\mathbb{C}: |w-c_n| < |d_ne_n|^{1/2} + |d_{n-1}e_{n-1}|^{1/2}\} .$$

Finally, let K_N^+ and K_N^- denote the closed convex hull of $\bigcup_{n=1}^{N} S_n^+$ and $\bigcup_{n=1}^{N} S_n^-$ respectively.

Then all eigenvalues of $A_2,\ldots,A_N$ are contained in $K_N^+ \cap K_N^- \cap (\bigcup_{n=1}^{N} \Delta_n)$.

Especially, if $c_n \in \mathbb{R}$ and $d_ne_n>0$ holds for $1\leq n\leq N$, then all eigenvalues of $A_2,\ldots,A_N$ are real.

References

1. M.G. de Bruin, E.B. Saff, R.S. Varga, On the zeros of generalized Bessel polynomials. I,II, Nederl.Akad.Wetensch. Indag.Math. 43 (1981), 1-25.

2. M.G. de Bruin, Generalized Padé tables and some algorithms therein. In: Proceedings of the 1st French-Polish meeting on Padé approximation and convergence acceleration techniques. CPT-81/PE. 1354, Centre de Physique Théorique Marseille, 1982

3. M.G. de Bruin, Zeros of polynomials generated by 4-term recurrence relations. In: Proceedings of a Conference on Rational Approximation and Interpolation, Springer-Verlag, Berlin-New York, 1984, 331-345

4. " , Simultaneous Padé approximation and orthogonality. In: Polynômes Orthogonaux et Applications, Proceedings, Bar-le-Duc 1984, Lecture Notes in Mathematics No. 1171, Springer-Verlag, Berlin-New York, 1985, 74-83

5. " , On the zeros of recursively defined polynomials, Analysis 6 (1986), 227-236

6. K. Dočev, On the generalized Bessel polynomials, Bulgar.Akad.Nauk. Izv. Mat.Inst. 6 (1962), 89-94 (in Bulgarian)

7. E.A. Van Doorn, A note on orthogonal polynomials and oscillation criteria for second order linear difference equations, Memorandum Nr. 522, Technische Hogeschool Twente, May 1985

8. " , Representations and bounds for zeros of orthogonal polynomials and eigenvalues of sign-symmetric tridiagonal matrices, Memorandum Nr. 525, Technische Hogeschool Twente, June 1985.

9. G. Frobenius, Über Relationen zwischen den Näherungsbrüchen von Potenzreihen, J. für Mathematik, 90 (1881), 1-17

10. J. Gilewicz, Approximants de Padé, Lecture Notes in Math.No.667, Springer-Verlag, Berlin, New-York, 1978

11. J. Gilewicz, E. Leopold, Location of the zeros of polynomials satisfying three-term recurrence relations. I. General case with complex coefficients, J. Approx. Theory 43 (1985), 1-14

12. J. Gilewicz, E. Leopold, Location of the zeros of polynomials satisfying three-term recurrence relations. II. General case with complex coefficients, in praparation

13. J. Gilewicz, Sur l'amélioration des théorèmes de localisation des zéros de polynômes, Congrès d'Analyse Numérique 1984, Bombannes (France)

14. J. Gilewicz, E. Leopold, Fine optimization of the zero-free region for the polynomials satisfying three-term recurrence relations, submitted to J. Approx. Theory

15. J. Gilewicz, E. Leopold, On the sharpness of results in the theory of location of zeros of polynomials defined three-term recurrence relations. In: Polynômes Orthogonaux et Applications, Proceedings, Bar-le-Duc 1984, Lecture Notes in Math. No. 1171, Springer, Berlin, 1985, 259-266.

16. E. Grosswald, Besselpolynomials, Lecture Notes in Math. No. 698, Springer-Verlag, Berlin, New-York, 1978

17. P. Henrici, Note on a theorem of Saff and Varga, Padé and Rational Approximation, Theory and Applications, Academic Press, New York, 1977, 157-161.

18. A.S. Householder, The theory of matrices in numerical analysis, Dover publications, Inc., New York, 1975

19. W.B. Jones, W.J. Thron, Continued fractions, Analytic Theory and Applications, Encycl. of Math. and its Applics. Vol. 11, London-Amsterdam - Don Mills, Ontario-Sydney-Tokyo, Addison-Wesley Publ. Co. 1980

20. E. Leopold, Approximants de Padé pour les fonctions de classes S, et localisation des zéros de certains polynômes, Thesis, Université d'Aix-Marseille I, 1982

21. E. Leopold, Location of the zeros of polynomials satisfying three-term recurrence relations. III Positive coefficients case, J. Approx. Theory 43 (1985), 15-24

22. E. Leopold, Location of the zeros of polynomials satisfying three-term recurrence relations. IV Application to some polynomials and to generalized Bessel polynomials, in preparation

23. D.J. Newman, T.J. Rivlin, The zeros of partial sums of the exponential function, J. Approx. Theory, 5 (1972), 405-412.

24. " , Correction: The zeros of the partial sums of the exponential function, J. Approx. Theory 16 (1976), 299-300.

25. O. Perron, Die Lehre von den Kettenbrüchen, 3 rd ed., vol.2, Teubner, Stuttgart, 1957

26. H.-J. Runckel, Zero-free parabolic regions for polynomials with complex coefficients, Proc. AMS 88 (1983), 299-304

27. " , Zero-free regions for polynomials with applications to Padé-approximants, In: Constructive Theory of Functions, Proceedings of the International Conference on constructive theory of functions, Varna 1984, Publishing house of the Bulgarian Academy of Sciences, Sofia 1984, 767-771.

28. " , Zeros of complex orthogonal polynomials, In: Polynômes, Orthogonaux et Applications, Proceedings, Bar-le-Duc 1984, Lecture Notes in Mathematics, No. 1171, Springer-Verlag, Berlin, New York, 1985, 278-282.

29. " , Pole- and zero-free regions for analytic continued fractions, Proc. AMS, (1986), Proc. AMS 97 (1986), 114-120.

30. " , Bounds for the eigenvalues of complex tridiagonal matrices, Analysis 6 (1986), 251-253.

31. E.B. Saff, R.S. Varga, On the zeros and poles of Padé approximants to e^z, Numer. Math. 25 (1975), 1-14

32. " , Zero-free parabolic regions for sequences of polynomials, SIAM J. Math. Anal. 7 (1976), 344-357

33. " , On the sharpness of theorems concerning zero-free regions for certain sequences of polynomials, Numer. Math. 26 (1976), 345-354

34. H.S. Wall, Analytic Theory of Continued Fractions, Van Nostrand, New York, 1948

RATIONAL APPROXIMATION AND INTERPOLATION OF FUNCTIONS BY BRANCHED CONTINUED FRACTIONS

Khristina I. Kuchminskaya
Institute of Applied Problems of Mechanics
and Mathematics, Ukrainian Acad. of Sciences
290047 Lvov, 3-b Naukova ul., USSR

Wojciech Siemaszko
Politechnika Rzeszowska
ul. Poznańska 2, P. O. Box 85
35-084 Rzeszów , Poland

1. INTRODUCTION.

It is well known that if we define a set of transformations of the form

$$T_0(z) = b_0 + z \quad , \quad T_k(z) = \frac{a_k}{b_k + z} \quad , \quad k=1,2,\ldots$$

then their compositions

$$\frac{P_0}{Q_0} = T_0(0) \ , \ \frac{P_1}{Q_1} = T_0(T_1(0)), \ \ldots \ , \ \frac{P_k}{Q_k} = T_0(T_1(\ldots T_k(0)\ldots)), \ \ldots$$

form a sequence of convergents of a continued fraction

$$b_0 + \cfrac{a_1}{b_1 + \cfrac{a_2}{b_2 + \ddots}}$$

The above construction will be easily generalized. Let b_0, $b_{i_1 i_2 \ldots i_k}$, $a_{i_1 i_2 \ldots i_k}$ be complex numbers and let $z_{i_1 \ldots i_k}$, $k=1,2,\ldots$ be complex variables. If we define now a set of transformations of the form

$$T_0(z_1, z_2, \ldots, z_N) = b_0 + z_1 + \ldots + z_N \ ,$$

$$T_{i_1}(z_{i_1 1}, z_{i_1 2}, \dots, z_{i_1 N}) = \frac{a_{i_1}}{b_{i_1} + z_{i_1 1} + z_{i_1 2} + \dots + z_{i_1 N}}, \dots$$

$$\dots, T_{i_1 i_2 \dots i_k}(z_{i_1 i_2 \dots i_k 1}, \dots, z_{i_1 i_2 \dots i_k N}) =$$

$$= \frac{a_{i_1 i_2 \dots i_k}}{b_{i_1 \dots i_k} + z_{i_1 \dots i_k 1} + \dots + z_{i_1 \dots i_k N}},$$

$i_k = 1, \dots, N$, $k = 1, 2, \dots$ Then we have

$$\frac{P_0}{Q_0} = T_0(0, 0, \dots, 0) = b_0$$

$$\frac{P_1}{Q_1} = T_0(T_1(0), T_2(0), \dots, T_N(0)) = b_0 + \sum_{i_1=1}^{N} \frac{a_{i_1}}{b_{i_1}},$$

$$\frac{P_2}{Q_2} = T_0(T_1(T_{11}(0), T_{12}(0), \dots, T_{1N}(0)), T_2(T_{21}(0), T_{22}(0), \dots,$$

$$\dots, T_{2N}(0)), \dots, T_N(T_{N1}(0), T_{N2}(0), \dots, T_{NN}(0))) =$$

$$= b_0 + \sum_{i_1=1}^{N} \frac{a_{i_1}}{b_{i_1} + \sum_{i_2=1}^{N} \frac{a_{i_1 i_2}}{b_{i_1 i_2}}}.$$

Continuing the above process k times we obtain an expression of the form

$$\frac{P_k}{Q_k} = b_0 + \sum_{i_1=1}^{N} \frac{a_{i_1}}{b_{i_1} + \sum_{i_2=1}^{N} \frac{a_{i_1 i_2}}{b_{i_1 i_2} + \ddots + \sum_{i_k=1}^{N} \frac{a_{i_1 \dots i_k}}{b_{i_1 \dots i_k}}}}.$$

In accordance with the theory of ordinary continued fractions we call P_k/Q_k , $k=0,1,\dots$ convergents of the continued fraction

$$b_0 + \sum_{i_1=1}^{N} \cfrac{a_{i_1}}{b_{i_1} + \sum_{i_2=1}^{N} \cfrac{a_{i_1 i_2}}{b_{i_1 i_2} + \ddots + \sum_{i_k=1}^{N} \cfrac{a_{i_1 \ldots i_k}}{b_{i_1 \ldots i_k} + \ddots}}}$$

This continued fraction will be written in a more compact form as

$$b_0 + \sum_{i_1=1}^{N} \frac{a_{i_1}}{b_{i_1}} + \sum_{i_2=1}^{N} \frac{a_{i_1 i_2}}{b_{i_1 i_2}} + \ldots + \sum_{i_k=1}^{N} \frac{a_{i_1 \ldots i_k}}{b_{i_1 \ldots i_k}} + \ldots$$

and because of its tree-like structure we will call it a branched continued fraction with N branches. We will write it shortly BCF.

The idea of constructing continued fractions of the above described form comes from V. Ya. Skorobogat'ko. It appeared for the first time in 1967 [17] . As the author writes in the preface to his second book on BCFs [18] , it was the result of generalization of Chaplygin method for solving differential equations.

During the last twenty years, theory of branched continued fractions has been intensively developed by V. Ya. Skorobogat'ko and his pupils. Their main results refer to :

a) general theory of BCFs - convergence criteria, value region problems etc. ,

b) applications of BCFs tree-like scheme for solutions of ordinary and partial differential equations,

c) applications of BCFs in the number theory (particularly for solving Diophante equations) ,

d) solution of systems of linear equations ,

e) the theory of Markov processes,

f) approximation and interpolation of multivariate functions.

These results are collected in three books written by V. Ya. Skorobogat'ko, P. I. Bodnarcuk and D. I. Bodnar [1] , [2] , [18] and for detailed references we refer to these books. In the next paragraphs we should like to present results concerning application of branched continued fractions to rational approximation and interpolation of multivariate functions. Some particular problems in this field were treated independently by J. Murphy and M. O'Donohe [14], and by A. Cuyt and B. Verdonk [3],[4].

2. BRANCHED CONTINUED FRACTIONS WITH POLYNOMIAL COEFFICIENTS.

In general, coefficients of BCF

$$b_0 + \sum_{i_1=1}^{N} \frac{a_{i_1}}{b_{i_1}} + \sum_{i_2=1}^{N} \frac{a_{i_1 i_2}}{b_{i_1 i_2}} + \dots \tag{2.1}$$

are real or complex functions defined in domain $D \subset \mathbb{R}^M$ or $D \subset \mathbb{C}^M$. If a's and b's coefficients are M-variable polynomials then convergents of this BCF are M-variable rational functions. Obviously, choosing various types of polynomials we will obtain various forms of BCFs not equivalent in the sense that there is no transformation of variables transforming one special form of BCF into another.

Four particular forms of our choice of corresponding polynomials seem to be of special interest. We are going to present these forms restricting our considerations to two-variable case (M=2) for the sake of simplicity of notations.

(a) The first interesting form of BCF with polynomial coefficients will be obtained from (2.1) if we put N=2 and

$$a_1 = x - x_0,\ a_2 = y - y_0,\ \dots,\ a_{i_1 \dots i_k 1} = x - x_k,\ a_{i_1 \dots i_k 2} = y - y_k \tag{2.2}$$

for $i_j = 1,2$, $j = 1,2,\dots$, $k = 1,2,\dots$. Here and further we assume that the sequences of points x_k and y_k, $k = 0,1,\dots$ are given. We will write BCF occuring in this case as ([2],[7])

$$B_1(x,y) = b_0 + \cfrac{x - x_0}{b_1 + \cfrac{x - x_1}{b_{11} + \cdots} + \cfrac{y - y_1}{b_{12} + \cdots}} + \cfrac{y - y_0}{b_2 + \cfrac{x - x_1}{b_{21} + \cdots} + \cfrac{y - y_1}{b_{22} + \cdots}} \tag{2.3}$$

(b) The second form of BCF considered here can be written as ([7], [14], [18]) :

$$B_2(x,y) = K_0 + \cfrac{(x - x_0)(y - y_0)}{K_1 + \cfrac{(x - x_1)(y - y_1)}{K_2 + \cdots}} \tag{2.4}$$

where

$$K_k = b_{kk} + \cfrac{x-x_k}{b_{k+1,k} + \cfrac{x-x_{k+1}}{b_{k+2,k} + \ddots}} + \cfrac{y-y_k}{b_{k,k+1} + \cfrac{y-y_{k+1}}{b_{k,k+2} + \ddots}} \qquad (2.5)$$

If we put $K_k(0)=b_{kk}$ and for $l=1,2,\dots$

$$K_k(l) = b_{kk} + \cfrac{x - x_k}{b_{k+1,k} + \cdots + \cfrac{x-x_{k+l-1}}{b_{k+l,k}}} + \cfrac{y - y_k}{b_{k,k+1} + \cdots + \cfrac{y-y_{k+l-1}}{b_{k,k+l}}} \qquad (2.6)$$

then the n-th convergent of BCF (2.4) will be written in a compact form as

$$\frac{P_n}{Q_n} = K_0(n) + \frac{(x-x_0)(y-y_0)}{K_1(n-2)} + \dots + \frac{(x-x_i)(y-y_i)}{K_{i+1}(n-2i-2)} + \dots$$
$$\dots + \frac{(x-x_{[\frac{n}{2}]-1})(y-y_{[\frac{n}{2}]-1})}{K_{[\frac{n}{2}]}(n-2[\frac{n}{2}])} \qquad (2.7)$$

where $[x]$ is an entier function.

(c) The third form of BCF can be defined as follows (15) :

$$B_3(x,y) = K_0 + \cfrac{x-x_0}{K_1 + \cfrac{x-x_1}{K_2 + \ddots}} + \cfrac{y-y_0}{L_1 + \cfrac{y-y_1}{L_2 + \ddots}} \qquad (2.8)$$

where for $k=0,1,\dots$

$$K_k = b_{k0} + \frac{(x-x_k)(y-y_k)}{b_{k+1,1}} + \frac{(x-x_{k+1})(y-y_{k+1})}{b_{k+2,2}} + \dots \qquad (2.9)$$

$$L_k = b_{0k} + \frac{(x-x_k)(y-y_k)}{b_{1,k+1}} + \frac{(x-x_{k+1})(y-y_{k+1})}{b_{k+2,2}} + \dots \qquad (2.10)$$

Denoting by $K_k(l)$ and $L_k(l)$, $l=0,1,\dots$ l-th convergents of ordinary continued fractions K_k and L_k we will define the n-th convergent of BCF $B_3(x,y)$ as

$$\frac{P_n}{Q_n} = K_0(\tfrac{n}{2}) + \left(\frac{x-x_0}{K_1([\frac{n-1}{2}])} + \dots + \frac{x-x_i}{K_{i+1}([\frac{n-i-1}{2}])} + \dots + \frac{x-x_{n-1}}{K_n(0)} \right) +$$

$$+ \left(\frac{y-y_0}{L_1([\frac{n-1}{2}])} + \dots + \frac{y-y_i}{L_{i+1}([\frac{n-i-1}{2}])} + \dots + \frac{y-y_{n-1}}{L_n(0)} \right) \qquad (2.11)$$

(d) The last useful form of BCF considered here is ([2], [18], [16])

$$B_4(x,y) = T_0 + \frac{x-x_0}{T_1} + \frac{x-x_1}{T_2} + \dots \qquad (2.12)$$

where

$$T_k = b_{k0} + \frac{y-y_0}{b_{k1}} + \frac{y-y_1}{b_{k2}} + \dots \qquad . \qquad (2.13)$$

Now the n-th convergent of this BCF is defined as

$$\frac{P_n}{Q_n} = T_0(n) + \frac{x-x_0}{T_1(n-1)} + \dots + \frac{x-x_i}{T_{i+1}(n-i-1)} + \dots + \frac{x-x_{n-1}}{T_n(0)}$$

where $T_k(l)$, $k,l=0,1,\dots$ is the l-th convergent of the ordinary continued fraction (2.13).

3. INTERPOLATION WITH BRANCHED CONTINUED FRACTIONS.

The following two-variable rational interpolation problem will be stated :

Let function $f(x,y)$ be defined in the domain D containing a triangular grid $G = \{(x_i,y_j) \in \mathbb{R}^2 : 0 \leq i+j \leq n\}$; we are seeking a rational two-variable function

$$R_n(x,y) = \frac{P_n(x,y)}{Q_n(x,y)} \qquad (3.1)$$

where $P_n(x,y)$ and $Q_n(x,y)$ are polynomials, and such that

$$f(x_i,y_j) - \frac{P_n(x_i,y_j)}{Q_n(x_i,y_j)} = 0 \qquad (3.2)$$

for all $(x_i,y_j) \in G$.

This problem cannot be solved so easily as it is done for one-variable functions. If we assume that $R_n(x,y)$ fulfilling (3.2) is uniquely determined by the same number of coefficients that is the number of points in the grid G , then the difficulty with prescribing degrees of polynomials $P_n(x,y)$ and $Q_n(x,y)$ arise.

Having in mind close connections between Thiele continued fraction and one-variable rational interpolation problem we will seek $R_n(x,y)$ as convergent of BCF with suitably chosen coefficients. Let us notice that convergents of (b), (c) and (d) branched continued fractions are uniquely determined by their $\frac{(n+1)(n+2)}{2}$ coefficients. This number is equal to the number of points occuring in the grid G.

Convergents of (b), (c) and (d) branched continued fractions posses one more useful property. Namely, if we arrange their coefficients in a triangle array

$$\begin{array}{lllll} b_{00} & b_{10} & \cdots & b_{n-1,0} & b_{n,0} \\ b_{01} & b_{11} & \cdots & b_{n-1,1} & \\ \vdots & \vdots & \cdot^{\cdot^{\cdot}} & & \\ b_{0,n-1} & b_{1,n-1} & & & \\ b_{0,n} & & & & \end{array}$$

then it is easy to see that value of the n-th convergent P_n/Q_n in the given point (x_k,y_l), $0 \leqslant k+l \leqslant n$ depends only on coefficients occupying a rectangular block

$$\begin{array}{lll} b_{00} & \cdots & b_{k0} \\ \vdots & & \vdots \\ b_{0l} & \cdots & b_{kl} \end{array}$$

Therefore, for the given data $f_{ij}=f(x_i,y_j)$, $0 \leqslant i+j \leqslant n$, if values of b_{ij}, $i \leqslant k$, $j \leqslant l$, except b_{kl} are known then we will find value of b_{kl}. Consecutive coefficients b_{kl} will be calculated according to the following recursive scheme

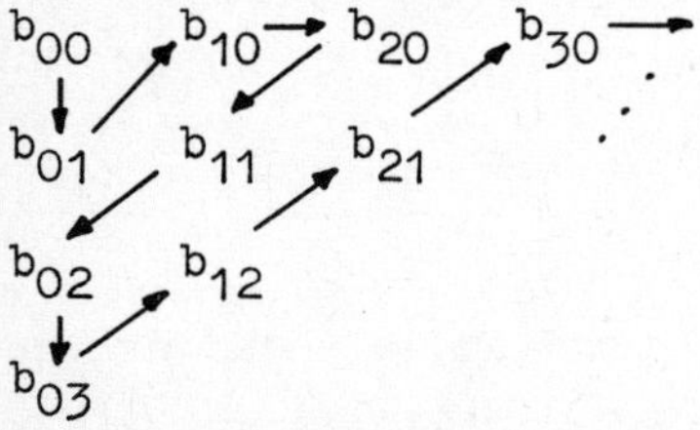

As an illustrative example, let us consider BCF of the form (b) and let all coefficients b_{ij}, $i \leqslant k$, $j \leqslant l$ be known except the value b_{kl}. If $k > l$ (cases when $k < l$ or $k=l$ are similar) then

$$f_{kl} = \frac{P_n(x_k,y_l)}{Q_n(x_k,y_l)} = K_0(x_k,y_l) + \frac{(x_k-x_0)(y_l-y_0)}{K_1(x_k,y_l)} + \dots + \qquad (3.3)$$

$$\dots + \frac{(x_k-x_{l-1})(y_l-y_{l-1})}{K_l(x_k,y_l)}$$

and

$$K_i(x_k,y_l) = b_{ii} + \left(\frac{x_k-x_i}{b_{i+1,i}} + \dots + \frac{x_k-x_{k-1}}{b_{ki}} \right) +$$

$$+ \left(\frac{y_l-y_i}{b_{i,i+1}} + \dots + \frac{y_l-y_{l-1}}{b_{i,l}} \right)$$

$i=0,1,\dots,l$. Thus

$$K_l(x_k,y_l) = b_{ll} + \frac{x_k-x_l}{b_{l+1,l}} + \dots + \frac{x_k-x_{k-1}}{b_{kl}} \qquad (3.4)$$

and from (3.3)

$$K_l(x_k,y_l) = \frac{(x_k-x_{l-1})(y_l-y_{l-1})}{-K_{l-1}(x_k,y_l)} + \frac{(x_k-x_{l-2})(y_l-y_{l-2})}{-K_{l-2}(x_k,y_l)} + \dots$$

$$\dots + \frac{(x_k-x_1)(y_l-y_1)}{-K_1(x_k,y_l)} + \frac{(x_k-x_0)(y_l-y_0)}{f_{kl}-K_0(x_k,y_l)} \qquad (3.5)$$

Now from (3.4) we have

$$b_{kl} = \frac{x_k-x_{k-1}}{-b_{k-1,l}} + \frac{x_k-x_{k-2}}{-b_{k-2,l}} + \dots + \frac{x_k-x_l}{K_l(x\ ,y\ ) - b_{ll}} \qquad (3.6)$$

Formula (3.6) will be simplified. Let us introduce the following notations

$$\varphi_{00}(x_i;y_j) = f_{ij}$$

$$\varphi_{10}(x_0,x;y_0) = \frac{x - x_0}{f(x,y_0) - f(x_0,y_0)}$$

$$\varphi_{01}(x\ ;y_0,y) = \frac{y - y_0}{f(x_0,y) - f(x_0,y_0)}$$

$$\varphi_{11}(x_0,x;y_0,y) = \frac{(x-x_0)(y-y_0)}{f(x,y) - f(x_0,y) + f(x_0,y_0) - f(x,y_0)}$$

$$\varphi_{ij}(x_0,x_1,\dots,x_{i-1};y_0,y_1,\dots,y_{j-1},y) =$$

$$= \begin{cases} (x-x_{i-1})\{\varphi_{i-1,j}(x_0,x_1,\dots,x_{i-2},x;y_0,y_1,\dots,y_{j-1},y) - \\ \quad - \varphi_{i-1,j}(x_0,x_1,\dots,x_{i-1};y_0,y_1,\dots,y_{j-1},y)\}^{-1} & \text{for } i>j \\ (y-y_{j-1})\{\varphi_{i,j-1}(x_0,x_1,\dots,x_{i-1},x;y_0,\dots,y_{j-2},y) - \\ \quad - \varphi_{i,j-1}(x_0,\dots,x_{i-1},x;y_0,\dots,y_{j-1})\}^{-1} & \text{for } i<j \\ (x-x_{i-1})(y-y_{i-1})\{\varphi_{i-1,i-1}(x_0,\dots,x_{i-2},x;y_0,\dots,y_{i-2},y) - \\ -\varphi_{i-1,i-1}(x_0,\dots,x_{i-1};y_0,\dots,y_{i-2},y) + \\ \quad +\varphi_{i-1,i-1}(x_0,\dots,x_{i-1};y_0,\dots,y_{i-1}) - \\ \quad -\varphi_{i-1,i-1}(x_0,\dots,x_{i-2},x;y_0,\dots,y_{i-1})\}^{-1} & \text{for } i=j \end{cases}$$

where $i,j=0,1,\dots$ and $\varphi_{-1,j}=\varphi_{i,-1}=\varphi_{-1,-1}=0$.

We have the following lemma ([7], [9], [18], [4]).

LEMMA. If all quantities $\varphi_{k,l}(x_0,x_1,\dots,x_k;y_0,y_1,\dots,y_l)$, $0\leqslant k+l\leqslant n$ exist then for BCF of the type (b) with coefficients equal

$$b_{kl} = \varphi_{k,l}(x_0,x_1,\dots,x_k;y_0,y_1,\dots,y_l) \tag{3.7}$$

equation (3.2) is fulfilled.

Quantities $\varphi_{k,l}$ are called partial reciprocal differences. The n-th convergent $P_n(x,y)/Q_n(x,y)$ of BCF of the type (b) with coefficients fulfilling (3.7) is now a rational function solving the problem of rational two-variable interpolation.

As we mentioned earlier, similar results will be obtained for other types of BCFs. Corresponding formulas for partial reciprocal differences for BCF of the type (d) are given in [7], [18], [16]. More complex situation occurs in the case of BCFs of the type (a). The n-th convergent of this fraction depends on $(n+1)^2$ coefficients and additional values of $f(x,y)$ are required. Originally, partial reciprocal differences for this BCF were defined for the set of values $\{f(x_i,y_j) : 0\leqslant i\leqslant n+1,\ 0\leqslant j\leqslant n+1\}$. Resulting convergent, however, interpolates $f(x,y)$ only on the triangular grid G defined at the beginning of this section.

Partial reciprocal differences are generalizations of reciprocal differences (Hildebrand [5], p.507). They appeared for the first time in [2] and were obtained by Kh. I. Kuchminskaya. Initially they were defined for branched continued fractions of the type (a). Modifications necessary for BCFs of the type (b),(d) were done also by Kh. I. Kuchminskaya in her Thesis [7].

In the end of this section let us notice that the following theorem will be proved.

THEOREM.([18], p.189, [16]).

Let $\frac{P_n(x,y)}{Q_n(x,y)}$ denote the n-th convergent of BCF of the types (a),(b), (c) or (d) interpolating given values

$$f_{ij} = f(x_i,y_j) , \quad 0 \leqslant i+j \leqslant n$$

at every point of the triangular grid $G = \{(x_i,y_j) : 0 \leqslant i+j \leqslant n\}$.
Then

(i) $\frac{P_n(x,y_0)}{Q_n(x,y_0)}$ is the n-th convergent of Thiele continued fraction interpolating values $f_{i,0} = f(x_i,y_0)$, $i=0,1,\dots,n$ at points $x_0,x_1,\dots,x_n$;

(ii) $\frac{P_n(x_0,y)}{Q_n(x_0,y)}$ is the n-th convergent of Thiele continued fraction interpolating values $f_{0,j} = f(x_0,y_j)$, $j=0,1,\dots,n$ at points $y_0,y_1,\dots,y_n$;

(iii) if $F(x,y)=f(x,y)Q_n(x,y) - P_n(x,y)$, $f \in \mathcal{C}^{n+m+2}(D)$, $G \subset D$ then for every compact set $K \subset D$, $(x,y) \in K$ we have

$$|f(x,y)Q_n(x,y) - P_n(x,y)| \quad \frac{1}{n+1} \sum_{k+l=0}^{n} \beta_{k,l} \tag{3.8}$$

where

$$\beta_{k,l} = \frac{h_x^{k+1}}{4(k+1)} \left\| \frac{\partial^{k+1} F}{\partial x^{k+1}} \right\|_K + \frac{h_y^{l+1}}{4(l+1)} \left\| \frac{\partial^{l+1} F}{\partial y^{l+1}} \right\|_K +$$

$$+ \frac{h_x^{k+1} \, h_y^{l+1}}{16(k+1)(l+1)} \left\| \frac{\partial^{k+l+2} F}{\partial x^{k+1} \partial y^{l+1}} \right\|_K ,$$

$_K$ is the sup norm on K and

$$h_x = \max \{|x_{i+1}-x_i| , \ i=0,1,\dots,(n-1)\} ,$$

$$h_y = \max \{|y_{j+1}-y_j| , \ j=0,1,\dots,(n-1)\} .$$

Inequality (3.8) will be used as an error formula for interpolation with convergents of branched continued fractions.

4. APPROXIMATION OF FUNCTIONS WITH BRANCHED CONTINUED FRACTIONS.

Since branched continued fractions are natural generalizations of ordinary continued fractions we will expect that they will be used for approximation of multivariate functions in the way similar to that realized by Padé approximants for one variable functions. Therefore, we will expect that for convergents $P_n(x,y)/Q_n(x,y)$ of BCFs with suitably chosen coefficients the following relation will be obtained

$$f(x,y)Q_n(x,y) - P_n(x,y) = \sum_{k+l>n} d_{kl}\, x^k y^l \tag{4.1}$$

if $f(x,y)$ is a double power series

$$f(x,y) = \sum_{k,l=0}^{\infty} c_{kl}\, x^k y^l \quad . \tag{4.2}$$

In fact, coefficients of BCFs approximating the given power series in the sense of relation (4.1) can be found. It will be done in two ways. The first one is similar to the Viscovatoff algorithm for expanding the given power series into corresponding continued fraction while the second one is obtained as the limit case of interpolation , when all knots of interpolation (x_i, y_j) , $i,j=0,1,\ldots$ tend to zero.

At the beginning we will show how Viscovatoff-like algorithm can be built for each type of BCF. Therefore, let us notice, that the given two-variable power series (4.2) will be decomposed as follows :

(a) $$f(x,y) = c_{00} + x\sum_{k,l=0}^{\infty} g_{kl}\, x^k y^l + y\sum_{k,l=0}^{\infty} h_{kl}\, x^k y^l \tag{4.3}$$

where $g_{k.l+1} + h_{k+1,l} = c_{k+1,l+1}$, $k,l=0,1,\ldots$;

(b) $$f(x,y) = \Big(c_{00} + \sum_{k=1}^{\infty} c_{k,0}\, x^k + \sum_{k=1}^{\infty} c_{0,k}\, y^k \Big) + $$

$$+ \ xy\sum_{k,l=0}^{\infty} g_{kl}\, x^k y^l \tag{4.4}$$

where $g_{kl} = c_{k+1,l+1}$, $k,l=0,1,\ldots$;

(c) $$f(x,y) = \sum_{k=0}^{\infty} c_{kk}(xy)^k + x\sum_{k\geq l} g_{kl}\, x^k y^l + y\sum_{k\leq l} h_{kl}\, x^k y^l \tag{4.5}$$

where $g_{kl} = c_{k+1,l}$, $h_{kl} = c_{k,l+1}$, $k,l=0,1,\ldots$;

(d) $$f(x,y) = \sum_{k=0} c_{0k}\, y^k + x\sum_{k,l=0} g_{kl}\, x^k y^l \tag{4.6}$$

where $g_{kl} = c_{k+1,l}$, $k,l=0,1,\ldots$.

In each case a new double power series appears and if only coefficients g_{00} and h_{00} of these series do not vanish, then we will find double power series that are reciprocal in the sense that we will calculate coefficients $\tilde{g}_{kl}$ and $\tilde{h}_{kl}$ such that

$$\left(\sum_{k,l=0}^{\infty} g_{kl} x^k y^l \right) \left(\sum_{k,l=0}^{\infty} \tilde{g}_{kl} x^k y^l \right) = 1$$

$$\left(\sum_{k,l=0}^{\infty} h_{kl} x^k y^l \right) \left(\sum_{k,l=0}^{\infty} \tilde{h}_{kl} x^k y^l \right) = 1 .$$

Therefore, if corresponding reciprocal series exist then we have :

in case of (a)

$$f(x,y) = c_{00} + \sum_{k,l=0}^{\infty} \frac{x}{\tilde{g}_{kl} x^k y^l} + \sum_{k,l=0}^{\infty} \frac{y}{\tilde{h}_{kl} x^k y^l} \quad ; \qquad (4.7)$$

in case of (b)

$$f(x,y) = \left(c_{00} + \sum_{k=0}^{\infty} c_{k0} x^k + \sum_{l=0}^{\infty} c_{0l} y^l \right) + \frac{xy}{\sum\limits_{k,l=0}^{\infty} \tilde{g}_{kl} x^k y^l} \qquad (4.8)$$

in case of (c)

$$f(x,y) = \sum_{k=0}^{\infty} c_{kk}(xy)^k + \frac{x}{\sum\limits_{k<l} \tilde{g}_{kl} x^k y^l} + \frac{y}{\sum\limits_{k>l} \tilde{h}_{kl} x^k y^l} \qquad (4.9)$$

in case of (d)

$$f(x,y) = \sum_{k=0}^{\infty} c_{0k} y^k + \frac{x}{\sum\limits_{k,l=0}^{\infty} g_{kl} x^k y^l} \quad . \qquad (4.10)$$

Repeating the above decompositions and expanding the resulting power series of variables x , y or $t=(xy)$ into corresponding ordinary continued fractions we obtain the expansins of the power series (4.2) into corresponding types of BCFs.

The above procedures work as far as suitable reciprocal series exist. However, it is difficult to find explicite formulas for coefficients of BCFs of each type approximating $f(x,y)$ in the sense of equation (4.1). Partial results are published in [10] , [14] , [15] , [18]. Calculation of coefficients of approximating BCFs as limits of partial reciprocal differences occuring for interpolating BCFs is much more efficient.

Let

$$\varphi^i_{k,l}(x_0,x_1,\dots,x_k;y_0,y_1,\dots,y_l) \quad , \ k,l=0,1,\dots \qquad (4.11)$$

denote the (k,l)-th partial reciprocal difference defined for a branched continued fraction of the type (b) for i=2, of the type (c) for i=3 and of the type (d) for i=4. If the following limits exist

$$\Phi^i_{k,l}(\bar{x},\bar{y}) = \qquad (4.12)$$

$$= \lim_{x_0,x_1,\dots,x_k\to\bar{x}} \ \lim_{y_0,y_1,\dots y_l\to\bar{y}} \left\{\varphi^i_{k,l}(x_0,x_1,\dots,x_k;y_0,\dots,y_l)\right\}$$

then they are called partial reciprocal derivatives at a point $(\bar{x},\bar{y})$.

If we put $x_i=0$, $y_i=0$, i=0,1,... for BCF (2.4), (2.8) or (2.12) and moreover if all corresponding partial reciprocal derivatives exist and

$$b_{kl} = \Phi^i_{k,l}(0,0) \quad , \ k,l=0,1,\dots \qquad (4.13)$$

for the arbitrary i , i=2,3,4, then the following lemma will be proved.

<u>LEMMA</u>. If f(x,y) is a function analytic in the neighbourhood of the origin and $P^i_n(x,y)/Q^i_n(x,y)$, i=2,3,4 denotes the n-th convergent of BCF of the type (b), (c) or (d) with coefficients given by (4.13) then condition (4.1) holds.

More complex but similar situation occurs for BCF of the type (a). Definition of corresponding partial reciprocal derivatives will be found in [2, p.235] and [18, p.191] .

Although definition of partial reciprocal derivatives seems to be not very easy for practical use, it is possible to find algorithms for calculating values of these derivatives. We will illustrate it on partial reciprocal derivatives for BCF of the type (d) ([2],[7],[18], [16]).

For BCF of the type (d) partial reciprocal differences are defined as :

$$\varphi_{0,0}(x_i;y_k) = f_{ik}$$

$$\varphi_{1,0}(x_i,x_j;y_k) = \frac{x_i-x_j}{\varphi_{0,0}(x_i;y_k) - \varphi_{0,0}(x_j;y_k)}$$

and for $s>1$

$$\varphi_{s+1,0}(x_{p_1},\dots,x_{p_s},x_i,x_j;y_k) =$$

$$= \frac{x_i - x_j}{\varphi_{s,0}(x_{p_1},\dots,x_{p_s},x_i;y_k) - \varphi_{s,0}(x_{p_1},\dots,x_{p_s},x_j;y_k)} \quad .$$

Similarly

$$\varphi_{s,1}(x_{p_0},\dots,x_{p_s};y_k,y_l) =$$

$$= \frac{y_k - y_l}{\varphi_{s,0}(x_{p_0},\dots,x_{p_s};y_k) - \varphi_{s,0}(x_{p_0},\dots,x_{p_s};y_l)}$$

and for $r > 1$

$$\varphi_{s,r+1}(x_{p_0},\dots,x_{p_s};y_{q_1},\dots,y_{q_r},y_k,y_l) =$$

$$= \frac{y_k - y_l}{\varphi_{s,r}(x_{p_0},\dots,x_{p_s};y_{q_0},\dots,y_{q_r},y_k) - \varphi_{s,r}(x_{p_0},\dots,x_{p_s};y_{q_0},\dots,y_{q_r},y_l)}$$

Values of partial reciprocal differences depend on the order of points x_i and y_j. We will symmetrize these differences by defining quantities $\psi_{k,l}$ as follows

$$\psi_{0,0}(x,y) = \varphi_{0,0}(x,y)$$

$$\psi_{1,0}(x_1,x_2;y) = \varphi_{1,0}(x_1,x_2;y)$$

$$\psi_{p,0}(x_0,\dots,x_p;y) = \varphi_{p,0}(x_0,\dots,x_p;y) - \psi_{p-2,0}(x\ ,\dots,x_{p-2};y)$$

$$\psi_{p,1}(x_0,\dots,x_p;y_1,y_2) = \frac{y_0 - y_1}{\psi_{p,0}(x_0,\dots,x_p;y\ ) - \psi_{p,0}(x_0,\dots,x_p;y_1)}$$

$$\psi_{p,r}(x_0,\dots,x_p;y_0,\dots,y_r) = \varphi_{p,r}(x_0,\dots,x_p;y_0,\dots,y_r) -$$

$$- \psi_{p,r-2}(x_0,\dots,x_p;y_0,\dots,y_{r-2})$$

$p \geqslant 0$, $r \geqslant 2$.

If we now define $\Psi_{k,l}(\bar{x},\bar{y})$ as

$$\Psi_{k,l}(\bar{x},\bar{y}) =$$

$$= \lim_{x_0,\dots x_k \to \bar{x}} \ \lim_{y_0,\dots,y_l \to \bar{y}} \left\{\psi_{k,l}(x_o,\dots,x_k;y_0,\dots,y_l)\right\}$$

then we have following relations

$$\Psi_{0,0}(\bar{x},\bar{y}) = f(\bar{x},\bar{y})$$

$$\Psi_{1,0}(\bar{x},\bar{y}) = \frac{1}{\frac{\partial f}{\partial x}(\bar{x},\bar{y})} \quad , \qquad {}_{0,1}(\bar{x},\bar{y}) = \frac{1}{\frac{\partial f}{\partial y}(\bar{x},\bar{y})}$$

and for $k > 1$, $l > 1$

$$\Psi_{k,0}(\bar{x},\bar{y}) = \Phi_{k,0}(\bar{x},\bar{y}) + \Psi_{k-2,0}(\bar{x},\bar{y})$$

$$\Phi_{k+1,0}(x,y) = \frac{k+1}{\frac{\partial}{\partial x}\Psi_{k,0}(\bar{x},\bar{y})}$$

$$\Psi_{k,l}(\bar{x},\bar{y}) = \Phi_{k,l}(\bar{x},\bar{y}) + \Psi_{k,l-2}(\bar{x},\bar{y})$$

$$\Phi_{k,l+1}(\bar{x},\bar{y}) = \frac{l+1}{\frac{\partial}{\partial y}\Psi_{l,k}(\bar{x},\bar{y})} \quad .$$

Consecutive values of $\Phi_{k,0}(\bar{x},\bar{y})$, $k=0,1,\ldots$ can be found from the following scheme

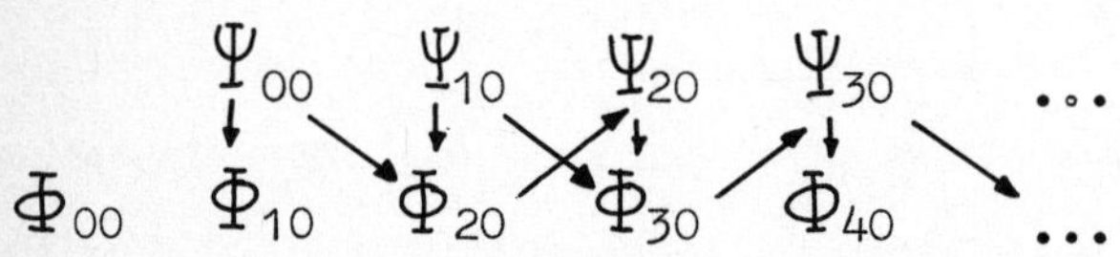

and then values of ${}_{k,l}(\bar{x},\bar{y})$, $l=0,1,\ldots$ from

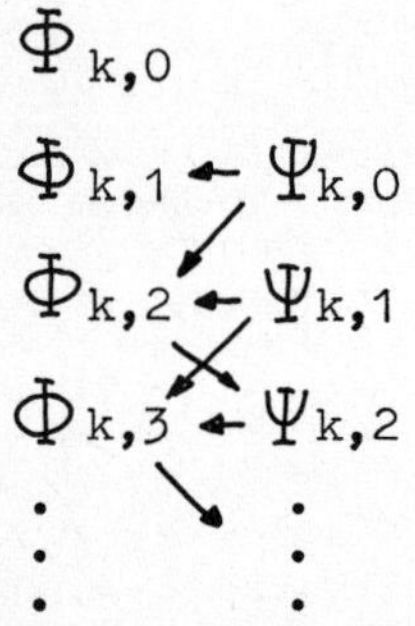

For example, if we consider the function $f(x,y)=(1+x+y)^{1/2}$ then we have

$$\Phi_{00}(x,y) = (1+x+y)^{1/2} \ , \quad \Phi_{k,0}(x,y) = 2(1+x+y)^{1/2}$$

$$\Phi_{k,2j-1}(x,y) = (1+x+y)^{1/2} \ , \quad \Phi_{k,2j}(x,y) = 4(1+x+y)^{1/2}$$

and BCF approximating $f(x,y)$ has the form

$$(1+x+y)^{1/2} = \left(1 + \frac{y}{2} + \frac{y}{2} + \ldots\right) + \frac{x}{\left(2 + \frac{y}{1} + \frac{y}{4} + \frac{y}{1} + \frac{y}{4} + \ldots\right) +}$$

$$+ \frac{x}{\left(2 + \frac{y}{1} + \frac{y}{4} + \frac{y}{1} + \frac{y}{4} + \ldots\right)} + \ldots$$

Finally, the following theorem can be proved.

THEOREM.

Let $f(x,y)$ be an analytic function in the neighbourhood of the origin. If $P_n(x,y)/Q_n(x,y)$ denotes the n-th convergent of BCF of the form (a),(b),(c) or (d) approximating $f(x,y)$ in the sense of (4.1) then

(i) $\dfrac{P_n(x,0)}{Q_n(x,0)}$ is the $\left(\left[\frac{n}{2}\right], \left[\frac{n}{2}\right]\right)$-th Padé approximant for the function $g(x)=f(x,0)$,

(ii) $\dfrac{P_n(0,y)}{Q_n(0,y)}$ is the $\left(\left[\frac{n}{2}\right], \left[\frac{n}{2}\right]\right)$-th Padé approximant for the function $h(y)=f(0,y)$,

(iii) if corresponding BCF is convergent in the neighbourhood of the origin than its values converge to values of $f(x,y)$.

Unfortunately, convergence properties of BCFs of the discussed types are still not known. We hope that results similar to convergence in Lebesgue measure of Padé approximants are obtainable.

5. FINAL CONCLUSIONS.

Branched continued fractions have approximately 10 times shorter history than ordinary continued fractions. Obviously, investigation of their properties is easier now, because we will rely on concepts created for investigation of ordinary continued fractions in their long and succesful life.

However, generalizations of known one dimensional properties are not only the reason why we hope that BCFs will have good future. Intuitively, nature of the Nature is "branched". Therefore, the structure of BCFs will be a tool for its description. But till now there are known applications of BCFs only for electrotechnical computations.

On the other hand, many numerical results show that BCFs will be used succesfuly for numerical computations. Here also only few steps have been made towards full description of their applications. We hope that the near future will bring answers to many questions which are now unsolved.

REFERENCES.

[1] D. I. Bodnar , Branched Continued Fractions, ed. Naukowaja Dumka Kiev, 1986 (in preparation) (in Russian)

[2] P. I. Bodnarcuk, V. Ya. Skorobogat'ko, Branched Continued Fractions and Applications,Naukovaja Dumka, Kiev (1974),pp.270 (in Ukrainian)

[3] A. Cuyt, A review of multivariate Padé approximation theory, J. Comp. Appl. Math. 12-13 (1985), 221-232

[4] A. Cuyt, B. Verdonk, Multivariate rational interpolation, Computing 34 (1985), 41-61

[5] F. B. Hildebrand, Introduction to Numerical Analysis, McGraw-Hill, New York (1974)

[6] Kh.I. Kuchminskaya, On interpolation formula for two-variable functions, in Continued Fractions and Applications ed.Naukovaja Dumka, Kiev (1976) 26-29 (in Russian)

[7] Kh.I. Kuchminskaya, Approximation and Interpolation of Functions by Continued and Branched Continued Fractions, Ph. D. Thesis, Ukrainian Acad. of Sciences, Lvov (1976) (in Russian)

[8] Kh.I. Kuchminskaya, On approximation of Functions by Continued and Branched Continued Fractions, Mat. Met. Fiz. Meh. Polya, 12 (1980) 3-10 (in Russian)

[9] Kh.I. Kuchminskaya, Interpolation of two-variable function by special type of branched continued fractions, in Mat. from 5-th Conference of Lvov Dept. Ukr. Acad. Sci., ed. AN USSR, Lvov (1978), 34-37 (in Russian)

[10] Kh.I. Kuchminskaya, Corresponding and associated branched continued fractions for double power series, Dokl. Akad. Nauk Ukr. SSR, Ser. A (1978) 7 , 614-617 (in Russian)

[11] Kh.I. Kuchminskaya, On sufficient conditions of the absolute convergence of two-dimensional continued fractions, Mat. Met. Fiz. Meh. Polya, 20 (1984) 19-23 (in Russian)

[12] Kh.I. Kuchminskaya, On the convergence of two-dimensional continued fractions,in "Constructive Theory of Functions", ed. Publishing House of the Bulg. Acad. of Sciences, Sofia (1984) 501-506

[13] Kh.I. Kuchminskaya, O. N. Sus, On approximation of functions by two-dimensional continued fractions, in VIII Shkola po Teorii Operatorow, Part II, Abstracts ,Riga (1983) 6-8 (in Russian)

[14] J. Murphy, M. O'Donohoe, A two-variable generalization of the Stieltjes-type continued fractions, Comp. Appl. Math. 4 (1978) 180-191

[15] W. Siemaszko, Branched continued fractions for double power series, J. Comp. Appl. Math. 9 (1980) 121-125

[16] W. Siemaszko, Thiele-type branched continued fractions for two-variable functions, J. Comp. Appl. Math. 9 (1983) 137-153

[17] V. Ya. Skorobogat'ko et al., Branched continued fractions, Dokl. Akad. Nauk Ukr. SSR ser.A No.2 (1966) 131-133 (in Russian)

[18] V. Ya. Skorobogat'ko, Branched Continued Fractions and their Applications in Computational Mathematics, ed. Nauka, Moskva (1983) pp. 311 (in Russian)

Polynomial condition of Leja

Wiesław Pleśniak

Instytut Matematyki UJ, ul. Reymonta 4, 30-059 Kraków, Poland

Abstract. A short review of recent results concerning a polynomial condition of Leja's type is given. Then some applications to the polynomial approximation are indicated.

1. Condition L^*.

Let E be a subset of the complex plane $\mathbb{C}$ and $a \in E$. We define a measure μ over E as follows. For $A \subset E$, $\mu(A) = m_1(\{t>0: \{|z-a|=t\} \cap A \neq \emptyset\})$, where m_1 is the Lebesgue linear measure. Then the famous polynomial lemma of Leja reads as follows.

1.1. Polynomial Lemma [4].

Suppose that for some $t_0 > 0$, $\mu(E \cap \{|z-a| \leq t_0\}) = t_0$. *Then every family* $\mathcal{F}$ *of polynomials with*

(1.1) $\mu(\{z \in E: \sup\{|f(z)|: f \in \mathcal{F}\} = \infty\}) = 0$

satisfies

(1.2) *For every* $b > 1$ *there exists a neighborhood* V *of* a *and a constant* $M > 0$ *such that for each* f *in* $\mathcal{F}$, $\sup |f|(V) \leq M b^{\deg f}$.

The Lemma appeared to be a very useful tool in the complex analysis of one or several variables including the case of infinitely many variables. One of its most beautiful applications is Leja's proof of Hartogs' theorem on separately analytic functions (see [5]). The Lemma has also been the starting point for studying the following condition L^* which is involved in many problems of polynomial approximation.

Let E be a subset of $\mathbb{C}^n$ and let μ be a non-negative function defined on the family of Borel subsets of E such that $\mu(\emptyset) = 0$. In what follows, such a function μ will simply be called a measure.

1.2. Definition. The pair (E, μ) is said to satisfy *condition* L^* at a point a belonging to the polynomially convex hull E of E if every family $\mathcal{F}$ of polynomials in $\mathbb{C}^n$ with property (1.1) satisfies (1.2). The pair (E, μ) is said to satisfy L^* if it satisfies L^* at every point $a \in E$.

Let us recall some known examples of pairs (E, μ) satisfying L^*. By the Polynomial Lemma 1.1 we get:

1.3. Example. Let E be a rectifiable Jordan arc in $\mathbb{C}$. Take μ to be the length measure over E. Then (E,μ) satisfies L*. In particular, if E is an interval and μ is the Lebesgue linear measure then (E,μ) satisfies L*.

By Example 1.3 and Fubini's theorem one easily obtains

1.4. Example. Let E be a subset of $\mathbb{R}^n$. Suppose that for a point a of the closure $\bar{E}$ of E there exists a non-singular affine mapping l in $\mathbb{R}^n$ such that $a \in l(J^n) \subset E \cup \{a\}$ where J^n is the Cartesian power of $J=[0,1]$. Let m_n denote the n-dimensional Lebesgue measure. Then (E,m_n) satisfies L* at the point a. In particular, for every bounded convex subset E of $\mathbb{R}^n$ or else for every bounded domain in $\mathbb{R}^n$ with Lipschitz boundary, the pair (E,m_n) satisfies L*.

The next example is closely related to the notion of L- *regularity* in $\mathbb{C}^n$ and the *complex Monge-Ampère operator* theory in $\mathbb{C}^n$. Let $\mathcal{L}$ denote the set of all plurisubharmonic functions u on $\mathbb{C}^n$ such that $u(z)=\log(1+|z|)+O(1)$ as $|z|\to\infty$, where $|.|$ is any norm in $\mathbb{C}^n$. To every subset E of $\mathbb{C}^n$ there corresponds a function $V_E(z)= \sup\{u(z): u\in\mathcal{L}, u\leqslant 0 \text{ on } E\}$ called the *extremal plurisubharmonic function* of E (see [14] , [13]). In general, V_E is not plurisubharmonic. If, however, E is not *pluripolar*, i.e. if there is no plurisubharmonic function u on $\mathbb{C}^n$, $u(z)\not\equiv-\infty$, such that $E\subset\{u(z)=-\infty\}$, then the upper regularization of V_E, $V^*_E(z) := \limsup_{w\to o} V_E(w)$, is a plurisubharmonic function belonging to the class $\mathcal{L}$ (see [13]).

E is said to be L- *regular at a point* $a\in E$ if $V_E(a)=0$. E is L- *regular* if it is L-regular at every point $a\in E$. Set $c(E)= \liminf_{|z|\to\infty} |z| \exp[-V^*_E(z)]$. The number c(E) is called the

L- *capacity* of E. If n=1, c(E) is known to be equal to the *logarithmic capacity* of E.

1.5. Example. Let E be an L-regular compact subset of $\mathbb{C}^n$. It is known that the extremal function V_E is then continuous in $\mathbb{C}^n$ (see [13]) and satisfies the complex Monge-Ampère equation $(dd^cV_E)^n=0$ in $\mathbb{C}^n\setminus E$, where $(dd^c)^n$ is the Monge-Ampère operator defined on the set of plurisubharmonic functions that are locally bounded in $\mathbb{C}^n$ (see [1] and previous articles by same authors). Moreover, $(dd^cV_E)^n$ is a Borel positive measure with support in E. It is known (see [15], [6], [2]) that $(E,(dd^cV_E)^n)$ satisfies L*. It is also seen that if (E,μ) satisfies L* and ν is a measure over E that dominates μ (i.e. $\nu(A)=0$ implies $\mu(A)=0$) then (E,ν) also satisfies L*. Hence, since the measure $(dd^cV_E)^n$ vanishes on every pluripolar set in $\mathbb{C}^n$ (see [1]), and since $F\subset\mathbb{C}^n$ is pluripolar if and only if $c(F)=0$ (see [13]), we can deduce that the pair (E,c) satisfies L^* (see also [3]).

Further examples of pairs (E,μ) satisfying L^* can be procured by the following

1.6. Proposition [7].

Let E *be a bounded subset of* $\mathbb{C}^n$ *and let* h *be a holomorphic mapping defined on a neighborhood* U *of* E *with values in* $\mathbb{C}^m$ $(m \leqslant n)$. *Let* μ *and* ν *be measures over* E *and* $h(E)$, *respectively, satisfying the following requirements:*

(i) *For each Borel subset* F *of* E *with* $\mu(E \setminus F)=0$, *the set* F *is not pluripolar.*

(ii) *For each Borel subset* H *of* $h(E)$, $\nu(H)=0$ *implies* $\mu(h^{-1}(H) \cap E)=0$.

If the pair (E,μ) *satisfies condition* L^* *at a point* $a \in E$ *and the mapping* h *is non-degenerate in a neighborhood of* a *(i.e. rank* $h := \sup \operatorname{rank}_z h = m$ *in a neighborhood of* a *), then the pair* $(h(E),\nu)$ *satisfies* L^* *at the point* $h(a)$.

We also notice a simple geometrical criterion of L^*-regularity.

1.7. Analytic accessibility criterion [11].

Let E *be a subset of* $\mathbb{C}^n$ *and* μ *a measure over* E. *Let* $a \in E$ *and suppose that there is an analytic curve* $h:[0,1] \to E$ *such that* $h(0)=a$ *and for each* $t \in (0,1]$, (E,μ) *satisfies* L^* *at* $h(t)$. *Then* (E,μ) *satisfies* L^* *at* a.

Actually, it is sufficient to assume h to be a semi-analytic arc (see [11]).

2. Polynomial approximation in L^2-norm

One of most important tools in polynomial approximation is the Bernstein-Markov inequality. We present here the version given by Nguyen Thanh Van and Zeriahi.

2.1. Proposition [8].

Suppose E *is an* L-*regular compact subset of* $\mathbb{C}^n$ *and* μ *is a positive measure over* E *such that* (E,μ) *satisfies* L^*. *Then for each polynomial* p,

$$\sup |p|(E) \leqslant M b^{\deg p} \left[\int_E |p(z)|^2 d\mu(z)\right]^{1/2} \qquad \text{(BM)}.$$

For subsets E of $\mathbb{R}^n$ a more general version of Proposition 2.1 was given in [9].

Let $\alpha:\mathbb{Z}_+ \to \mathbb{Z}_+^n$ be a bijection such that $|\alpha(k)| \leqslant |\alpha(k+1)|$ for each $k \in \mathbb{Z}_+$ where $|\alpha(k)| := \alpha_1(k)+\ldots+\alpha_n(k)$ is the length of the multiindex $\alpha(k)$. By (BM), the sequence

$\{z^{\alpha(k)}:=z^{\alpha_1(k)}\dots z^{\alpha_n(k)}\}_{k\in\mathbb{Z}_+}$ is linearly independent in the Hilbert space $L^2(E,\mu)$. Let $\{A_k\}_{k\in\mathbb{Z}_+}$ be the orthogonal sequence obtained from $\{z^{\alpha(k)}\}$ by the Hilbert-Schmidt procedure and let ν_k denote the L^2-norm of A_k. Then we have

2.2. Proposition [15], [8].

Let E *be an* L- *regular compact subset of* $\mathbb{C}^n$ *and let* μ *be a positive measure over* E *such that* (E,μ) *satisfies* L^*. *Then*

(i) $\lim_{k\to\infty} [\sup|A_k|(E)/\nu_k]^{1/|\alpha(k)|} = 1$

and (ii)

$$\limsup_{k\to\infty} [|A_k(z)|/\nu_k]^{1/|\alpha(k)|} = \exp V_E(z) \qquad \textit{for all } z\in\mathbb{C}^n\setminus E.$$

Suppose now that n=1. Given a function $f\in L^2(E,\mu)$, we consider the Fourier series of f with respect to the basis $\{A_k/\nu_k\}_{k\in\mathbb{Z}_+}$, $\sum a_k\nu_k^{-1}A_k$, where $a_k=(f,\nu_k^{-1}A_k)$.
Put $f_k=\sum_{l=0}^{\infty} a_l\nu_l^{-1}A_l$. Then we have

2.3. Proposition [10].

If no f_k *takes a fixed value* $a\in\mathbb{C}$ *in a neighborhood of* E *(independent of* k *) then* f *is the restriction* μ- *almost everywhere on* E *to a function* g *holomorphic in a neighorhood of* E.

2.4. Problem. Generalize Proposition 2.3 to the case n>1.

The polynomial condition L^* is also involved in polynomial approximation problems in Orlicz spaces of integrable functions that are natural generalizations of L^p-spaces. For details we refer the reader to [9] and [12].

References

[1] Bedford, E. and Taylor, B.A., A new capacity for plurisubharmonic functions, Acta Math. **149**, 1-40, 1982.

[2] Bedford, E. and Taylor, B.A., Fine topology, Shilov boundary, and $(dd^c)^n$, preprint 1985.

[3] Klimek, M., Extremal plurisubharmonic functions and L-regular sets in $\mathbb{C}^n$, Proc. Roy. Irish Acad. **82**$_A$, 217-230, 1982.

[4] Leja, F., Sur les suites de polynômes bornées presque partout sur la frontière d'un domaine, Math. Ann. **108**, 517-524, 1933.

[5] Leja, F., Teoria funkcji analitycznych, PWN Warszawa 1957.

[6] Levenberg, N., Monge-Ampère measures associated to extremal plurisubharmonic functions in $\mathbb{C}^n$, preprint 1984.

[7] Nguyen Thanh Van and Pleśniak, W., Invariance of L-regularity and Leja's polynomial condition under holomorphic mappings, Proc. Roy. Irish Acad. **84**$_A$, 111-115, 1984.

[8] Nguyen Thanh Van and Zeriahi, A., Familles de polynômes presque partout bornées, Bull.Soc.Math. France, **107**, 81-91, 1983.

[9] Pleśniak, W., Quasianalyticity in F-spaces of integrable functions, 558-571. In: *Approximation and Function Spaces*, ed. Z. Ciesielski, PWN Warszawa, North-Holland Publ. Comp. Amsterdam-New York-Oxford, 1981.

10] Pleśniak, W., On the distribution of zeros of the polynomials of best L^2-approximation to holomorphic functions, Zeszyty Nauk. Uniw. Jagiellon., Prace Mat. **22**, 29-35, 1981.

[11] Pleśniak, W., A criterion for polynomial conditions of Leja's type in $\mathbb{C}^n$, Univ. Iagiellon. Acta Math. **24**, 139-142, 1984.

[12] Pleśniak, W., Leja's type polynomial condition and polynomial approximation in Orlicz spaces, Ann. Polon. Math. **46**, 268-278, 1985.

[13] Siciak, J., Extremal plurisubharmonic functions in $\mathbb{C}^n$, Ann. Polon. Math. **39**, 175-211, 1981.

[14] Zaharjuta, V.P., Extremal plurisubharmonic functions, orthogonal polynomials and Bernstein-Walsh theorem for analytic functions of several variables, Ann. Polon. Math. **33**, 137-148, 1976 (Russian).

[15] Zeriahi, A., Capacité, constante de Tchebysheff et polynômes orthogonaux associés à un compact de $\mathbb{C}^n$, Bull. Soc. Math. France, **109**, 325-335, 1985.

BRANCHED CONTINUED FRACTIONS AND CONVERGENCE ACCELERATION PROBLEMS

V. Ya. Skorobogat'ko

Institute of Applied Problems
of Mechanics and Mathematics
Ukrainian Academy of Sciences
Lvov, 290047, U.S.S.R.

Acceleration of convergence of series, integral sums as well as of iterative methods will be based on the theory of branched continued fractions. Branched continued fraction is an expression of the form

$$a_0 + \sum_{i_1=1}^{N} \cfrac{a_{i_1}}{b_{i_1} + \sum_{i_2=1}^{N} \cfrac{a_{i_1 i_2}}{b_{i_1 i_2} + \cdots}} \tag{1}$$

The general theory of such continued fractions will be found in [1] . Roughly speaking, branched continued fractions will be treated as continued fractions for multivariate functions while ordinary continued fractions appear as one-dimensional case of such fractions. We are going to present some methods of acceleration of convergence of series.

Let us consider the series

$$S(t_1,\dots,t_m,\varepsilon_1,\dots,\varepsilon_p) = \sum_{k_1,\dots,k_p \geq 0} a_{k_1\dots k_p}(t_1,\dots,t_m)\, \varepsilon_1^{k_1} \dots \varepsilon_p^{k_p} \tag{2}$$

where ε_j, $j=1,2,\dots,p$ are parameters and t_i, $i=1,\dots,m$ are arguments.

Following methods of Kh. I. Kuchminskaya [1] we will find representation of the series (2) in a branched continued fraction form. In many cases succesive convergents of such branched continued fraction form a sequence accelerating convergence of this series. Similarly, the series (2) will be represented by a rational function of $\varepsilon_1, \dots, \varepsilon_p$ variables according to W. Siemaszko formulas. It will produce convergence acceleration procedure similar to Kh. I. Kuchminskaya one.

We will present now the procedure of acceleration of convergence for an iterative method and for a small parameter method. Moreover, we will find an estimation of the speed of acceleration of convergence.

For the sake of simplicity, we will consider only the case of one ordinary differential equation of the form

$$y' = f(x,y) \tag{3}$$

$$y(x_0) = y_0 \tag{4}$$

for which classical existence conditions of the Cauchy problem are fulfilled. It is known that the iteration series which appears in the contractive projection method, namely

$$\begin{gathered} y_0 + \sum_{i=1}^{\infty} (y_i - y_{i-1}) \\ y_n = y_0 + \int_{x_0}^{x} f(x, y_{n-1})\,dx \end{gathered} \tag{5}$$

converges uniformly to the solution of the problem (3)-(4).

Let us consider the power series

$$S(x,\lambda) = \sum_{k=0}^{\infty} c_k(x)\lambda^k \tag{6}$$

for which

$$y_k(x) - y_{k-1}(x) = c_k(x)\ .$$

We assume that if we choose x arbitrarily constant and belonging to the domain in which solution of the problem (3)-(4) exists, then the series (6) represents function for which analytic continuation is a meromorphic function of the parameter λ . It is obvious that for $\lambda=1$ the series (6) is equivalent to the iteration series (5).

Taking x constant and equal $\bar{x}$ in (6) we obtain ordinary power series for which we have the equality

$$\left| S(\bar{x},\lambda) - \frac{P_n(\bar{x},\lambda)}{P_m(\bar{x},\lambda)} \right| = \left| \left[\rho_{n+m}(\bar{x}) + O(\bar{x},\lambda^{n+m+1}) \right] R_{n+m}(\bar{x},\lambda) \right|$$

where $P_n(\bar{x},\lambda)$ and $P_m(\bar{x},\lambda)$ are polynomials. Coefficients of these polynomials are given in [2] and $\rho_{n+m}(\bar{x})$ has the form

$$\rho_{n+m}(\bar{x}) = (c_{n+m+1}(\bar{x}))^{-1} .$$

$$\cdot \begin{vmatrix} c_{n-m+1}(\bar{x}), c_{n-m+2}(\bar{x}), \ldots, c_{n+1}(\bar{x}) \\ c_{n-m+2}(\bar{x}), c_{n-m+3}(\bar{x}), \ldots, c_{n+2}(\bar{x}) \\ \vdots \quad \vdots \quad \vdots \\ c_{n+1}(\bar{x}), c_{n+2}(\bar{x}), \ldots, c_{n+m+1}(\bar{x}) \end{vmatrix} \begin{vmatrix} c_{n-m+1}(\bar{x}), \ldots, c_{n}(\bar{x}) \\ c_{n-m+2}(\bar{x}), \ldots, c_{n+1}(\bar{x}) \\ \vdots \quad \vdots \\ c_{n}(\bar{x}), \ldots, c_{n+m-1}(\bar{x}) \end{vmatrix}$$

It is known that $\lim_{n+m} \rho_{n+m}(\bar{x}) = 0$. If $\lambda = 1$ then $R_{n+m}(\bar{x})$ is the remainder of the iteration series (6) and since this series converges uniformly, we have

$$\left| S(x,\lambda) - P_n(x,1)/P_m(x,1) \right| \leqslant \max_{x_0 \leqslant x \leqslant \bar{x}} \left| \rho_{n+m}(x) \right| \cdot \varepsilon \qquad (8)$$

as $\left| R_{n+m}(x) \right| < \varepsilon$.

The above inequality shows that the use of rational functions $\frac{P_n(x,1)}{P_m(x,1)}$ leads to acceleration of the speed of convergence of iteration process. Similar estimations will be done for every component in the case of systems of equations.

Similar considerations will be done in a normed space. In this case we have to assume that the contractive projection method is defined and that the rational function $P_n(x,\lambda)/P_m(x,\lambda)$ is defined.

The above described method will be used also for acceleration of convergence of A. Poincaré small parameter method. Namely, let the component $y_k(x,\varepsilon)$ of a system of ordinary differential equations be represented in the form of convergent series

$$S(t,\varepsilon) = \sum_{k=0}^{\infty} f_k(t)\varepsilon^k$$

where ε is a parameter. Now the inequality (8) takes the form

$$\left| S(t,\varepsilon) - P_n(t,\varepsilon)/P_m(t,\varepsilon) \right| \leqslant \qquad (9)$$

$$\leqslant \max_{0 \leqslant t \leqslant T} \rho_{n+m}(t) \max_{0 \leqslant t \leqslant T} R_{n+m} .$$

If ρ_{n+m} tends to zero then the above inequality shows that the convergence of a small parameter method is accelerated.

The inequality (7) gives also an estimation of the speed of convergence of the method in integral norms. For example we have

$$\int_D \left| S(\bar{x},\lambda) - P_n(\bar{x},\lambda)/P_m(\bar{x},\lambda) \right| d\bar{x} \leqslant \tag{10}$$

$$\leqslant \left\{ \int_D \rho_{n+m}^2(\bar{x})\, d\bar{x} \right\}^{1/2} \cdot \left\{ \int_D R_{n+m}^2(\bar{x},\lambda)\, d\bar{x} \right\}^{1/2} +$$

$$+ \left\{ \int_D O^2(\bar{x},\lambda^{n+m+1}) d\bar{x} \right\}^{1/2} \cdot \left\{ \int_D R_{n+m}^2 (\bar{x},\lambda)\, d\bar{x} \right\}^{1/2}$$

where D is a domain in which the parameter $\bar{x}$ is defined. Let us notice that

$$\rho_{n+m}(\bar{x}) = x_{m+1}/c_{n+m+1}(\bar{x})$$

where x_{m+1} is the last component of the solution of the system of linear equations

$$\begin{bmatrix} c_{n-m+1}(\bar{x}), & c_{n-n+2}(\bar{x}), & \ldots, & c_n(\bar{x}) & , 0 \\ c_{n-m+2}(\bar{x}), & c_{n-m+3}(\bar{x}), & \ldots, & c_{n+1}(\bar{x}) & , 0 \\ \cdot & \cdot & & \cdot & \cdot \\ \cdot & \cdot & & \cdot & \cdot \\ \cdot & \cdot & & \cdot & \cdot \\ c_n(\bar{x}) & , c_{n+1}(\bar{x}) & , \ldots, & c_{n+m-1}(\bar{x}) & , 0 \\ c_{n+1}(\bar{x}) & , c_{n+2}(\bar{x}) & , \ldots, & c_{n+m}(\bar{x}) & , 1 \end{bmatrix} \begin{bmatrix} x_1 \\ x_2 \\ \cdot \\ \cdot \\ \cdot \\ x_m \\ x_{m+1} \end{bmatrix} = \begin{bmatrix} c_{n+1}(\bar{x}) \\ c_{n+2}(\bar{x}) \\ \cdot \\ \cdot \\ \cdot \\ c_{n+m}(\bar{x}) \\ c_{n+m+1}(\bar{x}) \end{bmatrix} \tag{11}$$

Quadrature formulas will be used for obtaining approximate values of integrals appearing in the inequality (10). Values of $\rho_{n+m}(x)$ on a grid covering domain D will be found by solving (11) in respectively many points.

Methods similar to the shown procedure will accelerate the practical realization of the small parameter method.

Similar estimations of the speed of acceleration of convergence will be obtained for the series (2) with the arbitrary but finite number of parameters. It is convenient to expand the obtained rational functions into the corresponding branched continued fraction since branched continued fractions are stable respectively to the round-off error of their components in practical computations.

REFERENCES.

1 V. Ya. Skorobogat'ko, "Theory of Branched Continued Fractions and their Applications in Computational Mathematics", ed. Nauka, Moskva (1983) pp. 311 (in Russian)

2 Z. I. Krupka, Approximation of meromorphic functions by rational functions and acceleration of convergence of power series, Dokl. Akad. Nauk Ukr. S. R., ser.A, No.10 (1982) 24-28

Two-point Padé-type and Padé approximants

in a non-commutative algebra.

André DRAUX

Université de Lille I
U.E.R I.E.E.A M3
59655 VILLENEUVE D'ASCQ - CEDEX
FRANCE

The Padé approximants can be deduced in the commutative case from the Padé type approximants in using the orthogonal polynomials (see [1] and [6]). The same method is used in the case of a non-commutative algebra in [2] and [3]. The case of two-point Padé approximants can be solved in a similar way, in particular in a non-commutative algebra.

1 - Two-point Padé type approximants.

Let K be a commutative field with a characteristic number 0 and A a non-commutative algebra on K with a unity element I.

K is assumed to have an absolute value denoted by $|.|$.

Let P be the set of the polynomials of a variable $x \in K$. The coefficients of the polynomials belong to A.

Let L and L^* be two formal power series :

$$L = \sum_{j=0}^{\infty} c_j x^j \text{ for } |x| \text{ small}$$

and

$$L^* = \sum_{j=1}^{\infty} c^*_{-j}\, x^{-j} \text{ for } |x| \text{ large, where } x \in K, \text{ and } c_j \text{ and } c^*_{-j} \in A.$$

Let v be an arbitrary polynomial of degree m ; $v \in P$.

$$v(x) = \sum_{i=0}^{m} b_{m-i}\, x^i.$$

v is assumed to be a quasi-monic polynomial (i.e. : b_0 has an inverse); v will be monic if $b_0 = I$.

Let us define $\tilde{v}(x)$ by :

$$\tilde{v}(x) = x^m\, v(x^{-1}).$$

Remark 1.

All the definitions and the properties in the sequel only concern the left two-point Padé type approximants. The case of right two-point Padé approximants can be deduced from the left approximants by inverting the order of all the products.

We look for the left function :

$$\overset{(\ell)}{f}_{k,m}(x) = \overset{(\ell)}{\tilde{w}}(x)(\tilde{v}(x))^{-1}$$

with deg $\overset{(\ell)}{\tilde{w}} = m-1$ such that :

$$L.\tilde{v}(x) - \overset{(\ell)}{\tilde{w}}(x) = O_+(x^k) \tag{1}$$

and

$$L^*\tilde{v}(x) - \overset{(\ell)}{\tilde{w}}(x) = O_-(x^{k-1}) \text{ with } o \le k \le m. \tag{2}$$

$O_+(x^k)$ is equivalent to the expression $x^k \sum_{j=o}^{\infty} e_j x^j$ and

$O_-(x^{k-1})$ is equivalent to $x^{k-1} \sum_{j=o}^{\infty} e_j^* x^{-j}$.

The function $\overset{(\ell)}{f}_{k,m}$ is called a left two-point Padé type approximant and is denoted by $\overset{(\ell)}{(k/m)}_{L,L^*}(x)$. The left two-point Padé type approximant is generally different from the right one.

The polynomial $\overset{(\ell)}{\tilde{w}}$ will be written :

$$\overset{(\ell)}{\tilde{w}}(x) = \sum_{j=o}^{m-1} \overset{(\ell)}{a_j} x^j.$$

v is called the generating polynomial of the approximant.

Theorem 2.

The left two-point Padé type approximant always exists and is unique.

Proof. The following relations

$$\sum_{i=o}^{m} c_{j-i} b_i - \overset{(\ell)}{a_j} = o, \forall j \in \mathbb{N} \text{ such that } o \le j \le k-1, \tag{3}$$

and

$$\sum_{i=o}^{m} c_{j-i}^* b_i - \overset{(\ell)}{a_j} = o, \forall j \in \mathbb{N} \text{ such that } k \le j \le m-1, \tag{4}$$

are easily deduced from the relations (1) and (2), and show that the coefficients $a_j^{(\ell)}$ always exist and are unique. □

A particular choice of the two formal power series L and L^* has been made. In fact it is possible to show that another choice gives a relation in which an approximant of the type $(k/m)_{L,L^*}$ (i.e : a polynomial of degree m-1 times the inverse of a polynomial of degree m) is used.

Let L and L^* be the two following formal power series :

$$L = \sum_{j=\nu}^{\infty} c_j x^j \text{ for } |x| \text{ small with } \nu \in \mathbb{Z}$$

and

$$L^* = \sum_{j=-\infty}^{\mu} c_j^* x^j \text{ for } |x| \text{ large with } \mu \in \mathbb{Z}$$

Then, from L and L^*, two other formal power series can be defined :

$$L_\ell = \sum_{j=o}^{\infty} (c_{j+\ell} - c_{j+\ell}^*)x^j \text{ for } |x| \text{ small,}$$

and

$$L_\ell^* = \sum_{j=-\infty}^{-1} (c_{j+\ell}^* - c_{j+\ell})x^j \text{ for } |x| \text{ large,}$$

with the convention that :

$$c_j^* = o \text{ if } j > \mu, \text{ and } c_j = o \text{ if } j < \nu.$$

We look for the two-point Padé type approximants of the series L and L^* which have the generating polynomial v :

$$\overset{(\ell)}{w}(x)\, x^{\nu}(\tilde{v}(x))^{-1} = \left(\sum_{j=o}^{m+\mu-\nu} a_j^{(\ell)} x^j\right) x^{\nu}\left(\sum_{j=o}^{m} b_j x^j\right)^{-1}$$

This approximant will be denoted by $(k+\ell+\overset{(\ell)}{r}/m+r)_{L,L^*}$, where $r = \max(0, -\nu)$, with $k \in \mathbb{N}$ such that $\nu - \mu - 1 \le k \le m$ and $\ell \in \mathbb{Z}$ such that $\min(\nu, \mu+1) \le \ell \le \mu+1$.

Property 3.

If the two approximants have the same generating polynomial, then :

$$(k+\ell+\overset{(\ell)}{r}/m+r)_{L,L^*} = \sum_{j=\nu}^{\ell-1} c_j x^j + \sum_{j=\ell}^{\mu} c_j^* x^j + x^{\ell}(k/m)^{(\ell)}_{L_\ell,L_\ell^*}(x) \qquad (5)$$

Proof : If the second member of (5) is denoted by B, and :

i) <u>if $\nu \le \mu+1$</u>, then it is easy to see that :

$$Lx^{-\nu}\tilde{v}(x) - Bx^{-\nu}\tilde{v}(x) = x^{\ell-\nu}(L_\ell \tilde{v}(x) - \overset{(\ell)}{\tilde{w}}(x)) = O_+(x^{k+\ell-\nu}) \qquad (6)$$

$$L^*x^{-\nu}\tilde{v}(x) - Bx^{-\nu}\tilde{v}(x) = x^{\ell-\nu}(L^*_\ell \tilde{v}(x) - \overset{(\ell)}{\tilde{w}}(x)) = O_-(x^{k+\ell-\nu-1}) \qquad (7)$$

where $\overset{(\ell)}{(k/m)}_{L_\ell,L^*_\ell} = \overset{(\ell)}{\tilde{w}}(x)(\tilde{v}(x))^{-1}$.

The relations (6) and (7) show that B is exactly $\overset{(\ell)}{\tilde{W}}(x)\, x^{\nu}(\tilde{v}(x))^{-1}$ with d deg $\overset{(\ell)}{W} = m+\mu-\nu$. From the property of uniqueness B is the approximant $\overset{(\ell)}{(k+\ell+r/m+r)}_{L,L^*}$.

ii) <u>if $\nu > \mu+1$</u>, then

$$L\,x^{-\nu}\tilde{v}(x) - Bx^{-\nu}\tilde{v}(x) = x^{1+\mu-\nu}(L_{\mu+1}\,\tilde{v}(x) - x^{\nu-\mu-1}\overset{(\ell)}{\tilde{w}}(x)) \qquad (8)$$

$$= O_+(x^{k+1+\mu-\nu})$$

$$L^*x^{-\nu}\tilde{v}(x) - Bx^{-\nu}\tilde{v}(x) = x^{1+\mu-\nu}(L^*_{\mu+1}\,\tilde{v}(x) - x^{\nu-\mu-1}\overset{(\ell)}{\tilde{w}}(x)) \qquad (9)$$

$$= O_-(x^{k+\mu-\nu}).$$

where $\overset{(\ell)}{(k/m)}_{L_{\mu+1},L^*_{\mu+1}} = x^{\nu-\mu-1}\overset{(\ell)}{\tilde{w}}(x)(\tilde{v}(x))^{-1}$.

The relations (8) and (9) also show that B is the approximant $(k+\mu+1+\overset{(\ell)}{r}/m+r)_{L_{\mu+1},L^*_{\mu+1}}$. □

As in [7] the numerator polynomial $\overset{(\ell)}{w}$ can be related to some associated polynomials with respect to some linear functionals. A left linear functional $\overset{(\ell)}{c}$ can be introduced from the moments c_i (coefficients of the formal power series L) thanks to the relations :

$$\overset{(\ell)}{c}(\lambda x^i) = c_i\,\lambda,\ \forall i \in \mathbb{N} \text{ and } \lambda \in A.$$

We also have another left linear functional $\overset{(\ell)}{c*}$ from the c^*_i's.

$$\overset{(\ell)}{c^*}(\lambda x^i) = c^*_{-i-1}\,\lambda,\ \forall i \in \mathbb{N} \text{ and } \lambda \in A.$$

The polynomials v and $\overset{(\ell)}{\tilde{w}}$ are shared in two parts as in [7].

$$v_{1,k}(x) = \sum_{i=o}^{k} b_i\, x^{k-i},$$

$$\tilde{v}_{2,m-k}(x) = \sum_{i=o}^{m-k} b_{k+i}\, x^{i},$$

$$w_{1,k}^{(\ell)}(x) = \sum_{i=o}^{k-1} a_i^{(\ell)}\, x^{k-1-i},$$

$$\tilde{w}_{2,m-k}^{(\ell)}(x) = \sum_{i=o}^{m-1-k} a_{k+i}^{(\ell)}\, x^{i}.$$

Property 4.

i)
$$w_{1,k}^{(\ell)}(t) = c^{(\ell)}\left((v_{1,k}(x) - v_{1,k}(t))(x-t)^{-1}\right) \tag{10}$$

and

$$\tilde{w}_{1,k}^{(\ell)}(\tilde{v}_{1,k})^{-1} = (k-1/k)_L^{(\ell)}, \tag{11}$$

which is the Padé type approximant of the formal power series L.

ii)
$$\tilde{w}_{2,m-k}^{(\ell)}(t) = c^{*(\ell)}\left((\tilde{v}_{2,m-k}(x) - \tilde{v}_{2,m-k}(t))(x-t)^{-1}\right) \tag{12}$$

If $\tilde{v}_{2,m-k}$ *is quasi-monic, then :*

$$\tilde{w}_{2,m-k}^{(\ell)}(\tilde{v}_{2,m-k})^{-1} = (m-1-k/m-k)_{\tilde{L}}^{(\ell)} \tag{13}$$

which is the Padé type approximant of the formal power series $\tilde{L}$ $(\tilde{L} = \sum_{i=o}^{\infty} c^*_{-i-1} t^i)$.

The linear functionals c *and* c^* *only act on the variable* x.

Proof. The relations (3) and (4) immediately give the relations (10) and (12). The relations (11) and (13) are easily deduced from the definition of a Padé type approximant (see [3]). □

The last property of the Padé type approximants which will be given in this paper, concerns the expression of the error.

Thanks to the coefficients c_i and c_i^*, other moments can be obtained :

$$d_i = c_i - c_i^*, \ \forall i \in \mathbb{Z} ,$$

with the convention that :

$$c_i = o \text{ if } i < o \text{ and } c_i^* = o \text{ if } i \geq o,$$

then some left linear functionals $\overset{(\ell)}{d}{}^{(j)}$ are defined from these moments by :

$$\overset{(\ell)}{d}{}^{(j)}(\lambda\, x^s) = d_{j+s}\, \lambda, \ \forall j \in \mathbb{Z} , \ \forall s \in \mathbb{N} \text{ and } \forall \lambda \in A.$$

The following results can be proved (see [4] for the proof):

Theorem 5.

$$L - \overset{(\ell)}{(k/m)}_{L,L^*}(x) = \Big(\sum_{j=k}^{\infty} \sum_{i=o}^{m} d_{j-i}\, b_i\, x^j \Big)(\tilde{v}(x))^{-1} \tag{14}$$

and, if $\tilde{v}$ *is quasi-monic :*

$$L^* - \overset{(\ell)}{(k/m)}_{L,L^*}(x) = - \Big(\sum_{j=k-1}^{\infty} \sum_{i=o}^{m} d_{j-i}\, b_i x^{j-m} \Big)(v(x^{-1}))^{-1} \tag{15}$$

This theorem is very important, since it permits to deduce the two-point Padé approximants (see the next section).

Other algebraic properties of two-point Padé type approximants can be found in [4].

2 - Two-point Padé approximants.

In the first section the generating polynomial v was chosen arbitrarily. Higher order approximants will be obtained for some particular choices of v. The highest order approximant is given by a generating polynomial such that :

$$\sum_{i=o}^{m} d_{j-i}\, b_i = o \ \forall j \in \mathbb{N} \text{ such that } k_1 - m \leq j \leq k_1 - 1.$$

Then v is completely determined.

In fact v is identical to the left orthogonal polynomial $\overset{(\ell)}{P}{}_m^{(k-2m)}$ with respect to the linear functional $\overset{(\ell)}{d}{}^{(k-2m)}$ (see [2]). In this case the numerator polynomial will be denoted by $Q^{(\ell)}_{k,m}$. Then the two-point Padé approximant is obtained and is denoted by $[k/m]^{(\ell)}_{L,L^*}$:

$$[k/m]^{(\ell)}_{L,L^*} = \tilde{Q}^{(\ell)}_{k,m} \, (\tilde{P}^{(\ell)(k-2m)}_{m})^{-1} \tag{16}$$

The approximation property is given by :

$$L - [k/m]^{(\ell)}_{L,L^*} = O_+(x^k),$$

and

$$L^* - [k/m]^{(\ell)}_{L,L^*} = O_-(x^{k-m-1}), \ \forall k \in \mathbb{N} \text{ such that } o \le k \le 2m .$$

The following theorems can be proved (see also [4] for the proofs):

Theorem 6.

i) If $m-1 < k \le 2m$, *then :*

$$Q^{(\ell)}_{k,m}(t) = c^{(\ell)} ((P^{(\ell)(k-2m)}_{m}(x) - P^{(\ell)(k-2m)}_{m}(t))(x-t)^{-1}).$$

ii) If $o \le k \le m-1$, *then :*

$$\tilde{Q}^{(\ell)}_{k,m}(t) = c^{*(\ell)} ((\tilde{P}^{(\ell)(k-2m)}_{m}(x) - \tilde{P}^{(\ell)(k-2m)}_{m}(t))(x-t)^{-1}).$$

Theorem 7.

The left and right two-point Padé approximantsare identical.

The existence and the uniqueness of two-point Padé approximant only depend on the existence and the uniqueness of the orthogonal polynomial, which depend on the Hankel matrix $M^{(k-2m)}_{m} = (d_{k-2m+i+j})^{m-1}_{i=j=o}$.

Theorem 8.

If $M^{(k-2m)}_{m}$ *has an inverse, then* $[k/m]_{L,L^*}$ *exists and is unique.*

Proof. If $M^{(k-2m)}_{m}$ has an inverse the existence and the uniqueness of $P^{(\ell)(k-2m)}_{m}$ has been proved in [2]. Then the result is obvious. □

The orthogonal polynomials $P^{(\ell)(k-2m)}_{m}$ are displayed in a two-dimensional array called the table P. Along a diagonal the superscript (k-2m) is constant, and along a column the index m is constant (see [2]). The numerator polynomials $Q_{k,m}^{(\ell)}$ are displayed in a table Q and are at the same place as the corresponding polynomial $P_m^{(k-2m)}$ in the table P.

The two-point Padé approximants are also displayed in a two-dimensional array, called two-point Padé table, as follows :

	[0/2]	...
[0/1]	[1/2]	...
[1/1]	[2/2]	...
[2/1]	[3/2]	...
	[4/2]	...

This table corresponds to the part of the table P in which all the orthogonal polynomials used by all the two-point Padé approximants are located.

In [4] some recurrence relations in the tables P and Q have been proved. The polynomials $P_m^{(k-2m)}$ and $Q_{k,m}$ verify the same three-term recurrence relations between three terms as follows.

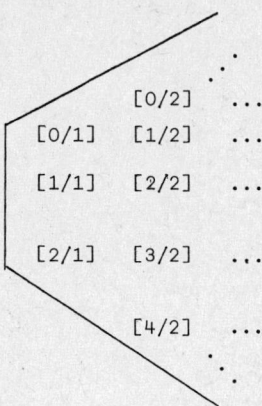

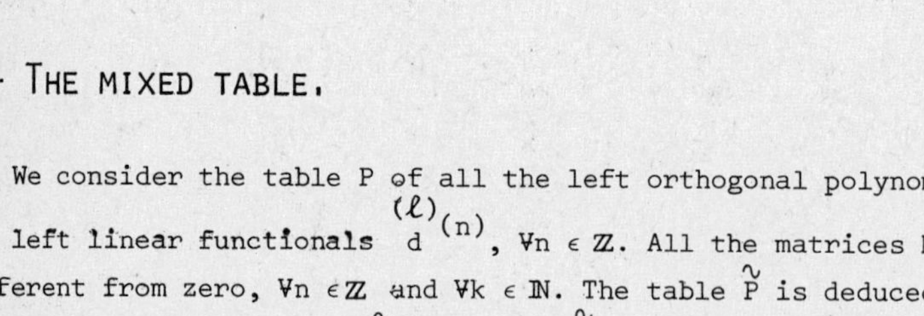

(17)

3 - THE MIXED TABLE.

We consider the table P of all the left orthogonal polynomials with respect to the left linear functionals $\overset{(\ell)}{d}{}^{(n)}$, $\forall n \in \mathbb{Z}$. All the matrices $M_k^{(n)}$ are assumed to be different from zero, $\forall n \in \mathbb{Z}$ and $\forall k \in \mathbb{N}$. The table $\tilde{P}$ is deduced from the table P by displaying the polynomials $\tilde{P}$. The table $\tilde{Z}$ contains the left following polynomials :

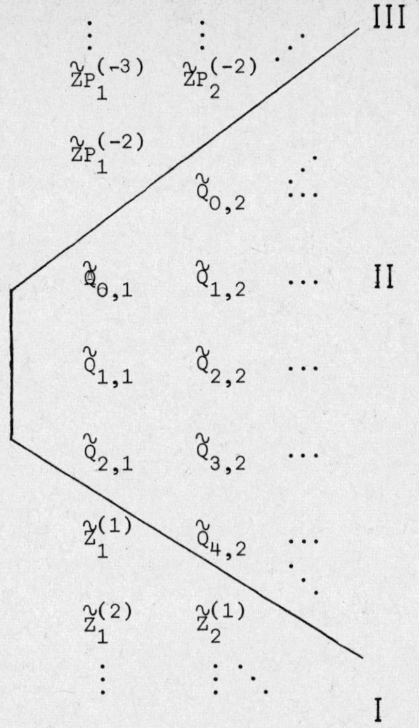

where $\tilde{Q}_{k,m}(x) = x^{m-1}\, Q_{k,m}(x^{-1})$, $\tilde{Z}_k^{(n)}$ (in the region I) is the numerator polynomial of the Padé approximant $[n-1+k/k]_L$.

Let us define the elements $\tilde{ZP}_i^{(-n)}$ in the region III.

The Padé approximant $[m-1-n/m]_{L^*}(t^{-1})$ can be written as (see [2] and [4]) :

$$\sum_{i=o}^{-n-1} c^*_{-i}\, t^{-i} + t^n \; \overset{(\ell)}{Q\tilde{W}}{}_m^{(n)}(t^{-1})\big(\overset{(\ell)}{\tilde{W}}{}_m^{(n)}(t^{-1})\big)^{-1}$$

$$= \overset{(\ell)}{Z\tilde{W}}{}_m^{(n)}(t^{-1})\big(\overset{(\ell)}{\tilde{W}}{}_m^{(n)}(t^{-1})\big)^{-1},$$

$\forall n \in \mathbb{Z}$ such that $n \leq -1$.

The polynomial $\overset{(\ell)}{W}{}_m^{(n)}$ is orthogonal with respect to the left linear functional $\overset{(\ell)}{\gamma}{}^{(m)}$ defined from the moments d_i such that :

$$\overset{(\ell)}{\gamma}{}^{(n)}(\lambda\, x^j) = d_{n-j}\, \lambda, \ \forall j \in \mathbb{N} \text{ and } \forall n \in \mathbb{Z}.$$

The polynomial $\overset{(\ell)}{\tilde{ZP}}{}_m^{(-j)}$ is defined by :

$$\overset{(\ell)}{\tilde{ZP}}{}_m^{(-j)}(x) = x^{m-1+j}\ \overset{(\ell)}{ZP}{}_m^{(-j)}(x^{-1}),\ \forall j \in \mathbb{N} \text{ such that } j \geq 1,$$

and

$$\overset{(\ell)}{ZP}{}_m^{(-j)}(x) = \overset{(\ell)}{Z\tilde{W}}{}_m^{(-j)}(x)\ \overset{(\ell)}{\lambda}{}_{m,m}^{(-j+1-2m)},$$

where

$$\overset{(\ell)}{\lambda}{}_{m,m}^{(-j+1-2m)} = \overset{(\ell)}{P}{}_m^{(-j+1-2m)}(0) = (\ \overset{(\ell)}{W}{}_m^{(-j)}(0))^{-1}.$$

Then the Padé approximant $[m-1-n/m]_{L^*}(t^{-1})$ also can be written as :

$$\overset{(\ell)}{ZP}{}_m^{(n)}(t^{-1})(\ \overset{(\ell)}{P}{}_m^{(n-2m+1)}(t^{-1}))^{-1} \text{ for } n \leq -1.$$

The table Z is obtained from the table $\tilde{Z}$ by using the polynomials $Z_k^{(n)}$ in the region I, $Q_{k,m}$ in the region II and $t^{-1}\ \overset{(\ell)}{ZP}{}_m^{(-j)}(t)$ in the region III.

The following property will be given without its very long proof (see our report [4] from the University of Lille I).

Property 9.

i) All the three-term recurrence relations of type (17) verified by the polynomial $\tilde{P}$ in the region I and III are also verified by the polynomials of the table $\tilde{Z}$ displayed at the same places.

ii) All the three-term recurrence relations of type (17) verified by the polynomial P in the regions II and III are also verified by the polynomials of the table Z displayed at the same places.

This property is very useful to compute the elements of the mixed Padé table where the following elements are displayed : in the region I the Padé approximants of L in t, in the region II the two-point Padé approximants of L and L^*, and in the region III the Padé approximants of L^* in t^{-1}.

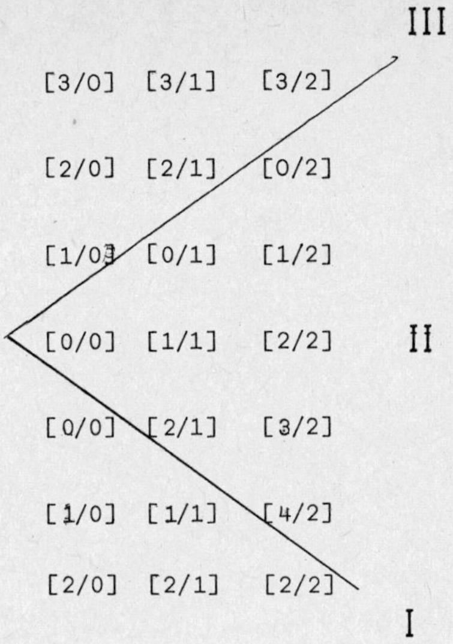

The initializations are

$$[p/0]_L(t) = \sum_{i=o}^{p} c_i t^i, \ \forall p \in \mathbb{N},$$

$$[0/0]_{L,L^*}(t) = 0,$$

$$[q/0]_{L^*}(t^{-1}) = \sum_{i=o}^{q} c_{-i}\, t^{-i}, \ \forall q \in \mathbb{N} \text{ such that } q \geq 1.$$

A last property gives the mean to compute the elements of the mixed Padé table.

Property 10.

The usual cross rule is verified in all the mixed Padé table, that is to say :

$$(N-C)^{-1} + (S-C)^{-1} = (W-C)^{-1} + (E-C)^{-1}$$

if the terms C, N, S, E and W are five elements of the mixed Padé table displayed as follows :

$$\begin{array}{ccc} & \cdot\, N & \\ \overset{W}{\cdot} & \cdot\, C & \overset{E}{\cdot} \\ & \cdot\, S & \end{array}$$

The proof of this property can be found in [3] and [4] .

BIBLIOGRAPHY.

1 BREZINSKI C.
Padé-type approximant and general orthogonal polynomials.
ISNM vol. 50, Birkhaüser Verlag, Basel, 1980.

2 DRAUX A.
Polynômes orthogonaux formels dans une algèbre non commutative.
Publication ANO 92 - Université de Lille I, 1982.

3 DRAUX A.
Approximants de type Padé et de Padé.
Publication ANO 96 - Université de Lille I, 1983.

4 DRAUX A.
Approximants de type Padé et de Padé en deux points.
Publication ANO 110 - Université de Lille I, 1983.

5 DRAUX A.
Bibliographie - Index.
Publication ANO 145 - Université de Lille I, 1984.

6 DRAUX A.
Polynômes orthogonaux formels - Applications.
Lecture Notes in Mathematics 974 - Springer Verlag, Heidelberg, 1983.

7 VAN ISEGHEM J.
Applications des approximants de type Padé.
Thèse 3ème cycle, Lille, 1982.

Existence of Chebyshev approximations by transformations of powered rationals

CHARLES B. DUNHAM

Computer Science Department
The University of Western Ontario
London, Canada N6A 5B7

1. Introduction

Consider a generalization of the approximation problem of (Dunham, 1977b, 1983) in which we do (real or complex) uniform approximation of a continuous function f on a topological space X by

$$F(A,x) := w(x,R(A,x))$$

where w is a function from $X\times\overline{R}$ (resp. $X\times\overline{C}$) to $\overline{R}$ (resp. $\overline{C}$) ($\overline{R}$ and $\overline{C}$ denote the extended real line and extended complex plane),

$$R(A,x) := [P(A,x)]^s / [Q(A,x)]^r \tag{1}$$

s and r are natural numbers, P,Q are linear approximating functions (let $\{\phi_1,\dots,\phi_n\},\{\psi_1,\dots,\psi_m\}$ be linearly independent subsets of $C(X)$ and define

$$P(A,x) := \sum_{k=1}^{n} a_k\phi_k(x) \qquad Q(A,x) := \sum_{k=1}^{m} a_{n+k}\psi_k(x) \quad)$$

such as in the previously cited papers or (Rice, 1969, p. 76), and parameters A are taken from a set P. Special cases were considered in most of the references with nonstandard (1) considered by Lau (1974), Kaufman and Taylor (1978), and Dunham (----).

For $\gamma>0$, if we multiply $P(A,.)$ by γ^r and $Q(A,.)$ by γ^s, we have multiplied numerator and denominator of (1) by γ^{rs}, getting the same ratio. Thus there is no loss of generality in normalizing Q, say by

$$\sum_{k=1}^{m} |a_{n+k}| = 1 \tag{2}$$

exactly as in (Dunham, 1977b). Let $\hat{P}$ be the set of parameters A satisfying (2).

Let W denote the set of x in X such that

$$|w(x,y)| \to \infty \qquad\qquad |y| \to \infty \tag{3}$$

We don't assume (3) for all x, since there are problems of practical interest eg. Taylor and Williams (1974) and Dunham (1976), where (3) fails.

For points x at which $Q(A,.)$ vanishes, a convention appears necessary to define the value of $F(A,x)$. It should be noted that in the classical case $F(A,x) = P(A,x) / Q(A,x)$, $Q(A,.) \geq 0$ on X, it was shown to be necessary that every such $F(A,.)$ be included for existence (Dunham, 1974) if we include all $F(A,.)$ with $Q(A,.) > 0$ (the admissible rationals).

2. Boehm-type Convention

DEFINITION: The pair (W,Q) have the dense bounded property if for any $Q(A,.) \not\equiv 0$ the set of points x at which $Q(A,.) \neq 0$ and $x \in W$ is dense in X.

In case $W = X$, this property reduces to Boehm's dense nonzero property (Boehm, 1965 : Rice, 1969, p. 84 : Dunham, 1977b, p. 285).

With the dense bounded property we can assign $F(A,.)$ a value for points x at which $Q(A,.)$ vanishes, a generalization of the convention of Boehm (Boehm, 1965 : Rice, 1969, p. 84). In the real case we define

$$F(A,x) = \limsup_{\substack{y \to x \\ Q(A,y) \neq 0}} F(A,y) \tag{4}$$

The definition for the complex case is a straightforward modification of that given by (Dunham, 1977b, p. 285). In particular the modulus of $F(A,x)$ is given by applying the modulus to $F(A,x)$ and $F(A,y)$ of (4).

3. Existence Theorem

THEOREM: Let f be continuous and bounded on X. Let (W,Q) have the dense bounded property. Let P be a non-empty closed subset of $\hat{P}$. Let $w(x,y)$ be continous from finite y to $\overline{R}$ (resp. $\overline{C}$) for all $x \in X$. Then there exists a best parameter X from P to f.

<u>Proof</u>: Let $\|f-F(A^k,.)\|$ be a decreasing sequence with limit $\rho(f) = \inf\{\| f-F(A,.)\| : A \varepsilon P\}$. If $\rho(f) = \infty$ the theorem is trivial, so we can assume without loss of generality that $\|f-F(A^1,.)\| < \infty$. If $\|F(A,.) - F(A^1,.)\| > 2\|f-F(A^1,.)\|$ then by the triangle inequality

$$\|f-F(A,.)\| > \|f-F(A^1,.)\| .$$

It follows that $\{\| F(A^k,.)\|\}$ is a bounded sequence. Let V be a finite subset of W on which $\{\phi_1,\dots,\phi_n\}$ are independent. Now by the inequality

$$|R(A,x)| = |P(A,x)|^s / |Q(A,x)|^r \geq |P(A,x)|^s / [\sum_{k=1}^{m} \|\psi_k\|]^r$$

$\|P(A^k,.)\|_V$ is bounded and by classical arguments (Rice, 1964, p. 25) the numerator coefficients of $\{A^k\}$ are bounded (the denominator coefficients were already bounded by the normalization (2)). Hence $\{A^k\}$ is a bounded sequence and has an accumulation point A^0 , assume without loss of generality $\{A^k\} \to A^0$. If $Q(A^0,x) \neq 0$, $R(A^k,x) \to R(A^0,x)$ (finite), $F(A^k,x) \to F(A^0,x)$ and

$$|f(x) - F(A^0,x)| = \lim_{k\to\infty} |f(x) - F(A^k,x)| \leq \rho(f) .$$

If $Q(A^0,x) = 0$,

$$|f(x) - F(A^0,x)| = \limsup_{\substack{y\to x \\ Q(A^0,y)\neq 0}} |f(y) - F(A^0,y)| \leq \rho(f) .$$

4. <u>Examples of Closed Parameter Sets</u>

Some examples of closed subsets of $\hat{P}$ for the complex case and $w(x,y) = \sigma(y)$ are given in Dunham (1977b). Examples of closed subsets for the real case are given below.

1. $\hat{P}$ is a closed non-empty set
2. $P_{GE}(Y) = \{A : Q(A,y) \geq 0$ on set $Y, A \varepsilon \hat{P}\}$ is a closed set if $\{\psi_1,\dots,\psi_m\}$ are continuous on Y . Restrictions on the range of Q can be generalized: see example 2 of Dunham (1977b)
3. Let $\mu,\nu \;\varepsilon\; C(X)$, $\mu \leq \nu$. The set of $A \varepsilon \hat{P}$,

$$\mu \leq F(A,.) \leq \nu \qquad (5)$$

is closed under the hypotheses of the preceding theorem.

Proof: By the convention we need only consider x at which $Q(A,x) \neq 0$. At such x, $R(A,x)$ is finite and $F(A^k,x) \to F(A,x)$. If (5) failed, it would also fail for A^k, k large.

With some care, the example could be extended to cover (5) holding on sets other than X.

5. Admissible Approximation

Consider real approximation. An approximant $F(A,.)$ is called admissible if $Q(A,.) > 0$ on X. It is of practical importance to know cases in which a best approximation by admissible functions is guaranteed to exist. Counter-examples from Taylor and Williams (1974) and Dunham (1976) suggest no such guarantee is easily obtainable if $|y| \to \infty$ doesn't imply $|w(x,y)| \to \infty$. Also, even in the case of approximation by ratios of power polynomials, existence of an admissible approximation is not guaranteed if X is not an interval (Dunham, 1983, remark p. 337).

THEOREM: Let $[\alpha,\beta]$ be a closed finite interval. Let $|y| \to \infty$ imply $|w(x,y)| \to \infty$ for all $x \in [\alpha,\beta]$. Let P and Q generate all power polynomials of degree $n-1$ (resp. $m-1$). Let $r = ts$, t a natural number. A best admissible approximation exists to all f continuous and bounded on $[\alpha,\beta]$.

Proof: There is a best approximation $F(A^0,.)$ with parameter from $P_{GE}[\alpha,\beta]$. Let $Q(A^0,.)$ have a zero c in $[\alpha,\beta]$. Then $(x-c)^{st}$ is a factor of the denominator of (1). If $(x-c)^{st}$ is not a factor of the numerator of (1), $|F(A^0,c)| = \infty$. But this is impossible unless $\rho(f) = \infty$ (in which case the theorem is trivial). Hence $(x-c)^{st}$ is a factor of the numerator of (1). We can, therefore, cancel $(x-c)^{st}$ from numerator and denominator of (1). (REMARK - after we have done this, the numerator is still a polynomial raised to the sth power). We repeat until the denominator has no zeros (and is, therefore, of one sign).

But in general we cannot cancel out common factors.

EXAMPLE: $F(A,x) = x^2/x$ in the case $s=2, r=1, X=[0,1]$,

EXAMPLE: $F(A,x) = x^3/(x)^2$ in the case $s=3, r=2, X=[0,1]$,

EXAMPLE: $F(A,x) = (x^2)^2/x^3$ in the case $s=2, r=3, X=[0,1]$.

A similar process can be used for infinite intervals or the complex case (in which we want denominators with no zeros on compact X).

References

Boehm, B.W. (1965). Existence of best rational Tchebycheff approximations. Pacific J. Math 15, 19-28.

Dunham, C.B. (1967). Transformed rational Chebyshev approximation. Numer. Math. 10, 147-152.

Dunham, C.B. (1974). Necessity of rationals with non-negative denominator. Mathematica 16, 251-253.

Dunham, C.B. (1976). Rational approximation with a vanishing weight function and with a fixed value at zero. Math. Comp. 30, 45-47.

Dunham, C.B. (1977a). Transformed rational Chebyshev approximation. J. Approximation Theory 19, 200-204.

Dunham, C.B. (1977b). Existence of transformed rational complex Chebyshev approximations. J. Approximation Theory 20, 284-287.

Dunham, C.B. (1983). Existence of transformed rational complex Chebyshev approximations, II. J. Approximation Theory 38, 334-337.

Dunham, C.B. (----). Rationals with repeated poles, in preparation.

Kaufman, E.H., jr. and Taylor, G.D. (1978). Uniform approximation with rational functions having negative poles. J. Approximation Theory 23, 364-378.

Lau, T.C. (1974). A class of approximations to the exponential function for the numerical solution of stiff differential equations. PhD thesis, University of Waterloo.

Rice, J.R. (1964). "The Approximation of Functions". Vol. 1, Addison-Wesley.

Rice, J.R. (1969). "The Approximation of Functions", Vol. 2, Addison-Wesley.

Schmidt, D. (1979). An existence theorem for Chebyshev approximation by interpolating rationals. J. Approximation Theory 27, 147-152.

Taylor, G.D. and Williams, J. (1974). Existence questions for the problem of Chebyshev approximation by interpolating rationals. Math. Comp. 28, 1097-1103.

BEST CHEBYSHEV RATIONAL APPROXIMANTS

AND POLES OF FUNCTIONS

R.K.Kovačeva

Institute of Mathematics

Bulgarian Academy of Sciences

1090 Sofia

Bulgaria

Abstracts: In this work, a theorem relating to best rational Chebyshev approximants with an unbounded number of the free poles is proved. This theorem provides a sufficient condition that a given function has a pole at a given point.

Let Δ be the real segment $[-1,1]$; let the function f be real and contimuos on Δ ($f \in C(\Delta)$). For each integer n ($n \in N$) we denote by r_n the class of the rational functions of order n: $r_n = \{p/q, q \neq 0, \deg p \leq n, \deg q \leq n\}$. Let R_n be the best Chebyshev approximant to f on Δ in the class r_n:

$$\|f - R_n\|_{\Delta} = \inf\{\|f - r\|_{\Delta}, r \in r_n\},$$

where $\|\dots\|_{\Delta}$ is the sup-norm on Δ. It is well known that the rational function R_n always exists and is uniquely determined by the alternation theorem of Chebyshev (see [1]). We set

$$R_n = P_n/Q_n,$$

where the polynomials P_n and Q_n have no a common divisor and Q_n is monic. The zeros $\xi_{n,1},\dots,\xi_{n,\mu_n}$ of Q_n are called free poles of R_n; $\mu_n \leq n$. For each $n \in N$ we denote by $\mathcal{P}_n$ the set of the poles of R_n in the extended complex plane $\bar{C}$ (the poles are counted with regard to their multiplicities). Let L be the set of the concentration points of $\mathcal{P}_n$, as $n \in N$, in $\bar{C}$ and l the set of the limit points.

The following theorem is found in [2]:

Theorem 1: Let $f \in C(\Delta)$. Suppose, $L = l$, $L \cap \Delta = \emptyset$ and L is finite. Then

a) f is holomorphic in the domain $\bar{C}-L$ ($f\in H(\bar{C}-L)$);

b) for any compact set K, $K\subset C-L$, the relation

(1) $$\lim \| f - R_n \|_K^{1/n} = 0$$

holds.

We shall use the following notation: for each $a\in L$ and $n\in N$ we shall renumber the free poles of R_n so that $|\xi_{n,k}(a) - a| \leq |\xi_{n,k+1}(a) - a|$ $k=1,\ldots,\mu_n$.

The result of the present paper is the following

Theorem 2. Let $f\in C(\Delta)$. Suppose, $L = 1$ is finite, $L\cap\Delta = \emptyset$ and there is a point a, $a\in L-\infty$ and an integer p such that

(2) $$\overline{\lim} |\xi_{n,k}(a) - a|^{1/n} < 1, \quad k = 1,\ldots,p$$

(3) $$\underline{\lim} |\xi_{n,p+1}(a) - a| > 0$$

Then the function f has a pole of order p at the point a.

2. Theorem 2 can be proved with the method of Gončar introduced in [3]. We shall give another idea of a proof.

We first suppose that f is not a rational function. In the opposite case $R_n \equiv f$ for all $n\in N$ sufficiently large.

We say that the integer n is normal if the number of the poles of R_n in $\bar{C}$ is equal to n. Let Λ be the set of the normal integers. The alternation theorem of Chebyshev implies:

a) if $n\in\Lambda$ and if (n',n'') is such a pair that $n',n''\in N$ and $n'<n<n''$; then $R_n \equiv R_{n'}$;

b) if $n_0\in\Lambda$ and if the order of the alternation of $f - R_{n_0}$ on Δ is equal to $2n_0 + 1 + m$ with $m>0$, then $R_{n_0+s} \equiv R_{n_0}$ for $s = 1,\ldots,m$.

Consequently Λ is infinite if and only if f is not a rational function.

We obtain, later, from the theorem of the alternation that for each pair (n',n'') of consistent normal integers (see [2])

(4) $$(R_{n'} - R_{n''})(z) = A_{n'} \frac{w_{n'+n''}(z)}{(Q_{n'}Q_{n''})(z)}, \quad A_{n'} \neq 0,$$

where $w_{n'+n''}$ is a polynomial of degree $= n'+n''$, it is monic, its zeros are simple and belong to Δ. $A_{n'}$ is given by the formula

(5) $$A_{n'} = (P_{n''}Q_{n'} - P_{n'}Q_{n''})(b)/w_{n'+n''}(b),$$

where b is an arbitrary complex number, $b\in C-\Delta$.

We shall assume in our further considerations that $\Lambda \equiv N$. It is clear that we don't lose the generality (see a) and b) in the stament above). In that case (4) and (5) are valid for each $n \in N$.

We set now (for n N, n is sufficiently large)

$\xi_{n,k}(a) = \xi_{n,k}$, k=1,...,p,

$q_n(z) = \prod_{k=1}^{p} (z - \xi_{n,k})$ and

$Q_n^* = Q_n/q_n$.

Denote by P_n^*/q_n the principal part of R_n at $\xi_{n,1},\ldots,\xi_{n,p}$. The calculation implies that

(6) $$P_n(\xi_{n,k}) = P_n^*(\xi_{n,k})Q_n^*(\xi_{n,k}).$$

The next lemma is the basic to our later conciderations:

Lemma: In the conditions of Theorem 2 the following assertions are equivalent:

1. The function f has a pole of order p at a;
2. For each k=1,...,p
$$\lim |P_n^*(\xi_{n,k})|^{1/n} = 1.$$

Proof of the lemma: We shall use the following notation: $U_r(\Gamma_r)$ is an open disk (circumference) of radius r centered at a. Select a positive number r such that $U_r \cap (L \cup \Delta) = \emptyset$. We set $U_r = U$ and $\Gamma_r = \Gamma$. It follows from Theorem 1 that

(7) $$\lim \| f - R_n \|_{\Gamma}^{1/n} = 0$$

Consequently the function $\varphi(z)$, given by $\varphi(z) = f(z)(z-a)^p$, is holomorphic on $\bar{U}$ (see (2) and (3)). Consequently, for $z \in \bar{U}-a$

$$f(z) = p_1(z)\,(z-a)^{-p} + f_1(z),$$

where p_1 is a polynomial of degree $\leq$ p-1 and $f_1 \in H(\bar{U})$. The function f has a pole of order p at a if and only if $p_1(a) \neq 0$; $f \in H(\bar{U})$ if and only if $p_1 \equiv 0$.

Using the results of [4], we obtain from (7)

$$\lim \|(P_n^* q_n^{-1})(z) - p_1(z)(z-a)^{-p}\|_{\Gamma}^{1/n} = 0$$

Consequently (see (2))

$$\lim \|P_n^* - p_1\|_{\bar{U}}^{1/n} < 1$$

The last inequality yields

$$\lim |P_n^*(\xi_{n,k}) - p_1(a)|^{1/n} < 1$$

This proves the lemma.

We note now that in the conditions of the theorem

(8) $$\overline{\lim} |Q_n^*(\xi_{n,k})(Q_n^*(\xi_{n+1,k}))^{-1} - 1|^{1/n} < 1, \quad K = 1,\ldots,p$$

Indeed, since $R_n \neq R_{n+1}$, we obtain from (5) that $\xi_{n,k} \neq \xi_{n,k+1}$. On the other hand, for arbitrary $\varepsilon, \varepsilon > 0$, there is $U_r(a) = U$ such that

(9) $\quad |Q_n^*(t)(Q_n^*(z))^{-1}| < (1+\varepsilon)^n$

holds for each $z, t \in \overline{U}$ and every $n \in \mathbb{N}$, n sufficiently large, $n \geq n_1(\varepsilon)$ (see [3]); $r = r(\varepsilon)$.

Using Gauchy's formula for $Q_n^*(\xi_{n,k})$ and $Q_n^*(\xi_{n+1,k})$, we obtain

$$\left|\frac{Q_n^*(\xi_{n,k}) - Q_n^*(\xi_{n+1,k})}{Q_n^*(\xi_{n+1,k})}\right| = \frac{1}{2\pi}\left|\int_\Gamma \frac{Q_n^*(t)(\xi_{n,k} - \xi_{n+1,k})\, dt}{(t-\xi_{n,k})(t-\xi_{n+1,k})Q_n^*(\xi_{n+1,k})}\right| \leq$$

$$\leq c_1 \frac{\|Q_n\|_{\overline{U}}}{Q_n^*(\xi_{n+1,k})} |\xi_{n+1,k} - \xi_{n,k}| \quad , \; n \geq n_2 \geq n_1(\varepsilon).$$

(c_1 is a positive constant).

The statement (8) follows now from (9) and from (2).

We prove in the same way that

(10) $\quad \overline{\lim} |Q_{n+1}^*(\xi_{n,k}) Q_{n+1}^*(\xi_{n+1,k}))^{-1} - 1|^{1/n} < 1 \qquad$ and

(11) $\quad \overline{\lim} | w_{2n+1}(\xi_{n,k}) w_{2n+1}^{-1}(\xi_{n+1,k}) - 1|^{1/n} < 1.$

We notice now that in the conditions of Theorem 2

(12) $\quad \lim |P_n^*(\xi_{n,k})|^{1/n} = 1$

To prove (12) we shall use an idea of Buslaev (see [5]), namely, we shall evaluate A_n by putting $b = \xi_{n,k}$ and $b = \xi_{n+1,k}$ in (5) and then multiply the results; we obtain (we remember that $\xi_{n,k} \neq \xi_{n+1,k}$)

$$\prod_{k=1}^{p} \frac{P_n(\xi_{n,k}) Q_{n+1}^*(\xi_{n,k})}{w_{2n+1}(\xi_{n,k})} = \prod_{k=1}^{p} \frac{P_{n+1}(\xi_{n,k}) Q_n^*(\xi_{n,k})}{w_{2n+1}(\xi_{n+1,k})}$$

From (6), we get

$$\prod_{k=1}^{p} \frac{P_n^*(\xi_{n,k})}{P_{n+1}^*(\xi_{n,k})} = \prod_{k=1}^{p} \frac{w_{2n+1}(\xi_{n,k})}{w_{2n+1}(\xi_{n+1,k})} \frac{Q_n^*(\xi_{n+1,k})}{Q_n^*(\xi_{n,k})} \frac{Q_{n+1}^*(\xi_{n+1,k})}{Q_{n+1}^*(\xi_{n,k})}$$

The statement (12) follows from (9), (10) and (11).

The validity of Theorem 2 follows now from the lemma and from (12).

References

1. N.I.Ahiezer , Approximationtheory, Moskow, 1965,(Russian).

2. K.N.Lungu, On properties of functions resulting from the assymptotik of the poles of rational best approximants, International Conferen-

ce on Constructive Function Theory, Varna, 1983, pp. 106-110 (Russian)

3. A.A.Gončar, On the convergence of diagonal Padé approximants in the spherical metrics, Papers dedicated to academician L.Iliev 70th birthday, p. 29-36, Publishing House of the Bulg.Acad. of Sciences, 1984.

A.A.Gončar, L.D.Grigoryan, Estimates of the norm of holomorphic functions, Mat.Sb. 99(1976), 634-638.

5. V.I.Buslaev, On the poles of the mth row in the Padé table, Mat. Sb. 117(1982), 435-441.

HYPERBOLIC APPROXIMATION OF MEROMORPHIC FUNCTIONS

K. Reczek

1. Consider a function f meromorphic in the unit disc D. Let f have $m < \infty$ poles in D (counted with their multiplicities). Denote

$$M_f(r) = \sup\{|f(z)|: |z| = r\}.$$

We define the order of f as follows:

$$\varrho(f) = \inf\{\mu > 0: \exists r_0 < 1: \forall r \in (r_0, 1)\ M_f(r) < \exp(1 - r)^{-\mu}\}.$$

If $m = 0$ then $\varrho(f)$ can be computed by means of the coefficients of the Maclaurin expansion of f. If f is a meromorphic function, then the Maclaurin coefficients have to be replaced by some coefficients of the Padé approximants (see [1], [5]). Our aim is to estimate the growth of f if the values $f(z_n)$ for some sequence (z_n) bounded in D are known.

2. Let (z_n) be a sequence of points such that $|z_n| \leq d < 1$ for each n. Denote

$$\omega_0(z) = 1,$$

$$\omega_{n+1}(z) = \omega_n(z) \cdot \left(\frac{z - z_{n+1}}{1 - \bar{z}_{n+1} z}\right),$$

$$\alpha_n(z) = (z - z_1) \cdot \ldots \cdot (z - z_n),$$

$$\beta_n(z) = (1 - \bar{z}_1 z) \cdot \ldots \cdot (1 - \bar{z}_n z).$$

The rational function P is called a hyperbolic polynomial (more precisely: a hyperbolic polynomial with respect to the sequence (z_n)) of degree k if

$$P(z) = \sum_{i=0}^{k} a_i \omega_i(z), \quad a_k \neq 0,$$

or, equivalently, $P(z) = \Pi(z)/\beta_n(z)$, where Π is an algebraic polynomial of degree k. Denote by $\mathcal{P}_n$ (resp. H_n) the space of all algebraic (resp. hyperbolic) polynomials of degree not greater than n.

Suppose now that f is holomorphic at each point z_i. We define the (k/l)-th hyperbolic Newton - Padé approximant of f as a function

$$f_{k/l} = \frac{P_{k/l}}{Q_{k/l}}$$

such that $P_{k/l} \in H_k$, $Q_{k/l} \in \mathcal{P}_l - \{0\}$ and the quotient function

$$\frac{Q_{k/l} \cdot f - P_{k/l}}{\alpha_{k+l+1}}$$

is holomorphic at z_i for $i = 1, 2, \ldots, k+l+1$. In the sequel we shall always assume that $Q_{k/l}(z) = z^{l'} + \ldots$, $l' \leq l$. The existence and uniqueness of $f_{k/l}$ can be proved in a similar way as for the Padé approximants. Note that the described method of approximation is related to the generalized Padé approximation, which has been led by Gončar [2].

3. First we shall formulate the following

Lemma 1. Let $\varphi_w(z) = \dfrac{z - w}{1 - \bar{w}z}$. For every $d < 1$ there exist positive numbers c_1, c_2 such that for every w, $|w| \leq d$, and for every $z \in D$

$$c_1(1 - |z|) \quad 1 - |\varphi_w(z)| \leq c_2(1 - |z|). \tag{1}$$

Proof. This lemma follows immediately from the mean value theorem by setting $c_1 = \inf\{|\varphi_w'(z)| : |z| = 1, |w| \leq d\}$ and $c_2 = \sup\{|\varphi_w'(z)| : |z| = 1, |w| \leq d\}$.

Theorem 1. By the above assumptions about f, the zeros of the approximants $f_{n/m}$, $n = 1, 2, \ldots$, tend to the poles of f and $\lim_{n\to\infty} f_{n/m}(z) = f(z)$ uniformly in every compact subset of D which does not contain any poles of f.

Proof. This theorem is a hyperbolic version of the well-known Montessus de Ballore theorem. Our proof is based on an idea related to the Shapiro proof (cf. [4]).

Denote by $\zeta_1, \ldots, \zeta_m$ the poles of f. Put

$$q(z) = (z - \zeta_1) \cdot \ldots \cdot (z - \zeta_m).$$

Then $f(z) = \dfrac{\Phi(z)}{q(z)}$, where Φ is holomorphic in D. Let K be a compact subset of $D - \{\zeta_1, \ldots, \zeta_m\}$. Suppose that there exists a number $\varepsilon > 0$ and a sequence of integers (n_k) such that $|f_{n_k}(x_k) - f(x_k)| \geq \varepsilon$ for some points $x_k \in K$. Since m does not change, we can simplify the notations: $f_k := f_{n_k/m} = P_k/Q_k$. Put

$$Q_k(z) = Q_{k,1}(z) \cdot Q_{k,2}(z),$$

where the zeros of $Q_{k,1}$ (resp. $Q_{k,2}$) lie inside the disc $\{z: |z| < 2\}$ (resp. out of the disc). We may assume that

$$\lim_{k\to\infty} Q_{k,1}(z) = Q(z). \tag{2}$$

Choose two numbers r and R such that $|z| < r < R < 1$ for $z \in K$ and $|\zeta_j| < r$ for $1 \leq j \leq m$. By the Hermite interpolation formula we get from the definition of f_k that

$$\phi(z)Q_{k,1}(z) - P_k(z)q(z) = \frac{\alpha_{n_k+m+1}(z)}{2\pi i\beta_{n_k}(z)Q_{k,2}(z)} \cdot$$

$$\int_{|t|=R} \frac{\phi(t)\beta_{n_k}(t)Q_k(t)}{(t-z)\alpha_{n_k+m+1}(t)}\, dt.$$

Hence, by (1),

$$|\phi(z)Q_{k,1}(z) - P_k(z)q(z)| \leq C_R \frac{|\omega_{n_k}(z)| \cdot \max\{|Q_k(t)| : |t| = R\}}{\min\{|\omega_{n_k} t| : |t| = R\} \cdot |Q_{k,2}(z)|}$$

$$\leq C_R' \left(\frac{1 - c_1(1-r)}{1 - c_2(1-R)}\right)^{n_k}$$

where C_R' does not depend on m. But R can be chosen as close to 1 as we wish. Hence, since (2) holds, we have

$$\lim_{k\to\infty} P_k(z)q(z) = \phi(z)Q(z) \tag{3}$$

uniformly for $z \in K$. Consequently, $Q(\zeta_j) = 0$ for $1 \leq j \leq m$. Thus, $Q = q$ and $Q_{k,2} = 1$ for large k. This implies that $\lim_{k\to\infty} Q_k(z) = q(z)$ and, since (3) holds, f_k tends to f uniformly on K.

<u>Theorem 2</u>. Let f be a function as in theorem 1. Put

$$f_n := f_{n/m} = \frac{P_n}{Q_n},$$

$$P_n(z) = \left(\sum_{k=0}^{n} p_{ni} z^i\right) / \beta_n(z).$$

Let f be of order ρ. Then

$$\limsup_{n \to \infty} \frac{\ln^+ \ln^+ |p_{nn} + \bar{z}_n p_{n-1,n-1}|}{\ln n} = \frac{\rho}{\rho+1}, \tag{4}$$

where $\ln^+ a = \ln \max(1, a)$.

Proof. Let $z \in D - \{\zeta_1, \ldots, \zeta_m\}$. Then, according to theorem 1,

$$f(z) = \lim_{n \to \infty} f_n(z) = f_{n_0}(z) + \sum_{n > n_0} \left[f_n(z) - f_{n-1}(z) \right] \tag{5}$$

It follows from the definition of f_n that the function

$$\frac{f_n(z) - f_{n-1}(z)}{\alpha_{n+m}(z)} = \frac{[f(z) - f_{n-1}(z)] - [f(z) - f_n(z)]}{\alpha_{n+m}(z)}$$

is holomorphic at z_j for $1 \leq j \leq n+m$. We can easily check that

$$f_n(z) - f_{n-1}(z) = \frac{(p_{nn} + \bar{z}_n p_{n-1,n-1}) \alpha_{n+m}(z)}{\beta_n(z) Q_{n-1}(z) Q_n(z)}. \tag{6}$$

Suppose that $|p_{nn} + \bar{z}_n p_{n-1,n-1}| \leq \exp n^{\frac{\mu}{\mu+1}}$ when n is large enough. Then it follows from (5) and (6) that, for $|z|$ close to 1,

$$|f(z)| \leq K \cdot \sum_{n > n_0} \exp \, n^{\mu/(\mu+1)} \cdot |\omega_n z|,$$

where K is a constant. Hence, by (1),

$$M_f(r) \leq K \cdot \sum_{n > n_0} [1 - c_1(1 - r)]^n \cdot \exp\left(n^{\mu/(\mu+1)}\right).$$

By a standard reasoning we check the inequality $\rho(f) \leq \mu$ and, as a consequence, we get that

$$\limsup_{n \to \infty} \frac{\ln^+ \ln^+ |p_{nn} + \bar{z}_n p_{n-1,n-1}|}{\ln n} \geq \frac{\rho}{\rho+1} \tag{7}$$

/cf. e.g. [3], proof of lemma 2.5, or [5], proof of theorem 2.1./

Suppose now that equality (4) does not hold. Then, according to (7), there esists such a number $\mu > \rho$, that

$$\limsup_{n \to \infty} \frac{\ln^+ \ln^+ |p_{nn} + \bar{z}_n p_{n-1,n-1}|}{\ln n} > \frac{\mu}{\mu+1}. \tag{8}$$

We shall make use of the following equality, which is a simple consequence of the definition of f_n and the Cauchy formula:

$$p_{nn} = \frac{1}{2\pi i} \int_{|t|=r} \frac{\phi(t) Q_n(t) \beta_n(t)}{\alpha_{n+m+1}(t)} \, dt$$

for r close to 1. Thus,

$$|p_{nn}| \leq k \cdot \exp\left((1-r)^{-\mu}\right) \cdot \left(1 - c_2(1-r)\right)^{-n},$$

where k depends neither on r nor on n. Put $r = 1 - \left(\frac{\mu}{n}\right)^{1/\mu+1}$. Then

$$\limsup_{n\to\infty} \frac{\ln^+ \ln^+ |p_{nn} + \bar{z}_n p_{n-1,n-1}|}{\ln n} \leq \limsup_{n\to\infty} \frac{\ln^+ \ln^+ |p_{nn}|}{\ln n} \leq \frac{\mu}{\mu+1}.$$

Last inequality contradicts (8). This ends the proof.

Before we shall state the next theorem, we formulate the following

Lemma 2. Let f be a function holomorphic in a domain containing the set $\overline{\{z_n : n = 1, 2, \ldots\}}$. Suppose that, by previous denotations, $\deg Q_n = m$ for almost every n and

$$\limsup_{n\to\infty} |p_{nn} + \bar{z}_n p_{n-1,n-1}|^{1/n} \leq 1. \tag{9}$$

Then f can be extended to a function meromorphic in D, with not more than m poles in D.

The proof of this lemma is essentially the same as the proof of [5, theorem 3.2], so we omit it.

Theorem 3. Let f be a function holomorphic in a domain containing the set $\overline{\{z_n : n = 1, 2, \ldots\}}$; suppose that there exists a compact set A disjoint with the unit circle, such that all the finite poles of the approximants f_n lie in A, and $\deg Q_n = m$ for n large enough. Assume that there exists a positive number ϱ such that the equality (4) is satisfied. Then f can be extended to a function meromorphic in D, with not more than m poles. Moreover, the order of this extension is equal to ϱ.

Proof. The condition (4) implies inequality (9). Thus, f can be regarded as a function meromorphic in D. Due to the assumptions about the poles of f_n, we can follow the proof of theorem 2, with only slight modifications.

References

[1] F. Beuermann, Wachstumsordnung, Koeffizientenwachstum und Nullstellendichte bei Potenzreihen mit endlichem Konvergenzkreis, Math. Zeit. 33 (1931), 98-108.

[2] A. A. Gončar, On the convergence of generalized Padé approximants to meromorphic functions /Russian/, Mat. Sb. 98 (140) (1975), 564-577.

[3] A. Janik, A characterization of the growth of analytic functions by means of polynomial approximation, Univ. Iag. Acta Math. 24 (1984), 295-319.

[4] J. Karlsson, H. Wallin, Rational approximation by interpolation procedure in several variables, in: Padé and rational approximation, A. P., New York - San Francisco - London, 1977, pp. 83-100.

[5] K. Reczek, Rational approximation and estimation of the growth of meromorphic functions, Zesz. Nauk. Akad. Górn.-Hutn. /Opuscula Mathematica/ 3 (1986)/to appear/.

Krzysztof Reczek
Institute of Mathematics
University of Mining and Metallurgy

Al. Mickiewicza 30,
30-059 Kraków,
Poland.

THREE DIFFERENT APPROACHES TO A PROOF OF CONVERGENCE FOR PADÉ APPROXIMANTS

Herbert Stahl *
TU Berlin/Sekr. FR 6-8
Franklinstr. 28/29
1000 Berlin 10 (FRG)

Abstract

Three different ways of proving the convergence of close-to-diagonal sequences of Padé approximants to functions with branch points are compared. It is assumed that the functions to be approximated have all their singularities in a compact set of capacity zero.

1. Introduction

In this lecture we will look at three different approaches to a proof of convergence for Padé approximants to functions $f(z)$ with branch points. It is assumed that all singularities of $f(z)$ are contained in a compact set $E \subseteq \hat{\mathbb{C}}$ of (logarithmic) capacity $cap(E) = 0$. ($\hat{\mathbb{C}}$ denotes the extended complex plane.) Since it is convenient to deal with Padé approximants expanded about infinity, we further assume that $f(z)$ is given by a function element with positive radius of convergence

(1.1) $$f(z) = \sum_{j=0}^{\infty} f_j z^{-j}.$$

Thus, the Padé approximant

(1.2) $$[m/n](z) := \frac{p_{mn}(\frac{1}{z})}{q_{mn}(\frac{1}{z})}, \quad m, n \in \mathbb{N},$$

is defined by the relation

(1.3) $$p_{mn}(\tfrac{1}{z}) - q_{mn}(\tfrac{1}{z}) f(z) = O(z^{-m-n-1})$$

---

* Research supported in part by Natural Sciences and Engineering Research Council Canada.

for $z \to \infty$, where $p_{mn} \in \Pi_m$ and $q_{mn} \in \Pi_n$. (Π_n, $n \in \mathbb{N}$, denotes the collection of all polynomials of degree not greater than n.)

The three approaches to be surveyed include Nuttall's method [1],[2] and two methods developed by the author, the first of which is contained in [3] and [4], the second one is new. While the first two of the three approaches are based on the investigation of orthogonal polynomials, the third one is totally different in this respect. No use is made of the well-known orthogonality property of the denominator polynomials of the Padé approximants. This special feature allows us to apply the method in other fields, in particular, we can use it for the investigation of Hermite-Padé polynomials of type I (the so-called Latin case), which represent an important and interesting generalization of Padé polynomials (cf. [5] of [6]). Only minor modifications are necessary to cover this subject. However, here we shall not pursue these further questions and confine ourselves to the classical Padé approximants.

The aim of this lecture is the description and discussion of the general line of argumentation underlying the three different approaches. Thereby we shall and avoid technical subtilities, which may obscure the understanding. Rigorous proofs can be found in the given references. For the third approach, which is new, a more detailed treatment is in preparation and will be published elsewhere.

The outline of the talk is as follows: In the next section we will start with two examples and then give a theorem about the unique existence of sets of minimal capacity, which are fundamental for the convergence theorems. In Section 3 the main convergence result is stated and discussed. After this in Section 4 more information is given about the structure of the set of minimal capacity. This enables us to formulate the somewhat more restrictive assumptions of Nuttall's result in Section 5. In Section 6 some auxiliary results are presented, and then in the last three sections we survey the different approaches to proofs of the convergence theorems.

The material in the last part of this lecture has been developed during a visit with John Nuttall at the University of Western Ontario. I am greatly indepted to him for the encouragement and stimulus he gave to my research work. He has drawn my attention to Hermite-Padé polynomials, which then led to the new approach presented here. I wish to express my gratitude to the Department of Physics of the University of Western Ontario and especially to John Nuttall and his family for the warm hospitality and kindness extended to me.

2. Sets of Minimal Capacity

It has been assumed that the function $f(z)$ has all its singularities in a compact set $E \subseteq \hat{\mathbb{C}}$. Hence, it can be analytically continued along any arc in $\hat{\mathbb{C}} \sim E$ issuing from infinity. Since $f(z)$ has branch points, this continuations are multi-valued. On the other hand rational functions are single-valued in the whole extented plane $\hat{\mathbb{C}}$. Therefore, it is not possible to expect the Padé approximants to converge in the full domain $\hat{\mathbb{C}} \sim E$.

Since the function $f(z)$ has to be single-valued in any domain of convergence, there exist cuts in $\hat{\mathbb{C}}$ connecting the branch points in such a way that in the complement $f(z)$ is single-valued. On these cuts the Padé approximants do not converge to $f(z)$. Of course, there may exist many possibilities for such cuts, and the question therefore is: which of them are associated with the Padé approximants? Or to reformulate this question more sceptically: Given a function $f(z)$ with branch points, exist there specific cuts chosen and marked by the convergence behaviour of the Padé approximants? We will illustrate the situation by two examples:

Example A - - Let the four points z_j, $j = 1,\dots,4$, be defined by $\pm 1.5 \pm i$. We consider the function

(2.1) $$f_1(z) := \left[\prod_{j=1}^{4}\left(1-\frac{z_j}{z}\right)\right]^{1/4}.$$

Example B - - Using again the four points from Example A, we now consider the function

(2.2) $$f_2(z) := \left[\prod_{j=1}^{4}\left(1-\frac{z_j}{z}\right)\right]^{1/2}.$$

In Example A any system of cuts that connects all four points $\{z_j\}$ as shown in Figure 1, will force the function $f_1(z)$ to be single-valued in the complement. Here the connection of the four points $\{z_j\}$ is a necessary and sufficient condition for the single-valuedness of $f_1(z)$.

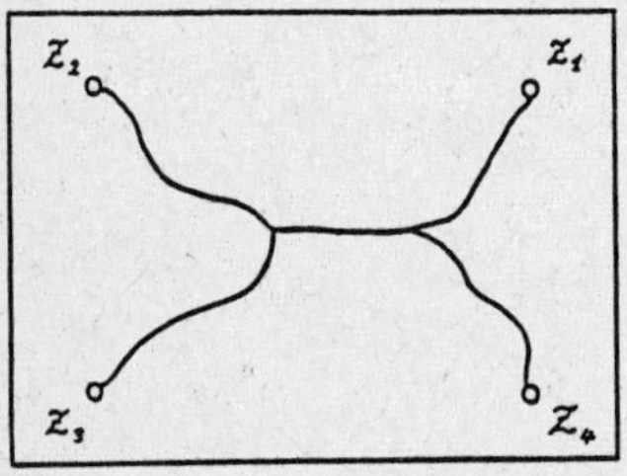

Figure 1

In Example B only pairs of points have to be connected to make $f_2(z)$ single-valued. So far we do not have criteria to decide which points have to be paired. The Figures 2a, b, and c show different possibilities.

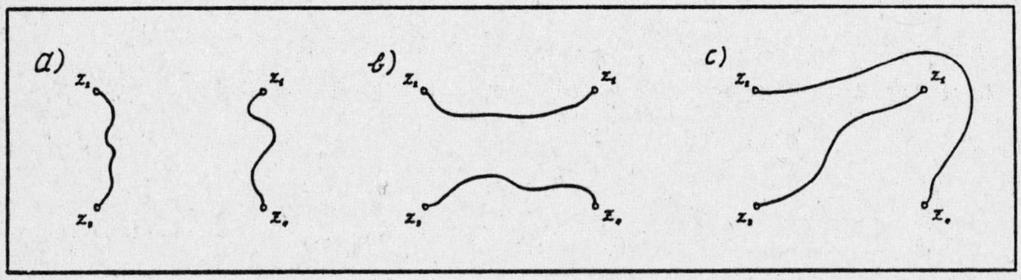

Figure 2

Example B shows that not only the exact location but also the connectivity of the cuts may allow different possibility. It turns out that the Padé approximants to $f(z)$ mark in a unique way certain cuts in and these cuts can be characterized by a property of minimal capacity. The unique existence of sets of minimal capacity is established in the next theorem, which has been proved in [8] (see also [7]).

Theorem 1 - - Let $f(z)$ be given by an analytic function element in a neighbourhood of infinity. There uniquely exists a compact set $K_0 \subseteq \mathbb{C}$ such that

(i) $D_0 := \hat{\mathbb{C}} \sim K_0$ is a domain in which $f(z)$ has a single-valued analytic continuation,

(ii) $\mathrm{cap}(K_0) = \inf_K \mathrm{cap}(K)$, where the infimum extends over all compact sets $K \subseteq \mathbb{C}$ satisfying (i), and

(iii) $K_0 \subseteq K$ for all compact sets $K \subseteq \mathbb{C}$ satisfying (i) and (ii).

Definition - - The set K_0 is called minimal set (for single-valued analytic continuation of $f(z)$) and the domain $D_0 \subseteq \hat{\mathbb{C}}$ extremal domain.

Remarks - - 1) Condition (iii) is of minor importance. If it is dropped, K_0 is uniquely determined only up to a set of capacity zero.

2) In Theorem 1 the assumption that $f(z)$ has all its singularities in a compact set $E \subseteq \mathbb{C}$ of capacity zero is not essential (cf. [8]).

3) Theorem 1 is true in the same way for analytic and meromorphic continuations. Since the poles of a meromorphic function are isolated, their capacity is zero and therefore the minimal set differs in both cases only by the points of polar singularities.

4) Below in Lemmas 1, 2, and 3, more information is given about the structure of the minimal set K_o.

5) If $f(z)$ has no branch points, it is reasonable to define E as the minimal set K_o. With this convention we have $cap(K_o) > 0$ if and only if $f(z)$ has branch points.

Let us look again at the two examples. In Figures 3 and 4, which are taken from [9], the minimal sets K_o are shown together with the poles of the Padé approximants [37/37] to the functions $f_1(z)$ and $f_2(z)$ of Exam-

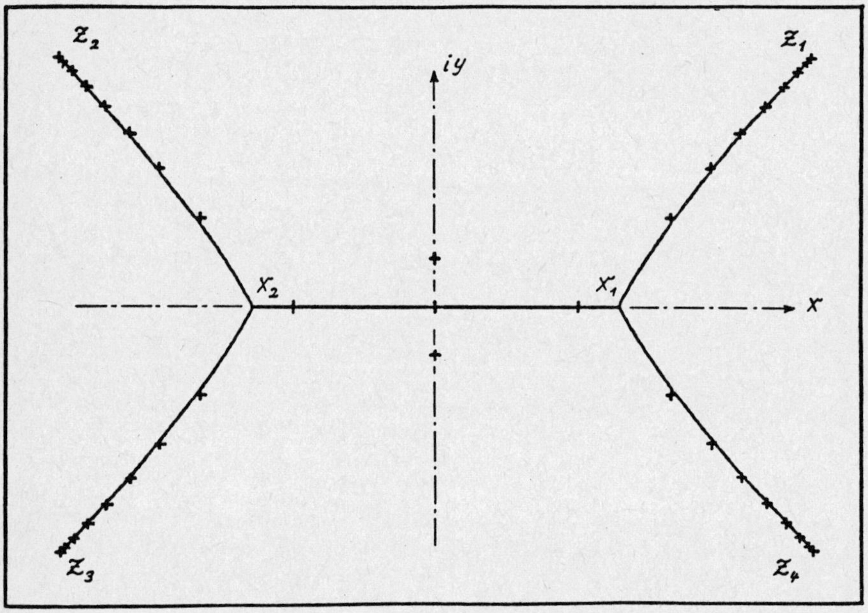

Figure 3

ple A and B, respectively. As it has already been indicated, in Example B the minimal set K_o consists of two separate cuts. In both cases a large majority of poles clusters on the cuts, but there are exceptions. In Example A these are the two poles on the imaginary axis, in Example B it is the single pole in the middle.

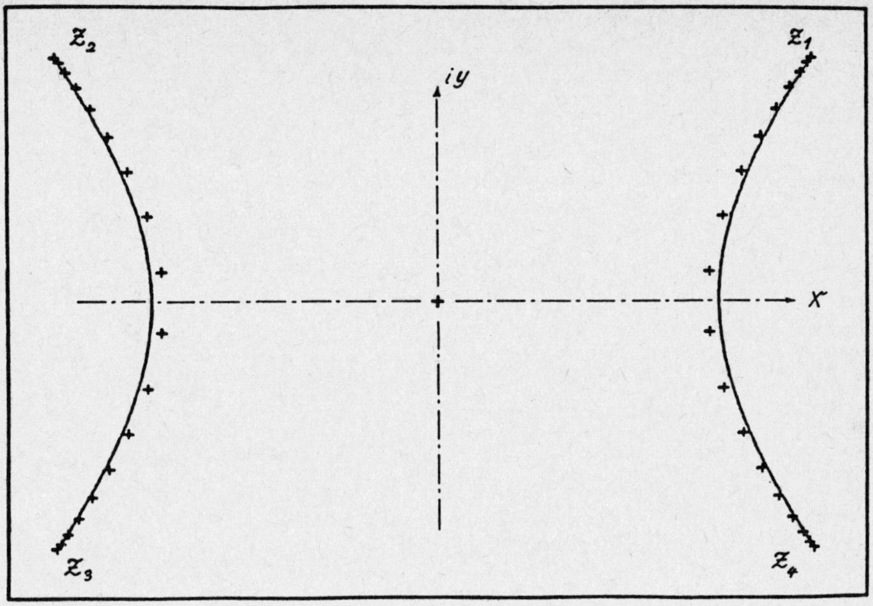

Figure 4

Unfortunately, such spurious poles may arise everywhere in $\mathbb{C}$, and they can only be excluded if the function $f(z)$ belongs to some special classes. This can be shown by rather simply structured examples (cf. [10]). However, the spurious poles appear together with nearby zeros and will therefore disturb convergence only on small sets. The possibility of spurious poles necessitates the use of a weaker concept of convergence than that of uniform or locally uniform convergence.

In Example A and B all four branch points play a role in the determination of the minimal set K_0, but this is not true in every case. It may happen that some of the branch points are hidden on another sheet of the Riemann surface defined by analytic continuation of $f(z)$. In general it is difficult to decide in advance, which of the singularities of $f(z)$ are contained in the minimal set K_0 and which will remain hidden. In [8] examples are given to illustrate this aspect of the problem.

3. The Convergence Result

Let the function $F(z)$ be defined by

(3.1) $$F(z) := \exp\{-g_{D_0}(z,\infty)\}, \quad z \in D_0,$$

where D_0 is the extremal domain for single-valued analytic continuation of $f(z)$ and $g_D(z,\infty)$ Green's function of the domain D with logarithmic singularity at infinity. We define $F(z) \equiv 0$ if $f(z)$ has no branch points. In order to make the definition of the logarithmic capacity independent of the special role of infinity we introduce

(3.2) $$\operatorname{cap}_x(V) := \operatorname{cap}\{z \in \mathbb{C};\ \tfrac{1}{z-x} \in V\}$$

for all capacitable sets $V \subseteq \hat{\mathbb{C}}$, $V \neq \hat{\mathbb{C}}$, and $x \notin V$. It is easy to see that for a sequence $V_n \subseteq V_0$, $n \in \mathbb{N}$, $\lim_{n\to\infty} \operatorname{cap}_x(V_n) = 0$ is equivalent to $\lim_{n\to\infty} \operatorname{cap}_y(V_n) = 0$ if $V_0 \subseteq \hat{\mathbb{C}}$ is a compact set and $x, y \in \hat{\mathbb{C}} \sim V_0$.

A sequence $\{(m,n)\} \subseteq \mathbb{N}^2$ is called close-to-diagonal or quasi-diagonal if

(3.3) $$\lim_{m+n\to\infty} \frac{m}{n} = 1.$$

We are now prepared to formulate the main result.

Theorem 2 - - Let the function $f(z)$ be defined by (1.1) and have all its singularities in a compact set $E \subseteq \hat{\mathbb{C}}$ of capacity zero. Then any close-to-diagonal sequence of Padé approximants $\{[m/n](z)\}$ to the function $f(z)$ converges in capacity to $f(z)$ in the extremal domain D_0. More precisely: For any compact set $V \subseteq D_0$, $\varepsilon > 0$, and $x \notin V$ we have

(3.4) $$\lim_{m+n\to\infty} \operatorname{cap}_X \{z \in V;\ |f(z)-[m/n](z)| > (F(z)+\varepsilon)^{m+n}\} = 0,$$

and

(3.5) $$\lim_{m+n\to\infty} \operatorname{cap}_X \{z \in V;\ |f(z)-[m/n](z)| < (F(z)-\varepsilon)^{m+n}\} = 0.$$

Remarks - - 1) The class of functions considered in Theorem 2 is rather large. It contains for instance all algebraic functions, but also functions with infinitely many branch points and essential singularities. What is not allowed is the existence of a natural boundary of positive capacity. For counterexamples see [11] or [12]. However, looking carefully at the proofs of Theorems 1 and 2 it can be verified that from all singularities only those on ∂D_0 have to be taken in consideration. If these singularities are contained in a compact set of capacity zero, Theorem 2 remains valid.

2) It follows from Theorem 2 that in Examples A and B any close-to-diagonal sequence of Padé approximants $\{[m/n](z)\}$ will converge in capacity outside of the minimal sets K_0 given in Figures 3 and 4.

3) As we have already mentioned in connection with the Padé approximants [37/37] to the functions $f_1(z)$ and $f_2(z)$ of Example A and B, respectively, there may be spurious poles everywhere in $\mathbb{C}$. Therefore it is necessary to use a convergence concept that allows for exceptional sets. Besides of the convergence in capacity also convergence in (planar Lebesque) measure has this ability. However, it has been shown in [13] by examples that certain pole -elimination procedures will not work satisfactory under the premise of convergence in measure, while they generate uniformly convergent approximants in case of convergence in capacity.

4) The result given in (3.4) can be called <u>geometric convergence in capacity</u>. The function $F(z)$ gives the convergence factor at the point $z \in D_0$.

5) The estimations of (3.4) and (3.5) together show that with the exception on certain sets of asymptotically vanishing capacity the $m+n$-th root of the approximation error is almost circular.

6) If the function $f(z)$ has no branch points, then the extremal domain D_0 is given by $\hat{\mathbb{C}} \sim E$ and $F(z) \equiv 0$. Hence, we have convergence faster than geometric. Theorem 2 has been proved for this case in [14]. Under this strong assumptions we get the result not only for close-to-diagonal, but also for essentially non-diagonal sequences satisfying

$$\overline{\lim_{m+n\to\infty}} \frac{|m-n|}{m+n} < 1. \tag{3.6}$$

7) If the function $f(z)$ has branch points, then essentially non-diagonal sequences of Padé approximants, i.e. sequences not satisfying (3.3), cannot converge in the whole domain D_0. There remains a non-empty domain of divergence, in which the Padé approximants tend to infinity or to zero depending on wether the sequence belongs to the upper or the lower triangle of the Padé table. In [15] and [16] these problems have been investigated for Hamburger and Stieltjes functions.

8) In general the Padé table of $f(z)$ will not be normal. However, it turns out that

$$\lim_{m+n\to\infty} \frac{\operatorname{degr}(p_{mn})}{m} = 1 , \qquad \lim_{m+n\to\infty} \frac{\operatorname{degr}(q_{mn})}{n} = 1 , \tag{3.7}$$

and also the size of blocks is asymptotically vanishing in relation to $m+n$. A more precise formulation of (3.7) is given in Theorem 4 in Section 8.

9) In Theorem 2 the only essential assumption is the requirement that the function $f(z)$ has all its singularities in a compact set of capacity zero. Nuttall's convergence theorem ([1], [2]) requires additional assumptions, which are stated and discussed in Section 5, when we have more informations about the structure of the minimal set K_0.

We close this section with a remark on the connection between Padé approximation and the property of minimal capacity. This connection is fundamental for our results, but nevertheless it may have been suprising at first sight. The following considerations can probably give some explanations and light up the underlying logic. Let $D \subseteq \hat{\mathbb{C}}$ be a domain with $\infty \in D$ and $f(z)$ having a single-valued analytic continuation in D. Further, let the function $F_D(z)$ be defined analogously to $F(z)$ in (3.1) except that we now use the domain D instead of the extremal domain D_0. The functions $F_D(z)$ and $F(z)$ have a zero of order 1 at infinity. From the definition of the logarithmic capacity it follows that

$$\frac{F(z)}{F_D(z)}\bigg|_{z=\infty} = \exp\left[\frac{cap(K_0)}{cap(\mathbb{C}\sim D)}\right] \leq 1. \tag{3.8}$$

In (3.8) the upper bound is attained if and only if the domain D satisfies assertion (ii) of Theorem 1. Since $F(z)$ is the convergence factor of the sequence of Padé approximants $\{[m/n]\}$, it follows from (3.4) and (3.8) that for $m+n \to \infty$ the Padé approximants behave like best rational approximants in small circular neighbourhoods of infinity if we neglect certain exceptional sets of asymptotically vanishing capacity.

At this point we would like to recall a classical result [17] by Walsh on the connection between best rational and Padé approximants (for a new treatment see [18]), which establishes a similar but more precisely formulated relation for fixed indices m and n.

4. More About Sets of Minimal Capacity

Figures 3 and 4 give the impression that the cuts constituting the minimal set K_0 are smooth or even analytic arcs. Indeed, it has been shown in [19] that this is true. Furthermore, Green's function $g_{D_0}(z,\infty)$ of the extremal domain D_0 possesses a certain symmetry property near the cuts, and the cuts themselves are trajectories of a quadratic differential. The last two properties are fundamental for the first two of the three approaches surveyed in this lecture. Relevant results are put together in the next three lemmas.

Lemma 1 ([19, Thm. 1]) - - We have

$$K_0 = E_0 \cup \bigcup_{\iota\in I} \mathcal{J}_\iota , \tag{4.1}$$

where $\{\mathcal{J}_\iota\}_{\iota\in I}$ is a family of open analytic pair-wise disjoint Jordan arcs, and $E_0 \subseteq \mathbb{C}$ a compact set of capacity zero. The set $E_0 \sim E$ consists of isolated points only.

Remark - - Without loss of generality we can assume that the arcs $\mathcal{J}_\iota, \iota\in I$, are extremal in the following sense: If $\mathcal{J}_\iota, \iota \in I$, belongs to $\{\mathcal{J}_\iota\}$, then no open proper subarc of $\mathcal{J}_\iota$ belongs to the family.

In Example A the family $\{\mathcal{J}_\iota\}_{\iota\in I}$ consists of the five arcs connecting the points $\{z_1, z_2, z_3, z_4, x_1, x_2\}$, which constitute the set E_0. The set $E_0 \sim E$ is equal to $\{x_1, x_2\}$ (cf. Figure 3). In Example B we only have two arcs and $E_0 \sim E = \phi$ (cf. Figure 4).

Taking into account the assumed extremality of the arcs $\mathcal{J}_\iota, \iota\in I$, we see that $E_0 \sim E$ consists of connecting points between arcs $\mathcal{J}_\iota$.

Lemma 2 (Symmetry Property, [19, Thm. 1, Cor.]) - - For any $\iota \in I$ and any $x \in J_\iota$ we have

$$\frac{\partial}{\partial n_1} g_{D_0}(x,\infty) = \frac{\partial}{\partial n_2} g_{D_0}(x,\infty), \tag{4.2}$$

where $\frac{\partial}{\partial n_1}$ and $\frac{\partial}{\partial n_2}$ denote the normal derivations to both sides of the Jordan arcs in $\{J_\iota\}_{\iota \in I}$.

Remark - - The symmetry (4.2) is a local condition for the minimality of K_0.

Lemma 3 ([19, Thm. 1, Lem. 5]) - - Let $\frac{\partial}{\partial z} := \frac{\partial}{\partial x} - i\frac{\partial}{\partial y}$, $z = x + iy$, denote the complex derivation. The function

$$G(z) := \left\{\frac{\partial}{\partial z} g_{D_0}(z,\infty)\right\}^2 \tag{4.3}$$

can be completed by analytic continuation to a function analytic in $\hat{\mathbb{C}} \sim E_0$. The Jordan arcs J_ι, $\iota \in I$, are trajectories of the quadratic differential $G(z)dz^2$. On every J_ι, $\iota \in I$, we have

$$G(z)dz^2 \leq 0. \tag{4.4}$$

If there are only finitely many, say n, branch points in K_0, then $G(z)$ is a rational function

$$G(z) = \frac{Y(z)}{X(z)} \tag{4.5}$$

with $X \in \mathbb{P}_n$ and $Y \in \Pi_{n-2}$.

Quadratic differentials have been introduced in connection with extremal problems in geometric function theory. For more information we refer to [20].

5. The Assumptions in Nuttall's Theorem

It has already been mentioned that in [1] and [2] assumptions are used which are more detailed than those of Theorem 2. A certain structure of the minimal set K_0 and a specific behaviour of the function $f(z)$ on K_0 is required. We shall state Nuttall's result as Theorem 3 after some preparatory definitions.

Let $\omega(z) := \{f_+(z) - f_-(z)\}$ be the discontinuity of $f(z)$ on arcs $J_\iota, \iota \in I,$ in K_0. (By + and - we denote the two sides or banks of the arcs $J_\iota, \iota \in I.$) Since K_0 is of minimal capacity, the discontinuity function ω cannot vanish identically on any arc $J_\iota, \iota \in I.$ It is analytic in a neighbourhood of every J_ι. This implies that its zeros on $J_\iota, \iota \in I,$ are isolated.

If there are only finitely many branch points, we know from Lemma 3 that the function $G(z)$ defined by (4.3) is rational. Its denominator polynomial X has simple zeros at every end-point of branches in K_0, and no zero elsewhere (cf. [19 , Thm. 1]). With X and ω we introduce

(5.1) $$\sigma(z) := \omega(z) X(z)^{\frac{1}{2}}$$

which is defined for $z \in J_\iota, \iota \in I,$ up to the sign, which may be different on the two sides of the arcs J_ι. The sign is determined by the root of X, which we suppose to be locally analytic in D_0. Given the appropriate behaviour of $f(z)$ near the end-points of $J_\iota, \iota \in I,$ which will be supposed in the next theorem for all $J_\iota, \iota \in I,$ the function $\sigma(z)$ can be extended to all $z \in \bar{J}_\iota$.

Based on the differential (4.4) we introduce a metric on $J_\iota, \iota \in I,$ by

(5.2) $$d(z', z'') := \left| \int_{z'}^{z''} \sqrt{G(\xi)}\, d\xi \right|, \quad z', z'' \in \bar{J}_\iota.$$

From (4.3) it follows that the equilibrium distribution γ on K_0 is uniformly distributed on every J_ι, $\iota \in I$, with respect to this metric. There also exists an immediate connection with the distribution of poles (and zeros) of the Padé approximants $[m/n]$ on and near K_0: In this new metric the poles are asymptotically almost equally spaced. Cf. Figures 3 and 4 as well as Theorem 4 below.

Theorem 3 ([2, Thm. 7.5]) - - Let the function $f(z)$ have $2N$ branch points $\{a_1, \ldots, a_{2N}\} \subseteq \mathbb{C}$. Using the notations of Lemma 1 it is assumed that

(i) $I = \{1, \ldots, N\}$, $E_0 = \{a_1, \ldots, a_{2N}\}$,

(ii) $K_0 = \bigcup_{\iota=1}^{n} \bar{J}_\iota$, and all $\bar{J}_\iota$ are disjoint,

(iii) there exist constants A, B, L, and $\lambda > 0$ with

(5.3) $0 < A \leq |G(z)| \leq B < \infty$ for all $z \in K_0$,

(iv) for all $\iota \in I$ and $z', z'' \in \bar{J}_\iota$

(5.4) $|G(z')^{-1} - G(z'')^{-1}| < L(\log d(z', z''))^{-1-\lambda}$.

Under these assumptions the sequence of Padé approximants $[m/n]$, $n \in \mathbb{N}$, converges in capacity to $f(z)$ in the extremal domain D_0.

Remarks - - 1) The first two conditions are concerned with the structure of the minimal set K_0, the last two with the required behaviour of $f(z)$ on K_0. Conditions (i) and (ii) imply that the branch points of $f(z)$ have to be pair-wise connected in K_0 by disjoint cuts. Thus, function $f_2(z)$ in Example B belongs to the considered class of functions, while function $f_1(z)$ in Example A does not.

2) From condition (iii) it follows that the discontinuity function $\omega(z)$ is not allowed to have zeros on the arcs $J_\iota, \iota \in I$.

3) Condition (iv) implies that the function $f(z)$ has to be of square root type at every branch point a_j, $j = 1, \dots, 2N$.

4) As a trade-off for the more restrictive assumptions in Theorem 3, results can be obtained that are in specific respects more precise than those of Theorem 2. For instance, it can be shown that the number of spurious poles of the Padé approximants $[n/n]$ is bounded by $N-1$. Or as another example, it is possible to prove estimates for the approximation error that are one degree more precise than those given in (3.4) (cf. [2, Thm. 6.9]). We shall return to these questions in Sections 7 and 8.

5) Conditions (i)-(iv) can be considered as being technical to a large extend. It may be conjectured that relaxations or modifications are possible. Indeed, steps in this direction have been undertaken in [21] and [22], where asymptotics of generalized Jacobi polynomials are investigated. In this problem K_0 consists of three connected arcs and the function $f(z)$ has non-square root singularities. However, in one or the other form conditions exceeding the assumptions of Theorem 2 are necessary if one wants to prove results of the more precise type we mentioned in the last remark.

6. Orthogonality and the Remainder Formula

In this section we state two lemmas containing an orthogonality relation for the denominators of Padé approximants and a rather general remainder formula for Padé approximants. But first let us begin with some notations.

The set of zeros of a polynomial P (taking into account multiplicities) is denoted by $Z(P)$. Hence, we have $card(Z(P)) = degr(P) \le n$ for any $P \in \Pi_n$, $n \in \mathbb{N}$. The counting measure $\chi(P)$ of $Z(P)$ is called zero distribution of P, i.e. we have $\chi(P)(B) = card(Z(P) \cap B)$ for every set $B \subseteq \mathbb{C}$ and $P \in \Pi_n$. The subset of polynomials $P \in \Pi_n$ satisfying the standardization

$$(6.1) \qquad P(z) = \prod_{x \in Z(P)} H(z,x)$$

is denoted by $\mathbb{P}_n$, $n \in \mathbb{N}$, where the function $H(z,x)$, $z, x \in \hat{\mathbb{C}}$, called standardized linear factor, is defined as

$$(6.2) \qquad H(z,x) := \begin{cases} z - x & \text{for } |x| \le 1 \\ (z-x)|x|^{-1} & \text{for } |x| > 1 \\ 1 & \text{for } x = \infty . \end{cases}$$

In $\mathbb{P}_n$ a polynomial is uniquely determined by its zero set or zero distribution. The particular advantage of the standardization (6.1) becomes apparent if some zeros of P tend to infinity.

The m,n-Padé approximant can be uniquely represented as

$$(6.3) \qquad [m/n](z) = \frac{P_{mn}(z)}{Q_{mn}(z)\, z^{m-n}} , \quad m, n \in \mathbb{N},$$

where the two polynomials $P_{mn} \in \Pi_m$, $Q_{mn} \in \mathbb{P}_n$ are supposed to be prime. It follows from (1.2) that up to a common factor the polynomials P_{mn} and Q_{mn} are given by $z^m p_{mn}(\frac{1}{z})$ and $z^n q_{mn}(\frac{1}{z})$, respectively.

From relation (1.3) together with Cauchy's formula and the analyticity of $f(z)$ in the extremal domain D_0 we get

Lemma 4 (Orthogonality Property, [3, Lem. 3.12]) -- For every $m, n \in \mathbb{N}$ the denominator polynomial $Q_{mn} \in \mathbb{P}_n$ in (6.3) satisfies the orthogonality relation

(6.4) $$\oint_C \zeta^{\ell} Q_{mn}(\zeta)\,\zeta^{m-n} f(\zeta)\,d\zeta = 0 \qquad \text{for} \quad \ell = 0,\dots,n-1,$$

where the integration path C has to be homotop in D_0 to a circle around infinitiy.

Remarks - - 1) If we move the integration path C to the boundary of D_0, then those parts of integral (6.4) extending over arcs contained in J_ι, $\iota \in I$, depend on the discontinuity function $\omega(z)$ only since the integration path has to run through J_ι twice, once forward and once backwards.

2) It can be shown that if a polynomial $Q \in \mathbb{P}_n$ satisfies orthogonality relation (6.4) there exists an associated polynomial $P \in \Pi_m$ such that the polynomials $q(\frac{1}{z}) := z^{-n}Q(z)$ and $p(\frac{1}{z}) := z^{-m}P(z)$ satisfy relation (1.3) for given $m, n \in \mathbb{N}$. The polynomial P is explicitely represented by the integral formula

(6.5) $$P(z) = \frac{z^{n_0-n}}{2\pi i} \oint_C \frac{Q(z)z^{m-n_0} - Q(\zeta)\zeta^{m-n_0}}{z-\zeta} f(\zeta)\,d\zeta$$

where $n_0 := \min(m,n)$ and the integration path C has to be the same as that in (6.7).

3) Contrary to the classical theory (cf. [21, Ch. III]) orthogonality relation (6.4) does not in general determine the polynomial $Q_{mn} \in \mathbb{P}_n$ uniquely. This phenomenon corresponds to the existence of blocks in the Padé table. (For a more detailed discussion cf. [12, Lem 2]).

The remainder formula, which will be given in the next lemma, shows there exists a one-to-one relationship between the asymptotic behaviour of the polynomials $Q_{mn} \in \mathbb{P}_n$ and the convergence behaviour of the Padé approximants $[m/n]$. This connection allows us to carry out investigations in one field by means of the other one.

Lemma 5 (Remainder Formula, [3, Lem. 3.12]) - - For $m, n \in \mathbb{N}$ and any $P \in \Pi_{n_0}$, $n_0 := \min(m,n)$, we have

$$f(z) - [m/n](z) =$$

(6.6)

$$= \frac{1}{2\pi i Q_{mn}(z) P(z) z^{m-n_0}} \oint_C \frac{Q_{mn}(\zeta) P(\zeta) \zeta^{m-n_0} f(\zeta)}{\zeta - z} d\zeta,$$

where the integration path C has to be homotop in D_0 to a negative oriented circle around infinity, and it has to separate $z \in D_0$ from K_0.

Remarks - - 1) It is important that the polynomial P in formula (6.6) is completely arbitrary, so it is possible to locate zeros at positions, where the function $f(z)$ or the set K_0 have irregularities which we may want to suppress in the integral in (6.6).

2) Since we can move the integration path C to the boundary of D_0, remainder formula (6.6) is defined for all $z \in D_0$.

7. Nuttall's Method

We now come to the description of the approach of [1] and [2]. It is based on the orthogonality of the polynomials Q_{mn} introduced in (6.3). Having established a formula for their asymptotics, Theorem 3 follows immediately by remainder formula (6.6) given in Lemma 5. Here we will outline only the basic ideas of the procedure. In the presentation of the material we follow [5, Sec. 2]. As in [1] and [2] we suppose that $m = n$.

The method applied to get asymptotics is a generalization of Szegö's [19, Ch. X II] treatment of this problem in case of polynomials orthogonal on [-1,1] with respect to a positive weight function. The proof proceeds in two steps: Firstly, special weight functions are considered which allow a rather explicit determination of the orthogonal polynomials.

Then, in a second step, the solution of the first step is used to approximate the general case.

<u>The First Step:</u>

From assumption (i) and (ii) in Theorem 3 it follows that the path of integration in orthogonality relation (6.4) can be moved to the N disjoint cuts $\bar{J}_l$, $l = 1,\ldots,N$. With the discontinuity function $\omega(z)$ defined in Section 4, relation (6.4) can be written as

$$\text{(7.1)} \qquad \sum_{l=1}^{N} \int_{\bar{J}_l} \zeta^{\ell} Q_{nn}(\zeta)\omega(\zeta)\,d\zeta = 0, \qquad \ell = 0,\ldots, n-1.$$

(In Szegö's treatment only one arc identical to the interval $[-1,1]$ exists.)

The first step is characterized by the assumption that the function $\sigma(z) = \chi(z)^{\frac{1}{2}}\omega(z)$, $z \in K_0$, defined by (5.1) is the reciprocal of a polynomial, i.e.

$$\text{(7.2)} \qquad \sigma(z) = \frac{1}{\rho(z)}, \qquad \rho \in \Pi_M, \; M \in \mathbb{N}.$$

In the sequel orthogonality relation (7.1) is studied under this special assumption for $n \geq N+M$. To distinguish the orthogonal polynomials from those of the general case, we denote them by $\tilde{Q}_n \in \mathbb{P}_n$.

Let $\mathcal{R}_2$ be the Riemann surface defined by a doubling of the extended complex plane $\hat{\mathbb{C}}$ cut along the arcs of K_0 and sticked together crosswise. Because of its symmetry property (cf. Lemma 2), Green's function $g_{D_0}(z,\infty)$ can be harmonically continued to the complete surface $\mathcal{R}_2$. We have $g(z^{(2)},\infty^{(2)}) = -g(z^{(1)},\infty^{(1)})$. With $z^{(j)}$ we denote the z-coordinate on the j-th sheet of the Riemann surface. The subscript D_0 of Green's function is suppressed.

On $\mathcal{R}_2$ there exists a meromorphic function F_n with a n-th order pole at $\infty^{(1)}$, a $n-M-N+1$-th order zero at $\infty^{(2)}$, the same M zeros as the polynomial ϱ on the second sheet of $\mathcal{R}_2$ (we know that $\varrho \neq 0$ on K_0), and with $N-1$ zeros α_i, $i = 1, \dots, N-1$, somewhere on $\mathcal{R}_2$. The last $N-1$ zeros cannot be chosen arbitrarily; they are determined by the solution of the Jacobi inversion problem [24, II Ch. 4.8].

The function $F_n(z^{(1)}) + F_n(z^{(2)})$ is single-valued on $\hat{\mathbb{C}}$, and therefore a polynomial of degree n. Next we show that this function satisfies relation (7.1).

As in Section 5 we denote by + and - the two opposite sides of the arcs J_ι, $\iota = 1, \dots, N$, in K_0. From the construction of $\mathcal{R}_2$ it follows that $F_{n,+}(z^{(1)}) = F_{n,-}(z^{(2)})$ for all $z^{(j)}$ lying above K_0. We have $X_+(z)^{-\frac{1}{2}} = -X_-(z)^{-\frac{1}{2}}$ for all $z \in K_0$ since under the assumptions of Theorem 3 $X(z)$ has a simple zero at every end-point of J_ι, $\iota = 1, \dots, N$. Inserting $F_n(z^{(1)}) + F_n(z^{(2)})$ in relation (7.1), with the last two identities we get

$$\sum_{\iota=1}^{N} \int_{J_\iota} \zeta^{\ell}[F_n(\zeta^{(1)}) + F_n(\zeta^{(2)})]\varrho(\zeta)^{-1}X(\zeta)^{-\frac{1}{2}}d\zeta = \tag{7.3}$$

$$= \oint_C \zeta^{\ell} F_n(\zeta)\varrho(\zeta)^{-1}X(\zeta)^{-\frac{1}{2}}d\zeta = 0, \quad \ell = 0, \dots, n-1,$$

where C is a closed path on the second sheet of $\mathcal{R}_2$ homotopic to the lifting of K_0 to $\mathcal{R}_2$. The first equality in (7.3) simply follows from the lifting of the integral from K_0 to $\mathcal{R}_2$. The second one is a consequence of the analyticity of $F_n(z^{(2)})\varrho(z)^{-1}$ on the second sheet of $\mathcal{R}_2$ and the fact that $F_n(z^{(2)})\varrho(z)^{-1}X(z)^{-\frac{1}{2}}$ has a zero of order $n+1$ at $\infty^{(2)}$. With this considerations it is proved that

$$\tilde{Q}_{nn}(z) := c[F_n(z^{(1)}) + F_n(z^{(2)})] \in \mathbb{P}_n, \quad c \neq 0, \tag{7.4}$$

satiesfies relation (7.1). Since the functions F_n are undetermined up to a constant factor, we can assume without loss of generality that $c = 1$.

The asymptotic form of $\tilde{Q}_{nn}$ for $n \to \infty$ can now be found by considering the function

$$(7.5) \qquad \chi_n(z) := F_n(z) \exp(-ng(z,\infty)), \quad z \in \mathcal{R}_2, \; n \in \mathbb{N}.$$

Because $g(z,\infty) \sim \log|z|$, and $\sim \log\frac{1}{|z|}$ near $\infty^{(1)}$, and $\infty^{(2)}$, respectively, the function χ_n has a pole of order $M+N-1$ at $\infty^{(2)}$ and no other pole elsewhere. It has the same zeros as the polynomial ρ on the second sheet, and $N-1$ zeros at α_i, $i = 1, \ldots, N-1$. It is not single-valued on $\mathcal{R}_2$. Its periods are caused by the second factor in (7.5), and therefore $\log \chi_n(z)$ has pure imaginary periods, which are equal to $n\gamma(\bar{J}_i) \bmod (2\pi)$, $i = 1, \ldots, N$, for every cut J_i, where γ denotes the equilibrium distribution on K_0. These facts together imply that the factor $\exp(ng(z,\infty))$ is the dominant part of the asymptotic form of F_n. With (7.4) we get

$$(7.6) \qquad \frac{\tilde{Q}_{nn}(z)}{\exp(n\,g(z,\infty))} = \chi_n(z) + o(1)$$

for $n \to \infty$ on every compact set in the extremal domain D_0.

The informations about the structure of the function χ_n, $n \in \mathbb{N}$, allow to give an asymptotic estimation. Theorem 3 now follows from (7.6) and the remainder formula of Lemma 5.

We remark that the zeros of χ_n are not restricted to a particular subset of $\hat{\mathbb{C}}$. For $n \to \infty$ they can cluster everywhere in $\hat{\mathbb{C}}$ (cf. [10]). However, we have seen that under the special assumptions of Theorem 3 (and so far also (7.2)) the number of zeros of $\tilde{Q}_{nn}$ which do not cluster on K_0

is bounded by $N-1$. It may be interesting to note that this result cannot be proved under the weaker assumptions of Theorem 2.

The Second Step:

The aim of this step is to overcome the auxiliary assumption (7.2). Just as in Szegö [19, Ch. X II] it is shown that the orthogonal polynomials Q_{nn}, $n \in \mathbb{N}$, of the general case satisfy the integral equation

(7.7a) $$Q_{nn}(z) = c_n \tilde{Q}_n(z) + \sum_{l=1}^{N} \int_{\bar{J}_l} Q_{nn}(\zeta) K_n(\zeta, z)\, d\zeta$$

where the kernel is given by

(7.7b) $$K_n(z,\zeta) = \frac{\tilde{Q}_n(z)\tilde{Q}_{n+1}(\zeta) - \tilde{Q}_{n+1}(z)\tilde{Q}_n(\zeta)}{\zeta - z} c_n \frac{\sigma(\zeta) - \varrho(\zeta)^{-1}}{X(\zeta)^{\frac{1}{2}}},$$

and $c_n \in \mathbb{R}$ is an appropriate constant. The function $\sigma(z) = X(z)^{\frac{1}{2}}\omega(z)$ is defined as in (5.1), and $\varrho \in \Pi_M$ is a polynomial approximating σ^{-1} on K_0. Under the assumptions of Theorem 3 it is possible to make the difference $\sigma - \varrho^{-1}$, and thereby the kernel (7.7b) as small as we like for large degrees M of the polynomial ϱ. It follows from (7.7a) that this implies a similar asymptotic behaviour for both sequences of polynomials $\tilde{Q}_n$ and Q_{nn}, $n \in \mathbb{N}$. Using again remainder formula (6.6) of Lemma 5 we get a complete proof of Theorem 3.

Looking back at the whole concept of this first appoach we see that the restrictive character of the assumptions of Theorem 3 is mainly caused by the techniques applied in the second step. In particular, conditions (iii) and (iv), which imply that the function $f(z)$ must be of square root type at each branch point, are a consequence of integral equation (7.7a) and the necessity to make kernel (7.7b) sufficiently small.

8. The Second Approach

The second approach is again based on the investigation of asymptotics for the orthogonal polynomials $Q_{mn} \in \mathbb{P}_n$, but the type of asymptotics is different. While Nuttall considered the ratio $Q_{nn}(z)F(z)^n$, we now concentrate on the n-th root $|Q_{mn}(z)|^{\frac{1}{n}}$ for $n \to \infty$, which gives a weaker description of the asymptotic behaviour. However, the results are sufficient for a proof of Theorem 2, and they have the advantage that we can get them under the rather general assumptions of this theorem. This gives us an extensive and in a certain sense naturally closed coverage of convergence problems in the field of Padé approximant.

It seems that Faber [25] was the first to investigate the n-th root type of asymptotics for orthogonal polynomials. Of course, he was concerned with the classical case of polynomials $Q_n \in \mathbb{P}_n$ orthogonal on the interval $[-1, 1]$ with respect to a non-negative weight function $\omega(x)$. He has proved that

(8.1) $$\lim_{n\to\infty} \sqrt[n]{Q_n(z)} = \frac{1}{2}(z + \sqrt{z^2-1})$$

locally uniformly in $\mathbb{C} \sim [-1, 1]$ if $\omega(x) \geq c > 0$ for all $x \in [-1,1]$. In his treatment Green's function and logarithmic potentials have already been used as essential instruments.

In this lecture we define the logarithmic potential $p(z,\mu)$ of a measure μ with support $S(\mu) \subseteq \hat{\mathbb{C}}$ by $p(z,\mu) := \int \log|H(z,x)|d\mu(x)$, where $H(z,x)$ is the standardized linear factor introduced in (6.2). With the equilibrium distribution γ of the minimal set K_0 we get the representation

(8.2) $$g_{D_0}(z,\infty) = c_0 + p(z,\gamma).$$

Since $p(z,\mu)$ differs by a constant depending on μ from the usual definition of the logarithmic potential (cf. [26, Ch. I]), $c_0 \in \mathbb{R}$ is identical to Robin's constant only if $K_0 \subseteq \{|z| \leq 1\}$. For a sequence $\{\mu_n\}$ of measures we denote the weak limit (if it exists) by $\lim \mu_n$ or $\mu_n \to \mu$. We call an assertion to be true almost everywhere on a set $S \subseteq \hat{\mathbb{C}}$, abbreviated as qua.e. on S, if it is true for all $x \in S$ with an exception on a set of (outer) capacity zero.

The basic result in the present section is the next theorem. We can describe only the main ideas of its proof. A complete treatment can be found in [3] or [27].

Theorem 4 ([3, Thm. 3]) - - Under the assumptions of Theorem 2 we have

(8.3a) $$\lim_{m+n\to\infty} \frac{1}{n}\chi(Q_{mn}) = \gamma,$$

(8.3b) $$\overline{\lim_{m+n\to\infty}} |Q_{mn}(z)|^{\frac{1}{n}} \leq e^{-c_0} F(z)^{-1}$$

locally uniform in D_0, and in (8.3b) we have equality qua. e. in $\hat{\mathbb{C}}$. For every compact set $V \subseteq D_0$, $\varepsilon > 0$, and $x \notin V$ we have

(8.4) $$\lim_{m+n\to\infty} \mathrm{cap}_x \{z \in V;\ |Q_{mn}(z)| < [e^{-c_0}F(z)^{-1} - \varepsilon]^n\} = 0.$$

Remark: - - The most important part of Theorem 4 is limit (8.3a). The orther two limits are immediate or merely potential-theoretic consequences of (8.3a) (cf. [3, Sec. 4]).

From Theorem 4 and remainder formula (6.6) in Lemma 5 we get limit (3.4) in Theorem 2, and thereby a proof of the convergence in capacity of Padé approximants. For a proof of limit (3.5), which represents the second half of Theorem 2, the results of Theorem 4 are not sufficient.

It is necessary to show that the integral in remainder formula (6.6) cannot have too many zeros in D_0. This result is formulated in the following theorem:

Theorem 5 ([3, Sec. 4] or [28, Thm. 2]) - - Let the integral $I_{mn}(z)$ be defined by

(8.5a) $$I_{mn}(z) := \frac{1}{2\pi i} \oint_C \frac{Q_{mn}(\xi)^2 \xi^{m-n} f(\xi)}{\xi - z} d\xi$$

for $z \in D_0$, where C is a path separating z from K_0. Under the assumptions of Theorem 2 for every compact set $V \subseteq D_0$, $\varepsilon > 0$, and $x \notin V$ we have

(8.5b) $$\lim_{m+n \to \infty} \operatorname{cap}_x \{ z \in V;\ |I_{mn}(z)| < (e^{-c_0} - \varepsilon)^{m+n} \} = 0.$$

We will not go into details of the proof of Theorem 5 and confine ourselves to the remark that the methods are similar to those applied in the proof of Theorem 4, with which we shall be concerned after some general remarks about the characteristic difficulties of the problem.

In Faber's treatment [25] it is essential that

(i) the weight function ω is positive, and

(ii) its support $\operatorname{supp}(\omega)$ is contained in $\mathbb{R}$.

Both conditions are not satisfied in case of orthogonality relation (6.6), which is the starting point of the investigations in [3] or [27] .

With condition (i) we lose one of the basic instruments for the investigation of orthogonal polynomials: The bilinear form defining orthogonality is no longer Hermitian, and as a consequence the minimality property of the integral $\int Q_n(x)^2 \omega(x) dx$ is no longer valid. The absence of the Hermitian property is the principle difficulty to be overcome in the proof of Theorem 4.

In [29] , [30] , and [31] it has been shown that the classical result (8.1) is also true for complex valued weight functions ω with $supp(\omega) = [-1,1]$ if these weight functions satisfy certain regularity properties. Hence, there is hope that the loss of condition (i) will not be insurmountable.

On the other hand, asymptotic formulas have also been proved for polynomials $Q_n \in \mathbb{P}_n$ orthogonal with respect to positive weight functions ω given on smooth Jordan arcs not contained in $\mathbb{R}$. An exellent and comprehensive treatment can be found in Widom [32]. However, in Widom's paper the orthogonality is defined with respect to the conjugate powers $1, \bar{z}, \ldots, \bar{z}^{n-1}$ instead of the ordinary powers $1, z, \ldots, z^{n-1}$ as it is the case in relation (6.6). This difference opens ways to different worlds. In the first case we have the Hermitian property if the weight function is positive, in the later case this can only be true if $supp(\omega) \subseteq \mathbb{R}$.

The condition $supp(\omega) \subseteq \mathbb{R}$ has been so helpful in [25] since it implies that any polynomial $P \in \Pi_n$ can be paired with its conjugate polynomial $\bar{P}(z) := \overline{P(\bar{z})} \in \Pi_n$ such that the product $P(x)\bar{P}(x) = |P(x)|^2$ is non-negative for every $x \in \mathbb{R}$. It turns out that the symmetry property shown in Lemma 3 gives us a local substitue (in neighbourhoods of $J_\iota, \iota \in I$) for the operation of conjugation. For this purpose we now define the so-called reflection function.

For every arc $J_\iota, \iota \in I$, there exists an open neighbourhood U_ι and a conformal mapping

(8.6) $$\varphi_\iota : U_\iota \longrightarrow \mathbb{C}$$

with

(8.7a) $$\varphi_\iota(\mathcal{J}_\iota) \subseteq \mathbb{R},$$

(8.7b) $$|\operatorname{Im} \varphi_\iota(z)| = g_{D_0}(z,\infty), \quad z \in \mathcal{U}_\iota .$$

Further, we can assume that all image domains $\varphi_\iota(\mathcal{U}_\iota)$ are symmetric to $\mathbb{R}$, and the sets $\mathcal{U}_\iota$, $\iota \in I$, are pair-wise disjoint. In the open set $\mathcal{U} := \bigcup_{\iota \in I} \mathcal{U}_\iota$, which contains the set $K_0 \sim E_0$, we define the <u>reflection function</u> ϕ by

(8.8) $$\phi(z) := \varphi_\iota^{-1}(\overline{\varphi_\iota(z)}) \;, \quad z \in \mathcal{U}_\iota .$$

It is an anti-analytic mapping of the left-hand side to the right-hand side and vice versa of every arc $\mathcal{J}_\iota$ in K_0.

By moving the integration path C to the boundary of D_0, orthogonality relation (6.6) can be transformed in

(8.9) $$\oint_{C_0} \zeta^l Q_{mn}(\zeta)\zeta^{m-n} f(\zeta)d\zeta + \sum_{\iota \in I} \int_{\mathcal{J}_\iota^*} \zeta^l Q_{mn}(\zeta)\zeta^{m-n}\omega(\zeta)d\zeta = 0$$

for $l = 0,\ldots, n-1$, where C_0 consists of closed curves surrounding the points of E_0 and being contained in a small neighbourhood of this set. The arcs $\mathcal{J}_\iota^*$, $\iota \in I$, are given by $\mathcal{J}_\iota^* := \mathcal{J}_\iota \sim \operatorname{Int}(C_0)$, and ω is the discontinuity of $f(z)$ on the arcs $\mathcal{J}_\iota$ in K_0. We note that $\operatorname{cap}(E_0) = 0$. Hence, it is possible to find polynomials of comparatively low degree that are small on C_0.

After these preparations we come to the actual proof of (8.3a), which is carried out indirectly, i.e. we assume that for at least a (infinite) subsequence $\{(m,n)\} = N_0 \subseteq \mathbb{N}^2$ satisfying (3.3) we have

(8.10) $$\lim_{N_0} \frac{1}{n}\chi(Q_{mn}) = \mu_0 \neq \gamma .$$

Because of the semicontinuity of logarithmic potentials there exists at least one point $x_0 \in K_0$ with

$$(8.11) \qquad p(x_0, \mu_0) = \sup_{x \in K_0} p(x, \mu_0).$$

For the sake of simplicity we assume that $x_0 \in K_0 \sim E_0$, $\omega(x_0) \neq 0$, and $x_0 \neq 0$. If one of these assumptions were not satisfied, we could add a small positive amount of mass to μ_0, which would be located at the points we want to avoid. The operation would force the maximum to be assumed outside of these points. Of course, such a correction must be taken into account in all following steps. We will skip the rather technical details of this aspect of the procedure. To keep notation simple we assume $x_0 \in \mathcal{J}_0$.

Using the so-called principle of descent of potential theory [26, Thm. 1.3] it is not difficult to see that there exists a subsequence $N_1 \subseteq N_0$ and points $x_{mn} \in \mathcal{J}_0$, $(m,n) \in N_1$ with $\lim_{N_1} x_{mn} = x_0$ and

$$(8.12) \qquad \lim_{N_1} \frac{1}{n} \log |Q_{mn}(x_{mn})| = p(x_0, \mu_0).$$

From the assumption that $\mu_0 \neq \gamma$ it follows that either $\|\mu_0\| < 1$ or $p(x_0, \mu_0) > -c_0$, where the constant c_0 is defined by (8.2). In both cases it is possible (but not easy) to show that an infinite sequence of polynomials $\tilde{Q}_{mn} \in \mathbb{P}_n$, $(m,n) \in N \subseteq N_1$, exists satisfying the following four conditions (cf. [3, Lem. 4.2]):

(i) $$\overline{\lim}_N \frac{1}{n} \operatorname{degr}(\tilde{Q}_{mn}) < 1.$$

(ii) For a small $\varepsilon > 0$ and for the open neighbourhood $\Delta_0 := \varphi_0^{-1}(\{|z - \varphi_0(x_0)| < \varepsilon\})$ of the point $x_0 \in \mathcal{J}_0$ we have $x_{mn} \in \mathcal{J}_0 \cap \Delta_0$ for all $(m,n) \in N$,

(8.13a) $$\lim_N \frac{1}{2n} \log |Q_{mn}(x_{mn})\tilde{Q}_{mn}(x_{mn})| > -c_0,$$

and

(8.13b) $$\overline{\lim}_N \sup_{x\in C_0\cup(K_0\sim\Delta_0)} \frac{1}{2n} \log |Q_{mn}(x)\tilde{Q}_{mn}(x)| \leq -c_0,$$

where C_0 is the integration path of the first integral in (8.9). We assume that $\overline{\Delta}_0 \cap \overline{Int(C_0)} = \phi$.

(iii) For $0 < \delta < \frac{\pi}{2}$ there exist $\theta_{mn} \in [0,2\pi]$ with

(8.14) $$|\arg(Q_{mn}(z)\tilde{Q}_{mn}(z) - \theta_{mn}| \leq \delta$$

for all $z \in [J_0 \cap \Delta_0] \sim Z(Q_{mn})$ and all $(m,n)\in N$

(iv) We have $Z(Q_{mn}) \cap J_0 \cap \Delta_0 = Z(\tilde{Q}_{mn}) \cap J_0 \cap \Delta_0$ for all $(m,n) \in N$.

Let us review the main ideas underlying the construction of the polynomials $\tilde{Q}_{mn}$. From (8.10) and (8.11) it follows that in a neighbourhood of x_0 the asymptotic density of the zeros of the polynomials Q_{mn}, $(m,n)\in N$ must be somewhat thinner than the density of the equilibrium distribution γ. This fact is basic for the simultanous satisfaction of conditions (i) and (ii). Conditions (iii) and (iv) can only be satisfied if the zeros of the polynomial $\tilde{Q}_{mn} Q_{mn}$ are symmetrically located with respect to the arc J_0 in a neighbourhood of x_0. In order to achieve the required symmetry we demand that in Δ_0 the zeros of the polynomials $\tilde{Q}_{mn}$ are the images of the zeros of the polynomials Q_{mn} under the reflection function $\phi(z)$. On $J_0 \cap \Delta_0$ both polynomials have identical zeros.

In (8.9) we used a special integration path C. Let this path without the two subarcs contained in $J_0 \cap \Delta_0$ be denoted by C_1. From (8.13b) in conditon (ii) it follows that

(8.15) $$\overline{\lim}_N \left[\int_{C_1} |\tilde{Q}_{mn}(\xi) Q_{mn}(\xi) \xi^{m-n}| \, |f(\xi)| \, d|\xi| \right]^{\frac{1}{2n}} \leq e^{-c_0}.$$

On the other hand from (8.13a) in condition (ii), the conditions (iii), (iv), and a little more argumentation using estimates for the derivatives of the polynomial $\tilde{Q}_{mn} Q_{mn}$ (cf. [3, Sec. 4]) we get

(8.16) $$\underline{\lim}_N \left| \int_{\gamma_0 \cap \Delta_0} Q_{mn}(\xi) Q_{mn}(\xi) \xi^{m-n} \omega(\xi) d\xi \right|^{\frac{1}{2n}} > e^{-c_0}$$

Thus, in all the integrals of (8.9) the dominant part is given by the integral (8.16). This implies that for $(m, n) \in N$ sufficiently large we have

(8.17) $$\oint_C \tilde{Q}_{mn}(\xi) Q_{mn}(\xi) \xi^{m-n} f(\xi) d\xi \neq 0.$$

However, from condition (i) it follows that (8.17) is a contradiction to the orthogonality (8.9) or (6.6). Thus, we have proved limit (8.3a) in Theorem 4. The other assertions of this theorem, as already mentioned, are purely potential-theoretic consequences of (8.3a). With this result we close our considerations of the second approach.

9. A Third Approach

While both approaches surveyed so far are very similar in their basic strategy of investigation, we now come to a completely different approach. In particular, no use is made of the orthogonality of the denominator polynomials. We shall go directly from the defining relation (1.3) of the Padé approximants to the convergence proof.

Relation (1.3) can be rewritten as

(9.1) $$P_{mn}(z) - Q_{mn}(z) f(z) =: R_{mn}(z) = O(z^{-n_0-1})$$

for $z \to \infty$ and $m,n \in \mathbb{N}$, where $P_{mn}(z) := z^{n_1} p_{mn}(\frac{1}{z}) \in \Pi_{n_1}$, $Q_{mn}(z) := z^{n_1} q_{mn}(\frac{1}{z}) \in \Pi_{n_1}$, $n_0 := \min(m,n)$, and $n_1 := \max(m,n)$. The polynomials $p_{mn} \in \Pi_m$ and $q_{mn} \in P_n$ are the same as those in relation (1.3), but the polynomials P_{mn} and Q_{mn} differ by the factors $const.* z^{m-n_1}$ and $const.* z^{n-n_1}$, respectively, from those introduced in (6.3). With the new definition of P_{mn} and Q_{mn} we have

(9.2) $$[m/n](z) = \frac{P_{mn}(z)}{Q_{mn}(z)}.$$

The ordinary and reciprocal approximation errors are given by

(9.3a) $$f(z) - [m/n](z) = \frac{R_{mn}(z)}{Q_{mn}(z)},$$

(9.3b) $$\frac{1}{f(z)} - \frac{1}{[m/n](z)} = \frac{R_{mn}(z)}{f(z)\, P_{mn}(z)}.$$

In the present section the function R_{mn} will be called <u>remainder</u>, although it is only the remainder of relation (9.1).

Let $(\mathcal{R},\pi)$ be the Riemann surface defined by meromorphic continuation of the function $f(z)$, where $\pi : \mathcal{R} \to \hat{\mathbb{C}}$ is the canonical projection. By $\mathcal{M}(D)$ we define the set of all function meromorphic in a domain D.

From relation (9.1) we learn that the Riemann surface $\mathcal{R}$ is the natural domain of definition for $f(z)$ and $R_{mn}(z)$, $m,n \in \mathbb{N}$, while for the polynomials P_{mn}, Q_{mn}, $m,n \in \mathbb{N}$, the extended complex plane $\hat{\mathbb{C}}$ should be considered as natural domain of definition. We will symbolize this particularity by writing

(9.4a) $$f, R_{mn} \in \mathcal{M}(\mathcal{R}), \quad m,n \in \mathbb{N},$$

(9.4b) $$P_{mn}, Q_{mn} \in \mathcal{M}(\hat{\mathbb{C}}), \quad m,n \in \mathbb{N}.$$

Of course, the polynomials can be lifted to $\mathcal{R}$ (and they have to be lifted if we want to analyze them), but it is important, and perhaps the basic idea of the whole approach, to notice that on $\mathcal{R}$ the polynomials P_{mn} and Q_{mn} repeat their values identically on every sheet, while this is impossible for the functions f and R_{mn}, $m,n \in \mathbb{N}$.

By $z^{(1)}, z^{(2)}, \ldots$ we denote points on $\mathcal{R}$ lying above the same basic point $x \in \hat{\mathbb{C}}$, i.e. $\pi(z^{(j)}) = x$ for $j = 1, 2, \ldots$ We assume that the Padé approximants are developed on the first sheet of $\mathcal{R}$. Hence, the last equality in (9.1) is supposed to be true in a neighbourhood of $\infty^{(1)}$.

Using the fact that polynomials have identical values on every sheet of $\mathcal{R}$ we can eliminate one of two polynomials P_{mn} and Q_{mn} if we simultaneously consider relation (9.1) on two sheets of $\mathcal{R}$. This gives us the next lemma:

Lemma 6 - - Let $z', z'' \in \mathcal{R}$ with $\pi(z') = \pi(z'')$ and $m, n \in \mathbb{N}$. Then we have

(9.5a) $$f(z') - f(z'') = \frac{R_{mn}(z')}{Q_{mn}(z')} - \frac{R_{mn}(z'')}{Q_{mn}(z'')},$$

(9.5b) $$\frac{1}{f(z')} - \frac{1}{f(z'')} = \frac{R_{mn}(z')}{f(z')P_{mn}(z')} - \frac{R_{mn}(z'')}{f(z'')P_{mn}(z'')}.$$

Remark: - - We note that the left-hand sides of (9.5a) and (9.5b) are independent of $m, n \in \mathbb{N}$ and cannot vanish identically if z' and z'' belong to different sheets. On the right-hand side we essentially have differences of the ordinary and reciprocal errors. Lemma 6 implies that these errors cannot become small simultaneously on two sheets of $\mathcal{R}$ if we disregard certain exceptional points, which may exist, but have to be isolated.

Since by our definition the polynomials P_{mn} and Q_{mn} can be multiplied by a non-zero constant, we may assume that

(9.6) $$\frac{1}{n_1}\log\max(|P_{mn}(z)|, |Q_{mn}(z)|) =: p(z,\mu_{mn}),$$

where the function on the right-hand side is the logarithmic potential of a certain measure μ_{mn}, which is positive since the maximum of subharmonic functions is again subharmonic [26, Ch. I], and $\|\mu_{mn}\| \leq 1$.

In the sequel we consider sequences of indices $\{(m,n)\} =: N_0 \subseteq \mathbb{N}^2$ satisfying (3.3). Hence, we have

(9.7) $$\lim_{N_0}\frac{n_1}{n} = \lim_{N_0}\frac{n_0}{n} = 1.$$

Because of the weak compactness of the unit ball in the space of positive measures, there exists a subsequence $N_1 \subseteq N_0$ with $\lim_{N_1}\mu_{mn} =: \mu_0$, $\lim_{N_1}\frac{1}{n_1}\chi(P_{mn}) =: \mu_1$, and $\lim_{N_1}\frac{1}{n_1}\chi(Q_{mn}) =: \mu_2$. By taking a subsequence again we further get

(9.8a) $$\overline{\lim}_{N_1}\, p(z,\mu_{mn}) = p(z,\mu_0),$$

(9.8b) $$\overline{\lim}_{N_1}\,\frac{1}{n_1}\log|P_{mn}(z)| =: p(z,\mu_1) + c_1,$$

(9.8c) $$\overline{\lim}_{N_1}\,\frac{1}{n_1}\log|Q_{mn}(z)| =: p(z,\mu_2) + c_2$$

for z qua.e. on $\mathbb{C}$ [26, Thm. 3.8]. Already at this stage it follows from (9.1) and (9.6) that $|c_j| < \infty$, $j = 1,2$.

We now want to get asymptotic representation similar to those in (9.8), but on $\mathcal{R}$. For this purpose we use Green's potentials: Let $G \subseteq \mathcal{R}$ be a domain possessing a Green's function $g(z,w;G)$, $z,w \in G$. For a measure ν in G we define Green's potential by

(9.9) $$g(z,\nu) := -\int g(z,w;G)\,d\nu(w), \quad z \in \mathcal{R}.$$

On $\mathcal{R} \sim G$ we have $g(z,\nu) \equiv 0$. For any domain G having a Green's function there exists a subsequence $N_2 \subseteq N_1$ such that the following limits exist and are equal to the given right-hand sides qua.e. in G:

(9.10a) $$\overline{\lim}_{N_2} \frac{1}{n_1} \log \max(|P_{mn}(z)|, |Q_{mn}(z)|) = $$
$$= g_0(z) := g(z,\nu_0) + h_0(z),$$

(9.10b) $$\overline{\lim}_{N_2} \frac{1}{n_1} \log |P_{mn}(z)| = g_1(z) := g(z,\nu_1) + h_1(z),$$

(9.10c) $$\overline{\lim}_{N_2} \frac{1}{n_1} \log |Q_{mn}(z)| = g_2(z) := g(z,\nu_2) + h_2(z),$$

(9.10d) $$\overline{\lim}_{N_2} \frac{1}{n_0} \log |R_{mn}(z)| = g_3(z) := g(z,\nu_3) + h_3(z).$$

The functions $h_j(z)$, $j = 0,1,2,3$, are harmonic in $G \subseteq \mathcal{R}$ and ν_j, $j = 0,1,2,3$, denote measures with carriers on G. While the functions h_j depend on the domain G, the measures ν_j are independent of G in the following sense: For two domains $G_1, G_2 \subseteq \mathcal{R}$ the associated measures coincide on $G_1 \cap G_2$. In the sequel we will assume that $G \subseteq \mathcal{R}$ has been selected appropiately. The limits (9.10a-c) differ from (9.8a-c) only by the fact that the considered functions are lifted from $\hat{\mathbb{C}}$ to the Riemann surface $\mathcal{R}$.

<u>Lemma</u> 7 - - <u>With the sets</u> $I_0 := \{\infty^{(1)}\}$, $I_1 := \pi^{-1}(\infty)$, <u>and</u> $I_2 := I_1 \sim I_0$, <u>we have</u>

(i) $\nu_3(I_0) \geq 1$,

(ii) $\nu_3(B) \geq 0$ <u>for all Borel sets</u> $B \subseteq \mathcal{R} \sim I_2$.

(iii) <u>If</u> $G \subseteq \mathcal{R} \sim I_1$ <u>is a domain in which</u> π <u>is univalent, then for</u> $j = 0,1,2$ <u>we have</u>

(9.11a) $$0 \leq \nu_j(G) \leq |\nu_j(I_0)| \leq 1,$$

(9.11b) $$0 \leq v_j(\bar{G}) \leq 2|v_j(I_0)| - v_j(G).$$

Assertion (i) follows from the second equality in (9.1). The remainder R_{mn} is meromorphic on $\mathcal{R}$, and all polar singularities in $\mathcal{R} \sim I_1$ are generated by $f(z)$. Hence, their order is bounded by that of $f(z)$, which implies assertion (ii). The inequalities in (9.11a) are a consequence of the fact that a polynomial has the same number of zeros and poles on $\hat{\mathbb{C}}$. Finally, we get (9.11b) from (9.11a) if we regard that the projection $\pi: \partial G \to \mathbb{C}$ can be only bivalent if we disregard certain exceptional points.

We can refine the last reasoning to get the following lemma:

<u>Lemma</u> 8 - - <u>If the domain</u> G <u>satisfies the assumptions of assertion</u> (iii) <u>in Lemma</u> 7 and

(9.12) $$v_0(G) = 2|v_0(I_0)|,$$

<u>then there exists a set</u> $E_2 \subseteq \partial G$ <u>of points isolated in</u> $\mathcal{R}$ <u>such that the projection</u> $\pi: \partial G \to \mathbb{C}$ <u>is bivalent on</u> $\partial G \sim E_2$. <u>Further, we have</u> $\overline{\pi(G)} = \hat{\mathbb{C}}$.

Our next major step is the investigation of the function

(9.13) $$d(z) := g_3(z) - g_0(z), \quad z \in \mathcal{R}.$$

It follows from the inequalities

(9.14a) $$|f(z) - [m/n](z)| \leq \frac{|R_{mn}(z)|}{|Q_{mn}(z)|},$$

(9.14b) $$\left| \frac{1}{f(z)} - \frac{1}{[m/n](z)} \right| \leq \frac{|R_{mn}(z)|}{|f(z)||P_{mn}(z)|}$$

that in a certain sense $d(z)$ is an asymptotic upper estimate of either the ordinary or the reciprocal approximation error. However, it is not an asymptotic estimate in the full sense since in (9.10a) and (9.10b) only upper limits have been proved. Later we shall see that in the domain of convergence the density of the zeros of the polynomials Q_{mn} and P_{mn} will asymptotically vanish, which has consequences for the limits (9.10a) and (9.10b) and will allow to prove convergence in capacity.

From (9.1) and (9.6) we get

$$\text{(9.15)} \qquad d(z) \leq 0 \quad \text{for all} \quad z \in \mathcal{R}.$$

Up to now we only know that the function $d(z)$ is the difference of two functions subharmonic in $\mathcal{R} \sim \pi^{-1}(\infty)$. This gives us some informations about the topological nature of $d(z)$, but unfortunately this informations are not very precise. For instance, in general $d(z)$ will not be continuous; we know only that its discontinuities can be covered by an open set of arbitrary small capacity (cf. [26, Thm. 3.6]). In order to avoid technical subtilities and to keep the exposition simple we assume that the set

$$\text{(9.16)} \qquad \tilde{D} := \{ z \in \mathcal{R};\ d(z) < 0 \}$$

is open, which is obviously not true in general. It is not difficult to show that $\infty^{(1)} \in \tilde{D}$. By D we denote the component of $\tilde{D}$ containing $\infty^{(1)}$. In the sequel this domain D will play a prominent role. At the end of the section we will see that up to isolated point $\pi(D)$ is equal to the extremal domain D_0 introduced in Section 2, and therefore it will be the domain of convergence of the sequence $\{ [m/n](z);\ (m,n) \in N_0 \}$.

From Lemma 6 and our considerations after (9.14a, b) we get the next lemma:

Lemma 9 - - We have

(i) $\infty^{(1)} \in D$, and

(ii) the projection π is univalent in D.

For the study of logarithmic potentials it often is helpful to use the potential-theoretic notion of flux: The positive part of the defining measure of a logarithmic potential represents the springs and the mass of the measure gives the flux of these springs, while on the other hand the negative part of the defining measure describes the sinks. The point ∞ must be included in these considerations. In case of $d(z)$ the springs are given by the positive part of ν_3 and the negative part of ν_0, while the sinks are given by the negative part of ν_3 and the positive part of ν_0. At this stage of our investigation we cannot exclude the possibility of cancelations.

On a compact surface, as for instance on $\hat{\mathbb{C}}$, the total flux must be zero. On an arbitrary domain $G \subseteq \mathcal{R}$ this will in general not be true since there may be flux across the boundary ∂G. In case of the domain D it follows from $d(z) = 0$, $z \in \partial D$, together with (9.15) that the total flux on $\bar{D}$ must be non-positive. This gives us the following fundamental inequality

(9.17) $$\nu_3(\bar{D}) - \nu_0(\bar{D}) \leq 0.$$

With $I_0 := \{\infty^{(1)}\}$ it follows from (9.11b) in Lemma 7 and assertion (ii) in Lemma 9 that

(9.18) $$\nu_0(\bar{D} \sim I_0) \leq 2\,|\nu_0(I_0)| - \nu_0(D \sim I_0),$$

and further with (9.16) and (9.17) that

(9.19) $\quad v_3(I_0) + |v_0(I_0)| + v_3(\bar{D}\sim I_0) \leq v_0(\bar{D}\sim I_0) \leq 2|v_0(I_0)| - v_0(\bar{D}\sim I_0).$

Since we know by Lemma 7 that $v_3(I_0) \geq 1$, $v_3|_{\bar{D}\sim I_0} \geq 0$, and $v_0|_{\bar{D}\sim I_0} \geq 0$, from (9.19) we readily get

(9.20a) $\quad v_3(I_0) = 1,$

(9.20b) $\quad v_3|_{\bar{D}\sim I_0} = 0$

(9.20c) $\quad v_0(I_0) = -1,$

(9.20d) $\quad v_0(\partial D) = 2|v_0(I_0)| = 2,$

(9.20e) $\quad v_0|_{D\sim I_0} = 0.$

Here, we will pause a little before the final step and show that assertions (9.20a-c) already tell us a lot about the asymptotics of the Padé approximants and Padé polynomials: From (9.20a) and (9.1) it follows that the size of blocks in the Padé table is asymptotically vanishing in relation to n or m (cf. Remark 8 to Theorem 2). From (9.20b) it follows that the number of zeros of the remainder R_{mn} on any compact set in D is asymptotically vanishing in relation n or m. This result corresponds to Theorem 5 in Section 8. From (9.20c) it follows that

(9.21) $$\lim_{N_2} \frac{1}{n_1} \max(\mathrm{degr}(P_{mn}), \mathrm{degr}(Q_{mn})) = 1,$$

and soon we will see that from (9.21) we also get (3.7) of Remark 8 to Theorem 2.

We come back to the main line of our investigation: Equality (9.20d) together with assertion (ii) in Lemma 9 shows that the conclusions of Lemma 8 are valid for the domain D. In particular, we have $\overline{\pi(D)} = \hat{\mathbb{C}}$. Using Lemma 6 again we see that

(9.22) $\quad d(z) = 0 \qquad$ for $z \in \mathcal{R} \sim D.$

Let $G \subseteq \mathcal{R} \sim \pi^{-1}(\infty)$ be a domain containing $\bar{D}$. Because of (9.22) we have $\nu_0|_{G\sim\bar{D}} = \nu_3|_{G\sim\bar{D}}$, and with (9.20a) and (9.20e) further

(9.23) $$\nu_3|_G = \delta_{\infty^{(1)}},$$

where $\delta_{\infty^{(1)}}$ is Dirac's measure for the point $\infty^{(1)}$. From (9.20b), (9.20e), and (9.22) we get the important result

(9.24) $$d(z) = -2g(z,\infty^{(1)};G) \quad \text{for} \quad z \in \mathcal{R}.$$

Let now the image domain $\pi(D)$ be denoted by D_1. The function $f(z)$ is meromorphic in D_1, and $\infty \in D_1$. Let the set $E_1 \subseteq \partial D_1$ be the union of all non-polar singularities of $f(z)$ on ∂D_1 together with the image $\pi(E_2)$ of the set $E_2 \subseteq \partial D$ introduced in Lemma 8. Since π is univalent in D we have

(9.25) $$g_{D_1}(z,\infty) = -\tfrac{1}{2}\, d(\pi^{-1}(z)) \quad \text{for} \quad z \in D_1.$$

From Lemma 8 we know that π is exactly bivalent on $\partial D \sim E_2$. For arbitrary $z \in \partial D_1 \sim E_1$ let $z_1, z_2 \in \partial D \sim E_2$, $z_1 \neq z_2$, be the two uniquely existing points above z, and $\mathcal{U} \subseteq \mathbb{C} \sim E_1$, $\mathcal{U}_1, \mathcal{U}_2 \subseteq \mathcal{R}$ open neighbourhoods of z, z_1, and z_2, respectively, with $\pi(\mathcal{U}_1) = \pi(\mathcal{U}_2) = \mathcal{U}$. The inverse mappings from $\mathcal{U}$ on $\mathcal{U}_j$ are denoted by π_j^{-1}, $j = 1,2$. Since the measure ν_0 is identical on every sheet of $\mathcal{R}$, it follows from (9.23) that the function

(9.26a) $$h(z) := d(\pi_1^{-1}(z)) - d(\pi_2^{-1}(z))$$

is harmonic in $\mathcal{U}$. Because of (9.22) we have

(9.26b) $$h(z) = 0 \qquad \text{for} \quad z \in \partial D_1 \cap \mathcal{U}.$$

Hence, $\partial D_1 \cap \mathcal{U}$ is an analytic arc. It divides $\mathcal{U}$ in two subdomains, in one of which $h(z)$ is equal to $2g_{D_1}(z,\infty)$ and in the other one equal to $-2g_{D_1}(z,\infty)$. These results are summed up in the next lemma:

Lemma 10 - - The function $f(z)$ is single-valued in $D_1 \subseteq \hat{\mathbb{C}}$, and $\infty \in D_1$. The boundary ∂D_1 consists of open analytic arcs and a compact set E_1, which is the union of a subset of E and points isolated in $\hat{\mathbb{C}} \sim E$. The complement of $\overline{D}_1$ is empty. Green's function $g_{D_1}(z,\infty)$ possesses the symmetry property (4.2) stated in Lemma 2 with respect to the extremal domain D_0.

The properties attributed in Lemma 10 to the domain D_1 are practically the same as those which we got in Section 4 as characteristic consequences of the extremality of the domain D_0 and the set K_0. In the other direction it can be shown that D_1 is uniquely determined by the properties stated in Lemma 10, and further that the set $K_1 := \hat{\mathbb{C}} \sim D_1$ is of minimal capacity. The only difference between D_1 and D_0 is given by the polar singularities of $f(z)$ in D_1, which do not belong to D_0.

The proof of uniqueness of D_1 can be obtained much easier than that given in [8]. This follows from the comparatively precise knowledge we have of the topological structure of K_1. Such a proof can be based on a comparison of the flux of Green's function belonging two different domains. Or another possibility is given by Grötsch's classical method [33]. We will not go deeper into this subject here.

Next we show that the limits (9.8a-c) are identical. Let us assume that there exists a set $S \subseteq D$ of positive capacity with $g_1(z) < g_2(z)$ for all $z \in S$. From (9.10a-d), (9.13), and the inequality (9.14a) it follows by taking subsequences if necessery that $f(z) = 0$ qua.e. on S. This would imply that $f(z) \equiv 0$ on D. Hence, we have proved $g_1 \geq g_2$

qua.e. on D. Analogously, we can prove that $g_1 \leq g_2$ qua.e. on D if we use inequality (9.14b) instead of (9.14a). From the equality of g_1 and g_2 we get in (9.8a-c) the identities

(9.27) $$\mu_0 = \mu_1 = \mu_2 \text{ , and } c_1 = c_2 = 0.$$

With the information so far collected we are able to finish the proof of Theorem 2. From (9.8a-c), (9.13), (9.23), (9.25), and (9.27) we get

(9.28) $$\lim_{N_2} \frac{1}{n_1} \chi(P_{mn}) = \lim_{N_2} \frac{1}{n_1} \chi(Q_{mn}) = \gamma,$$

where γ is the equilibrium distribution on K_0 or K_1. The second part of (9.28) proves Theorem 4, while the first one shows that this theorem is also true for the numerator polynomials of Padê approximants.

The limits (3.4) and (3.5) in Theorem 2 follow from (9.25), (9.13), (9.20b), and (9.28) for the subsequence $N_2 \subseteq N_1$. For this conclusion it is fundamental that on any compact set $V \subseteq D_1$ we have

(9.29) $$\lim_{N_2} \frac{1}{n_1} \operatorname{card}(Z(Q_{mn}) \cap V) = 0.$$

Because of the uniqueness of the domain D_1, the proof extends to the original sequence N.

With the proof of Theorem 2 we have finished the description of the third approach. The material presented in this section is new. Of course, proofs have only been sketched. This is partly caused by the limitation of time for the lecture, but it is also intensional. The omission of technical, mainly potential-theoretical details may have improved the understandability of the main ideas of the approach. At least it is hoped that this effect could be achieved.

10. Final Remarks

For the last approach it is characteristic that only analytic properties of the investigated objects are used as long as possible, while the more arithmetical side is avoided. Even remainder formula (6.6) of Lemma 5, which can in one or the other form be considered as a standard tool for convergence proofs, has not been emploid.

Another remarkable feature is the fact that we get the structure and the properties of the convergence domains D_0 or D_1 as by-products of our analysis and have not to derive them from an extremality principle as in the two other approaches. In the second approach these derived properties, especially the symmetry of Green's function $g_{D_0}(z,\infty)$, are essential prerequisites of the analysis.

In case of the Hermite-Padé polynomials the last two approaches split up. The last one is suitable for the analysis of Hermite-Padé polynomials of type I (Latin case), while the second approach seems to be adequate for the type II polynomials (so called German case). At least Gonchar and Rahmanov's result 34 can be generalized by the second method since it is based on orthogonality. The crucial condition for the immediate generalization is the disjointness of the cuts arising in the new problem.

References

[1] Nuttall, J. (1977). The convergence of Padé approximants to functions with branch points, in "Padé and Rationa Approximation" (E.B. Saff and R.S. Varga, Eds.), Academic Press, New York, pp. 101 - 109.

[2] Nuttall, J. and Singh, S.R. (1978): Orthogonal polynomials and Padé approximants associated with a system of arcs, J. Approx. Theory, 21, pp. 1 - 42.

[3] Stahl, H. (1982): The convergence of Padé approximants to functions with branch points, submitted to J. Approx. Theory.

[4] Stahl, H. (1985): The convergence of generalized Padé approximants, will appear in J. Constr. Approx.

[5] Nuttall, J. (1982): The convergence of Padé approximants and their generalizations, Springer Lect. Notes Math., 925, pp. 246 - 257.

[6] Nuttall, J. (1984): Asymptotics of diagonal Hermite-Padé polynomials, J. Approx. Theory, 42, pp. 299 - 386.

[7] Nuttall, J. (1985): On sets of minimum capacity, in "Lecture Notes in Pure and Applied Mathematics" Dekker, New York.

[8] Stahl, H. (1985): Extremal domains associated with an analytic function I and II, Complex Variables, 4, pp. 311 - 324, 325 - 338.

[9] Nuttall, J. (1980): Sets of minimum capacity, Padé approximants and the bubble problem, in "Bifurcation Phenomena in Mathematical Physics and Related Topics" (C.Bardos and D.Bessis, Eds.), Reidel, Dordrecht, pp. 185 - 201.

[10] Stahl, H. (1985): On the divergence of certain Padé approximants and the behaviour of the associated orthogonal polynomials, in "Poly- Springer Lect. Notes Math., 1171, pp. 321 - 330.

[11] Rahmanov, E.A. (1980): On the convergence of Padé approximants in classes of holomorphic functions, Matem. Sbornik, 112 (154). English translation: Math. USSR Sb., 40 (1981), pp. 149 - 155.

[12] Stahl, H. (1985): Divergence of diagonal Padé approximants and the asymptotic behavior of orthogonal polynomials associated with non-positive measures, Constr. Approx., 1, pp. 249 - 270.

[13] Stahl, H. (1984): Convergence in capacity and uniform convergence, will appear in J. Approx. Theory.

[14] Pommerenke, Ch. (1973): Padé approximants and convergence in capacity, J. Math. Anal. Appl., 41, pp. 775 - 780.

[15] Stahl, H. (1976): Beiträge zum Problem der Konvergenz von Padéapproximierenden, doctorial dissertation, Technical University Berlin.

[16] Graves-Morris, P.R. (1981): The convergence of ray sequences of Padé approximants of Stiltjes functions, J. Comp. Appl. Math., 7, pp. 191 - 201.

[17] Walsh, J.L. (1964): Padé approximants as limits of rational functions of best approximation, J. Math. Mech., 13, pp. 305 - 312.

[18] Trefethen, L.N. and Gutknecht, M.H. (1985): On convergence and degeneration in rational Padé and Chebyshev approximation, SIAM J. Math. Anal., 16, pp. 198 - 210.

[19] Stahl, H. (1985): The structure of extremal domain associated with an analytic function, Complex Variables, 4, pp. 339 - 354.

[20] Jensen, G. (1975): Quadratic Differentials, Chap. 8 in "Univalent Functions" by Ch. Pommerenke, Vandenhoeck and Ruprecht, Göttingen.

[21] Gammel, J.L. and Nuttall, J. (1982): Note on generalized Jacobi polynomials, Springer Lect. Notes Math., 925, pp. 258 - 270.

[22] Nuttall, J. (1985): Asymptotics of generalized Jacobi polynomials, will appear in Constr. Approx.

[23] Szegö, G. (1959): Orthogonal Polynomials, American Mathematical Society, New York.

[24] Siegel, C.L. (1971): Topics in Complex Function Theory, Interscience, New York.

[25] Faber, G. (1922): Über nach Polynomen fortschreitende Reihen, Sitzungsberichte der Bayrischen Akademie der Wissenschaften, pp. 157 - 178.

[26] Landkof, N.S. (1972): Foundations of Modern Potential Theory, Springer Verlag, Berlin.

[27] Stahl, H. (1985): Orthogonal polynomials with complex valued weight functions I, will appear in Constr. Approx.

[28] Stahl, H. (1985): Orthogonal polynomials with complex valued weight Functions II, will appear in Constr. Approx.

[29] Baxter, G (1961): A convergence equivalence related to polynomials orthogonal on the unit circle., Trans. Amer. Math. Soc., 99, pp. 471 - 487.

[30] Nuttall, J. (unpublished manuscript): Orthogonal polynomials for complex weight functions and the convergence of related Padé approximants.

[31] Magnus, A (1986): Toeplitz matrix techniques and convergence of complex weight Padé approximants, will appear in J. Comp. Appl. Math.

[32] Widom, H. (1969): Extremal polynomials associated with a system of curves in the complex plane, Advances in Math., 3, pp. 127 - 232.

[33] Grötzsch, C. (1930): Über ein Variationsproblem der konformen Abbildung, Berichte der Sächsischen Akademie der Wissenschaften, Math.-Phys. Kl., 82, pp. 251 - 263.

[34] Goncar, A.A. and Rahmanov, E.A. (1981): On the convergence of simultaneous Padé approximants for systems of functions of Markov type, Proc. Steklov Math. Inst., 157, pp. 31 - 48.

On the continuity properties of the multivariate Padé–Operator $T_{m,n}$

Helmut Werner, Bonn

1. Introduction

In this paper we will continue the investigation that was begun in [6] concerning the dependence of the Padé approximation on the approximated function F. It is known from the univariate case that this dependence will be continuous iff F is itself a rational function of the degrees (m, n) or if F is normal, that is its Padé approximation is a nondegenerate rational function of degree (m, n). A very detailed report of the different notions of continuity in case of one real or complex independent variable is found in the recently appeared paper by Trefethen and Gutknecht [4]. In the multivariate case the situation is complicated for several reasons. First of all it is by no means clear what is meant by the Padé approximation of degree (m, n). Second there may be common zeros of numerator and denominator at one point, but they cannot be factored out and cancelled as in the univariate case.

With respect to the first point we adopt the definition given by A. Cuyt [1] which arises quite naturally if one solves the homogeneous linear system of equations for the terms of the denominator by means of determinantal relations. This was pointed out in [6] and we adopt the notations used there. A further consequence of the determinantal relations is that one has to allow for the "shift" $s = m \cdot n$, i.e. the denominator is of the form

$$Q(x) = q_0(x) + q_1(x) + \ldots + q_n(x), \quad \text{where the degree of } q_j, \text{ written } \partial q_j, \text{ is } s + j,$$

that is the terms are homogeneous forms and n is the degree or more intuitive the "length" of the polynomial Q. In contrast to other definitions this approach poses no difficulties with respect to the algebraic solvability of the Padé approximation, i.e. the question of existence.

We briefly summarize the necessary definitions (compare [6]). Given m, n and $N = m + n$. To allow for intuition assume $x = (x_1, x_2)$ to be two-dimensional. We consider the formal power series

$$(1) \qquad F(x) = c_0 + c_1(x) + c_2(x) + \ldots, \qquad \text{where } c_j \text{ is a form of degree } j.$$

We restrict our attention to the functions with $c_0 \neq 0$ and assume without loss of generality $c_0 = 1$ to simplify notations. Numerator P and denominator Q should be polynomials of degree (length) m resp. n with the said shift $s = m \cdot n$.

$$
(2) \qquad \begin{aligned} P(x) &= p_0(x) + p_1(x) + \ldots + p_m(x), \qquad \partial p_j = s + j, \\ Q(x) &= q_0(x) + q_1(x) + \ldots + q_n(x), \qquad \partial q_j = s + j. \end{aligned}
$$

Then $\frac{P}{Q}$ is called the Padé approximation $T_{m,n}F$ of F if the relation

$$
(3) \qquad (P - F \cdot Q)(x) = \mathcal{O}\left(|x|^{s+N+1}\right)
$$

holds.

This definition causes no algebraic problems but its numerical use is a matter of more concern (compare [1], [3]). If Q is positive (or negative) definite in the neighborhood of $x = 0$ the numerics may be handled by an appropriate computer arithmetic. Otherwise there will be critical directions under which curves arise (emanate) from the origin onto which $Q(x) = 0$ holds. Although there is a curve where $P(x) = 0$ osculating each zero curve of the denominator these two curves need not coincide and therefore may not cancel. This was the second difficulty mentioned before. To overcome it one may cut out sectors around the critical directions to retain a region R_0 around the origin on which Q will be nonzero. Details are given in [3]. On R_0 we may ask for the continuity behavior of $T_{m,n}$ as a function of F. To be specific we will use the semi–norm

$$
\|F\|_N := \max_{0 \le j \le N} [\max_{R_0} |c_j(x)|]
$$

to define neighborhoods of F. We will call a rational function $R = \frac{P}{Q}$ strongly degenerate in $\mathcal{R}_{m,n}$, where P, Q relative prime, if the defect

$$
d = \min(m - \operatorname{length} P, n - \operatorname{length} Q) > 0.
$$

Then $R = \frac{P}{Q}$ is defined as the Padé approximation of F, if

$$
(4) \qquad (P - F \cdot Q)(x) = \mathcal{O}\left(|x|^{s_0+N+1-d}\right), \quad s_0 = \partial q_0, \quad P \text{ and } Q \text{ relative prime.}
$$

This is in accordance with the above definition if P and Q are multiplied by an appropriate common factor.

We call F strongly non-normal if $T_{m,n}$F is strongly degenerate, F is called normal if $T_{m,n}F$ is such that $\partial q_0 = s$ and not degenerate.

It is seen in [2] that for normal functions F the statement holds:

$$
F_\varepsilon \to F \text{ in the semi–norm implies uniform convergence } T_{m,n}F_\varepsilon \Longrightarrow T_{m,n}F \text{ on } R_0.
$$

In this paper we show that $T_{m,n}$ is discontinuous at strongly non-normal functions F.

The proof we give is slightly different in its setup from those used in the univariate Tschebyscheff and Padé case (compare [5], [7]). In the univariate case a sequence of rational functions was constructed that was nondegenerate and had a pole moving closer and closer to the origin, the limiting function being degenerate because the pole had cancelled. The pole could, however, be chosen so that the corresponding power series F_ε existed and converged to F in the N–semi–norm. That is one had to analyse very carefully the behavior of the terms $c_{j,\varepsilon}(x)$ in the expansion F_ε.

In the multivariate case we dispense with the explicit construction of the singularities but modify the expansion of F to yield F_ε. Then we show that this can be done in such a way that the associate Padé approximations $T_{m,n}F_\varepsilon$ will have singularities close to the origin, depending upon the size of ε, right in the interior of the region R_0 corresponding to $T_{m,n}F$, hence there cannot be uniform convergence in R_0.

A comprehensive report of multivariate Padé approximation concerning its analytical behaviour in particular will soon be available from the Sonderforschungsbereich 72 der Universität Bonn by T. Schebiella and the author. It will also contain a more comprehensive list of references.

2. The case m=n=1

As an introduction we consider the example $m = n = 1$. Then

$$F(x) = 1 + c_1(x) + c_2(x) + \ldots, \qquad c_2 \not\equiv 0,$$

has to be approximated by $\frac{P}{Q}$, where $P = p_0 + p_1$ and $Q = q_0 + q_1$. The approximant is subjected to the equations

$$\begin{pmatrix} p_0 \\ p_1 \\ 0 \end{pmatrix} = \begin{pmatrix} 1 & 0 & 0 \\ c_1 & 1 & 0 \\ c_2 & c_1 & 1 \end{pmatrix} \begin{pmatrix} q_0 \\ q_1 \\ 0 \end{pmatrix}$$

corresponding to

$$(P - F \cdot Q)(x) = \mathcal{O}\left(|x|^{s_0+N+1}\right), \quad s_0 = \partial q_0 = m \cdot n = 1,$$

with the solution

$$q_0 = c_1, \quad q_1 = -c_2,$$

$$p_0 = c_1, \quad p_1 = c_1^2 - c_2$$

and therefore

$$\frac{P}{Q}(x) = 1 + \frac{c_1^2(x)}{c_1(x) - c_2(x)}.$$

Obviously $\frac{P}{Q}$ reduces to 1 if and only if $c_1 \equiv 0$. This means that we have $c_1 \equiv 0$ as a necessary and sufficient condition for reducibility in the case $m = n = 1$ (strong degeneracy of Padé aproximation). The factor of length $d = 1$ that may be cancelled is $T = c_1 - c_2$, so $P = T \cdot P^*$, $Q = T \cdot Q^*$. The equations that still hold are

$$\begin{pmatrix} p_0^* \\ p_1^* \end{pmatrix} = \begin{pmatrix} 1 & 0 \\ 0 & 1 \end{pmatrix} \begin{pmatrix} q_0^* \\ q_1^* \end{pmatrix}$$

or

$$(P^* - F \cdot Q^*)(x) = \mathcal{O}\left(|x|^{s_0^*+N+1-d}\right), \quad s_0^* = \partial q_0^* = 0.$$

We may define a sequence of functions converging to F in the semi-norm $\|\cdot\|_2$, a strongly degenerate function, characterized by $c_1 \equiv 0$, in the following way:

$$F_\varepsilon(x) = 1 + \varepsilon \cdot a(x) + c_2(x), \qquad \partial a = 1,$$

i.e. a is a fixed form of degree one.

It is clear that

$$\|F - F_\varepsilon\|_2 \longrightarrow 0 \qquad \text{for } \varepsilon \to 0.$$

On the other hand, the Padé approximant for $m = n = 1$ of F_ε cannot be reducible because of $\partial a = 1$, $a \not\equiv 0$, and

$$\frac{P_\varepsilon}{Q_\varepsilon} = 1 + \frac{\varepsilon^2 a^2}{\varepsilon a - c_2}$$

is singular along the curve given by $\varepsilon a - c_2 = 0$ if a does not divide c_2.
Suppose we choose a so that it is relative prime to c_2. Introduce polar coordinates

$$\begin{aligned} a(x) &= r \cdot \hat{a}(\varphi) \\ c_2(x) &= r^2 \cdot \hat{c}_2(\varphi). \end{aligned}$$

Then

$$\varepsilon a - c_2 = \varepsilon \cdot r \cdot \hat{a}(\varphi) - r^2 \cdot \hat{c}_2(\varphi) = 0$$

has the solutions $r = 0$ and $r_\varepsilon = \varepsilon \cdot \frac{\hat{a}(\varphi)}{\hat{c}_2(\varphi)}$.

Since we have a large freedom in choosing a we may arrange for the last expression to be positive for some region of φ and we see that for $\varepsilon \to 0$ we have singularities of $\frac{P_\varepsilon}{Q_\varepsilon}$ converging to zero. This shows that there is no convergence of the Padé approximants; that is the Padé–Operator is discontinuous. If $c_2 \equiv 0$ a second pertubation of the coefficients may be used to achieve

$$r_\varepsilon = \frac{\varepsilon}{\varepsilon_1} \cdot \frac{\hat{a}}{\hat{b}}, \quad c_2 = \varepsilon_1 \hat{b}, \quad \partial \hat{b} = 2,$$

for some angular region, and having $\varepsilon, \varepsilon_1$ going to zero in an appropriate way.

3. The behavior of the Padé–Operator

In this section we will show that strong degeneracy implies discontinuity of the Padé–Operator. For simplicity we assume that there is a factor T of length less or equal 1 such that

$$P = T \cdot P^*, \quad Q = T \cdot Q^*$$

and that the defect is not larger than one, i.e. we have precisely $d = 1$. Furthermore assume that F is not itself a rational function of the class (m, n).

Lemma 1 *Let $F(x) = c_0 + c_1(x) + \ldots$, $c_0 = 1$, be strongly non-normal. Then*

$$\begin{vmatrix} c_m(x) & \ldots & c_{m-n+1}(x) \\ \vdots & \vdots & \vdots \\ c_{m+n-1}(x) & \ldots & c_m(x) \end{vmatrix} = 0.$$

<u>Proof:</u> Q solves the system of linear equations

$$\text{(5)} \qquad \begin{pmatrix} c_{m+1}(x) & c_m(x) & \ldots & c_{m+1-n}(x) \\ c_{m+2}(x) & c_{m+1}(x) & \ldots & c_{m+2-n}(x) \\ \vdots & \vdots & & \vdots \\ c_{m+n}(x) & c_{m+n-1}(x) & \ldots & c_m(x) \end{pmatrix} \begin{pmatrix} q_0(x) \\ q_1(x) \\ \vdots \\ q_n(x) \end{pmatrix} = 0.$$

Since $d > 0$ the length of P^* resp. Q^* is at most $m-1$ resp. $n-1$. From

$$(P^* - F \cdot Q^*)(x) = \mathcal{O}\left(|x|^{s_0^*+N+1-d}\right) \tag{6}$$

and $p_m^* \equiv 0$ we conclude that

$$p_m^*(x) = c_m(x) \cdot q_0^*(x) + \ldots + c_{m-n+1}(x) \cdot q_{n-1}^*(x) = 0.$$

Together with the equations from (5), i.e.

$$\begin{array}{lr} c_{m+1}(x) \cdot q_0^*(x) \quad + \ldots + c_{m+2-n}(x) \cdot q_{n-1}^*(x) & = 0 \\ \vdots & \vdots \\ c_{m+n-1}(x) \cdot q_0^*(x) + \ldots + c_m(x) \cdot q_{n-1}^*(x) & = 0 \end{array} \tag{7}$$

and the existence of a nontrivial solution of these n homogeneous linear equations the result is immediate. ∎

Obviously, if the said determinant vanishes there is a nontrivial solution Q^* which leads to the relation (6) with $d > 0$.

Hence a <u>necessary and sufficient criterion for strong degeneracy</u> is

$$C_{m,n} := \begin{vmatrix} c_m & \ldots & c_{m-n+1} \\ \vdots & \vdots & \vdots \\ c_{m+n-1} & \ldots & c_m \end{vmatrix} \equiv 0.$$

Lemma 2 *If $d = 1$ then*

$$C_{m-1,n-1} = \begin{vmatrix} c_{m-1} & \ldots & c_{m-n+1} \\ \vdots & \vdots & \vdots \\ c_{m+n-3} & \ldots & c_{m-1} \end{vmatrix} \not\equiv 0.$$

<u>Proof:</u> Observe that q_0^*, as defined above, needs to be different from zero, because

$$q_0^* \equiv 0 \Longrightarrow p_0^* \equiv c_0 \cdot q_0^* \equiv 0,$$

hence $d > 1$, in contradiction to the assumption made. The determinant given is equal to q_0^* up to homogeneous factors. ∎

A special role is played by rational functions F and in particular by those that are degenerate. Of course the crude semi–norm $\|\cdot\|_N$ is not strong enough to identify rational functions but we can try to do the best possible within this framework.

We could say that F is a degenerate <u>quasi rational</u> (and omit the prefix quasi again) if there is a rational function $\frac{P}{Q}$ such that $\partial P < m$, $\partial Q < n$ and

$$(P - F \cdot Q)(x) = \mathcal{O}\left(|x|^{s_0+N+1}\right)$$

which is of order $d = 1$ higher than could be expected from the parameters.

Lemma 3 *If F is a degenerate quasi rational the rank of the matrix*

$$\mathbf{C}_{m,n+1}(x) := \begin{pmatrix} c_m(x) & \ldots & c_{m-n}(x) \\ \vdots & \vdots & \vdots \\ c_{m+n}(x) & \ldots & c_m(x) \end{pmatrix} \qquad (c_{m-n} \equiv 0 \text{ if } m-n < 0)$$

is less than $n+1$.

Proof: Due to the relation

$$(P - F \cdot Q)(x) = \mathcal{O}\left(|x|^{s_0+N+1}\right)$$

the above equations (7) are satisfied by the terms of Q without any $\mathcal{O}$–terms on the right hand side, together with the equation

$$c_{m+n}(x) \cdot q_0(x) + \ldots + c_m(x) \cdot q_n(x) = 0.$$

Since F is assumed to be degenerate ($d > 0$) the homogeneous form q_n will be zero. That is the dimension of the kernel of this n–dimensional homogeneous linear system is larger than 0, hence the rank of $\mathbf{C}_{m,n+1}$ ought to be less than $n + 1$. ∎

After these preparations we construct a sequence of functions F_ε ($\varepsilon \to 0$) to show the discontinuity of the Padé–Operator $T_{m,n}$. Assume F to be strongly non-normal and the length of Q to be $n - 1$, then $C_{m,n} \equiv 0$. Because $C_{m-1,n-1} \not\equiv 0$ we may modify c_{m+n-1} to

$$c_{m+n-1,\varepsilon}(x) := c_{m+n-1}(x) + \varepsilon \cdot a(x),$$

where a is a form of degree $\partial c_{m+n-1} = m + n - 1$ such that

$$C_{m,n,\varepsilon} \not\equiv 0.$$

Hence F_ε is not strongly non-normal, i.e. P_ε and Q_ε are relative prime up to homogeneous forms. Obviously

$$\|F - F_\varepsilon\|_N \to 0 \qquad \text{for } \varepsilon \to 0.$$

Consider the Padé approximants of F_ε. The denominator Q_ε is obtained from

$$\begin{pmatrix} c_{m+1}(x) & c_m(x) & \ldots & c_{m+1-n}(x) \\ \vdots & \vdots & & \vdots \\ c_{m+n-1,\varepsilon}(x) & c_{m+n-2}(x) & \ldots & c_{m+1}(x) \\ c_{m+n}(x) & c_{m+n-1,\varepsilon}(x) & \ldots & c_m(x) \end{pmatrix} \begin{pmatrix} q_{0,\varepsilon}(x) \\ q_{1,\varepsilon}(x) \\ \vdots \\ q_{n,\varepsilon}(x) \end{pmatrix} = 0.$$

In particular

$$q_{0,\varepsilon} = \begin{vmatrix} c_m & c_{m-1} & \ldots & c_{m+1-n} \\ \vdots & \vdots & \vdots & \vdots \\ c_{m+n-2} & c_{m+n-3} & \ldots & c_{m-1} \\ c_{m+n-1,\varepsilon} & c_{m+n-2} & \ldots & c_m \end{vmatrix}$$

$$= \underbrace{C_{m,n}}_{\equiv 0 \text{ by Lemma 1}} + (-1)^{1+n} \cdot \varepsilon \cdot a \cdot \underbrace{C_{m-1,n-1}}_{\not\equiv 0 \text{ by Lemma 2}}$$

and

$$q_{1,\varepsilon} = -\begin{vmatrix} c_{m+1} & c_{m-1} & \dots & c_{m+1-n} \\ \vdots & \vdots & \vdots & \vdots \\ c_{m+n-1,\varepsilon} & c_{m+n-3} & \dots & c_{m-1} \\ c_{m+n} & c_{m+n-2} & \dots & c_m \end{vmatrix}$$

$$= q_{1,0} + (-1)^{1+(n-1)} \cdot \varepsilon \cdot a \cdot \begin{vmatrix} c_{m-1} & c_{m-2} & \dots & c_{m+1-n} \\ \vdots & \vdots & \vdots & \vdots \\ c_{m+n-4} & c_{m+n-5} & \dots & c_{m-2} \\ c_{m+n-2} & c_{m+n-3} & \dots & c_m \end{vmatrix}$$

$$= q_{1,0} + \mathcal{O}\left(\varepsilon \cdot |x|^{s+1}\right),$$

in general

$$q_{j,\varepsilon} = \mathcal{O}\left(|x|^{s+j}\right) + \mathcal{O}\left(\varepsilon \cdot |x|^{s+j}\right) + \mathcal{O}\left(\varepsilon^2 \cdot |x|^{s+j}\right).$$

Since $q_{0,0}$ is equal to zero it would contradict to the length $n-1$ of Q if $q_{1,0}$ would also vanish. Therefore

$$Q_\varepsilon = (-1)^{1+n} \cdot \varepsilon \cdot a \cdot C_{m-1,n-1} + q_{1,0} + \mathcal{O}\left(\varepsilon \cdot |x|^{s+1}\right) + \mathcal{O}\left(|x|^{s+2}\right)$$

$$= r^s \cdot \varepsilon \cdot A(\varphi) + r^{s+1} \cdot \hat{q}_1(\varphi) + \mathcal{O}\left(r^{s+2}\right) + \mathcal{O}\left(\varepsilon \cdot r^{s+1}\right), \quad r = |x|,$$

after introducing polar coordinates and using the notation

$$A(\varphi) := (-1)^{1+n} \cdot r^{-s} \cdot a \cdot C_{m-1,n-1} \not\equiv 0,$$

if the factor a is properly chosen, and $\hat{q}_1(\varphi) = r^{-s-1} \cdot q_{1,0}$.
From $Q_\varepsilon(x) = 0$ we see that there is an s–fold root $r_\varepsilon = 0$ and a further root satisfying

$$r_\varepsilon = -\varepsilon \cdot \frac{A(\varphi)}{\hat{q}_1(\varphi)} + \mathcal{O}\left(r_\varepsilon^2\right) + \mathcal{O}\left(\varepsilon r_\varepsilon\right)$$

for every fixed φ with $\hat{q}_1(\varphi) \neq 0$ in a region for φ. We may select $\hat{a}(\varphi) := r^{1-N} \cdot a(x)$ so that $A(\varphi)$ is different from zero and so that $A(\varphi)$ and $\hat{q}_1(\varphi)$ have different signs. Hence there is a positive root r_ε with $r_\varepsilon \to 0$ for $\varepsilon \to 0$. The points (r_ε, φ) apart from some exceptions will not be zeros of P_ε, since there is no common factor. Hence there is a sequence of points where $\frac{P_\varepsilon}{Q_\varepsilon}$ is infinite, while $\frac{P}{Q}$ is finite. This shows that

$$\frac{P_\varepsilon}{Q_\varepsilon} \text{ cannot converge uniformly to } \frac{P}{Q}$$

on every region of uniform boundedness of $\frac{P}{Q}$. This establishes the discontinuity of the Padé–Operator $T_{m,n}$ at F if F is strongly non-normal.

We do not elaborate on the case $d > 1$ but instead we conclude with an example in which all previous arguments may be seen explicitely. Take $m = n = 2$ and denote by ax, bx first order forms, so e.g. $ax = a_1x_1 + a_2x_2$, not identically equal to zero.
Assume that the first terms of F are given by

$$c_0 = 1,\ c_1(x) = ax,\ c_2(x) = ax \cdot bx,\ c_3(x) = ax \cdot (bx)^2,\ c_4 \text{ arbitrary}, \ \dots.$$

By the above criterion F is strongly non-normal and P^*, Q^* may immediately be calculated from the reduced set of linear equations. For instance Q^* is obtained from the reduced system

$$c_2 \cdot q_0^* + c_1 \cdot q_1^* = 0$$

with $q_0^*(x) = c_1(x) = ax$ and $q_1^*(x) = -c_2(x) = -ax \cdot bx$ as

$$Q^*(x) = ax \cdot (1 - bx).$$

Correspondingly P^* also has the factor ax and one obtains

$$P^*(x) = ax \cdot (1 + ax - bx),$$

hence

$$T_{2,2}F = 1 + \frac{ax}{1 - bx},$$

a function regular in the neighborhood of $x = 0$. The shift is zero after cancellation of ax but the length of numerator and denominator remains 1.

The approximation functions F_ε. We perturb the third term

$$c_{3,\varepsilon}(x) := c_3(x) + \varepsilon c(x) \quad \text{with } \partial c = 3, \text{ a form } c \text{ to be specified later.}$$

In this case the denominator Q_ε must be calculated from the full system of linear equations

$$\begin{pmatrix} c_{3,\varepsilon} & c_2 & c_1 \\ c_4 & c_{3,\varepsilon} & c_2 \end{pmatrix} \begin{pmatrix} q_{0,\varepsilon} \\ q_{1,\varepsilon} \\ q_{2,\varepsilon} \end{pmatrix} = 0.$$

The result is

$$q_{0,\varepsilon} = C_{2,2} - \varepsilon c \cdot c_1 = -\varepsilon c \cdot c_1 \quad \text{in view of lemma 1,}$$

$$q_{1,\varepsilon} = q_{1,0} - \varepsilon c \cdot c_2$$

and $q_{1,0} \not\equiv 0$ if $d = 1$ holds,

$$q_{2,\varepsilon} = q_{2,0} + 2\varepsilon c \cdot c_3 + \varepsilon^2 c^2$$

and the already given special form

$$Q_\varepsilon = -\varepsilon c \cdot c_1 + q_{1,0} + q_{2,0} + \varepsilon c \cdot (c_2 - 2c_3) - \varepsilon^2 c^2.$$

Since $q_{1,0}$ does not vanish identically we can find angular regions where it is different from zero. c_1 being also not identically equal to zero, we can choose c so that $c \cdot c_1$ is of opposite sign to $q_{1,0}$ in at least part of this angular region. For fixed angle φ (after introducing polar coordinates and using the previous notations) it is seen that $Q_\varepsilon(x) = 0$ produces a three–fold zero at $r = 0$ and that one zero r_ε converges to 0 for $\varepsilon \to 0$. Its existence being established by the implicit function theorem, e.g. its derivative at the origin is

$$\frac{\partial r_\varepsilon}{\partial \varepsilon} = -\frac{\hat{c} \cdot \hat{c}_1}{\hat{q}_{1,0}} > 0 \quad \text{by construction.}$$

It is left to the reader to verify that P_ε and Q_ε have no common factor of length greater than zero and hence almost no zero of Q_ε is compensated by a zero of P_ε. Hence $T_{2,2}F_\varepsilon$

is singular close to the origin at places right in the interior of a region of boundedness for $T_{2,2}F$.

This would be even more transparent if ax, bx were assigned specific values, e.g. $ax := x_1, bx := x_2$. Again it is left to the reader to study the resulting curves of singularities. He will from the previous discussion obtain some feeling on how complex the structure of the multivariate Padé–Operator considered in dependence on F can be.

References

[1] A. Cuyt: *Abstract Padé Approximants for Operators: Theory and Application,* Lecture Notes in Mathematics, Springer, Berlin 1984

[2] A. Cuyt, H. Werner, L. Wuytack: *On the Continuity of the Multivariate Padé Operator, J. of Computational and Applied Mathematics* **11** (1984), pp. 95–102

[3] T. Schebiella: *Multivariate Padé Approximation,* Diplomarbeit, Bonn 1985

[4] L. Trefethen, M. Gutknecht: *On the Convergence and Degeneracy in Rational Padé and Chebyshev Approximation, SIAM J. Math. Analysis* **16** (1985), pp. 198–210

[5] H. Werner: *On the rational Tschebyscheff–Operator, Math. Zeitschr.,* **86** (1964), pp. 317–326

[6] H. Werner: *Multivariate Padé Approximation, Numer. Math.,* to appear

[7] H. Werner, L. Wuytack: *On the Continuity of the Padé Operator, SIAM J. Num. Analysis* **20** (1983), pp. 1273–1280

Professor Dr. H. Werner
Institut für Angewandte Mathematik
Universität Bonn
Wegeler Straße 6
D–5300 Bonn 1 (Germany)

THE MARCHAUD INEQUALITY FOR GENERALIZED MODULI OF SMOOTHNESS

Z. Wronicz

Institute of Mathematics, Stanisław Staszic

Academy of Mining and Metallurgy,

Cracow, Poland

1. Introduction. The natural generalisation of algebraic polynomials are polynomials associated with extended complete Chebyshev systems. An important tool for estimating the best approximation of a continuous function f by algebraic polynomials of degree at most n-1 is the nth order modulus of smoothness of the function f $\omega_n(f,h)$. It satisfies the following property: $\omega_n(P_{n-1},h) = 0$ for any polynomial of degree at most n-1. This property does not hold true for generalized polynomials. Therefore the modulus of smoothness $\omega_n(f,h)$ cannot be a good tool for estimating the best approximation by generalized polynomials. Because of this we have defined a modulus of smoothness associated with an extended complete Chebyshev system $U = \{u_i\}_{i=0}^{n-1}$ in [13]. This modulus satisfies the above property. We have also proved basic properties of generalized moduli of smoothness, a generalization of the Whitney theorem and applied those facts to approximation by L-splines.

The purpose of this paper is to prove the Marchaud inequality and a few further properties of generalized moduli of smoothness.

2. Basic properties of generalized divided differences. The system $U = \{u_i\}_{i=0}^{n-1}$ of the functions u_i of class C^n in the interval $I = [0,1]$ is called an extended complete Chebyshev system (ECT - system) in I if,

for any points $0 \leqslant t_o \leqslant t_1 \leqslant \ldots \leqslant t_k \leqslant 1$, $k = 0,\ldots,n-1$

$$D\begin{pmatrix} u_o,\ldots,u_k \\ t_o,\ldots,t_k \end{pmatrix} = \det\left[D^{d_j}u_i(t_j)\right]_{i,j=o}^{k} > 0 ,$$

where $d_j = \max\{l : t_j = t_{j-1} = \ldots = t_{j-l}\}$, $j = 0,\ldots,k$, and D is the differentiation operator.

The system U is a basis of the null space N_{L_U} of a linear differential operator L_U of the form: $L_U = D^n + \sum_{i=o}^{n-1} a_i(t)D^i$. Conversely, for every operator L of the above form there exists $\delta > 0$ such that, for every subinterval $J \subset I$, with the length $|J| < \delta$ its null space N_L has a basis $\{u_i^J\}_{i=o}^{n-1}$, which is an ECT - system in the subinterval J (see [4,7,9]).

An ECT - system U admits the representation

(1)
$$u_o(t) = w_o(t) ,$$
$$u_i(t) = w_o(t)\int_0^t w_1(\tau_1)\int_0^{\tau_1} w_2(\tau_2)\ldots\int_0^{\tau_{i-1}} w_i(\tau_i)\,d\tau_i\ldots d\tau_1 ,$$

$i = 1,\ldots,n-1$ $(\tau_o = t)$, where $w_i \in C^{n-i}(I)$, $w_i > 0$ for $t \in I$, $i = 0,\ldots,n-1$. The adjoint system $V = \{v_i\}_{i=o}^{n-1}$ is defined as follows:

(2)
$$v_o(t) = 1 ,$$
$$v_i(t) = \int_0^t w_{n-1}(\tau_1)\int_0^{\tau_1} w_{n-2}(\tau_2)\ldots\int_0^{\tau_{i-1}} w_{n-i}(\tau_i)\,d\tau_i\ldots d\tau_1 ,$$

$i = 1,\ldots,n-1$ $(\tau_o = t)$.

Define

$$D_j f(t) = \frac{d}{dt}\frac{f(t)}{w_j(t)} , \quad D_j^* f(t) = \frac{1}{w_j(t)}\frac{d}{dt} f(t) , \quad j = 0,\ldots,n-1,$$

$$L_k f = D_{k-1}\ldots D_o f, \quad Lf = L_n f, \quad L^* f = D_o^*\ldots D_{n-1}^* f.$$

Systems (1) and (2) span the null spaces of the differential operators L and L^*, respectively. Clearly $N_L = N_{L_U}$.

We define the <u>divided difference</u> of a function f at the points $t_o \leqslant \ldots \leqslant t_n$, $t_o < t_n$ w.r.t. the operator L (the system U) by

$$(3) \qquad [t_o,\ldots,t_n;f]_L = \frac{D\begin{pmatrix}u_o,\ldots,u_{n-1},f\\ t_o,\ldots,t_{n-1},t_n\end{pmatrix}}{D\begin{pmatrix}u_o,\ldots,u_{n-1},u_n\\ t_o,\ldots,t_{n-1},t_n\end{pmatrix}},$$

where u_n is any function satisfying the equation $Lu = 1$.

We may put $w_n = 1$ and define u_n by (1) (see [7,9,12]). It follows from the definition that the divided difference depends neither on the choice of a basis of the space N_L nor on the order of the points t_j, $j = 0,\ldots,n$.

Let M_i be the ith L^*B-spline (basic spline w.r.t. the system V and the partition $\Delta_i = \{t_i,\ldots,t_{i+n}\}$) i.e. the function satisfying the following conditions: 1° M_i is an L^*-spline w.r.t. the system V and the partition Δ_i, 2° $\operatorname{supp} M_i = [t_i,t_{i+n}]$ and 3° $\int_0^1 M_i(t)dt = 1$ (see [7,9,12]). Then

$$[t_i,\ldots,t_{i+n};f]_L = \int_{t_i}^{t_{i+n}} Lf(t)M_i(t)dt .$$

<u>Example 1.</u> $U = \{t^i\}_{i=o}^{n-1}$, $w_o = 1$, $w_i = i$ for $i \geqslant 1$, $Lf = \frac{1}{(n-1)!} D^n f$, $[t_j,\ldots,t_{j+n};f]_L = n[t_j,\ldots,t_{j+n};f]$, $I = [a,b]$, $-\infty < a < b < \infty$, where the last expression is the divided difference in the algebraic case (see [2,3]).

<u>Example 2.</u> $U = \{e^{\lambda_i t}\}_{i=o}^{n-1}$, $\lambda_i < \lambda_{i+1}$, $i = 0,\ldots,n-2$, $I = [a,b]$, $-\infty < a < b < \infty$, $L_U = (D-\lambda_o)\cdot\ldots\cdot(D-\lambda_{n-1})$.

<u>Example 3.</u> $L_U = D(D^2+1)\cdot\ldots\cdot(D^2+m^2)$, $N_{L_U} = T_m$ - the space of trigonometric polynomials of degree at most m (see [8]).

<u>Example 4.</u> (see [4]). $U = \left\{\frac{1}{\lambda_i+t}\right\}_{i=o}^{n-1}$, where $0 < \lambda_o < \lambda_1 < \ldots < \lambda_{n-1}$, $I = [a,b]$, $0 < a < b < \infty$.

Put

$$\begin{bmatrix}u_o,\ldots,u_j\\ t_o,\ldots,t_j\end{bmatrix} f\Bigg] = \frac{D\begin{pmatrix}u_o,\ldots,u_{j-1},f\\ t_o,\ldots,t_{j-1},t_j\end{pmatrix}}{D\begin{pmatrix}u_o,\ldots,u_j\\ t_o,\ldots,t_j\end{pmatrix}}, \quad j = 1,\ldots,n .$$

Further, we need the following theorems:

Theorem 1. (Mühlbach[5]). Let $\{u_o,\dots,u_n\}$, $\{u_o,\dots,u_{n-1}\}$ and $\{u_o,\dots,u_{n-2}\}$ be Chebyshev systems over I. Consider n+1 different points $t_i \in I$, $i = 0,\dots,n$. Then

$$\begin{bmatrix} u_o,\dots,u_n \\ t_o,\dots,t_n \end{bmatrix} f \Bigg] = \frac{\begin{bmatrix} u_o,\dots,u_{n-1} \\ t_1,\dots,t_n \end{bmatrix} f \Bigg] - \begin{bmatrix} u_o,\dots,u_{n-1} \\ t_o,\dots,t_{n-1} \end{bmatrix} f \Bigg]}{\begin{bmatrix} u_o,\dots,u_{n-1} \\ t_1,\dots,t_n \end{bmatrix} u_n \Bigg] - \begin{bmatrix} u_o,\dots,u_{n-1} \\ t_o,\dots,t_{n-1} \end{bmatrix} u_n \Bigg]} .$$

Theorem 2. ([13]). Let $\Delta = \{0 \leqslant t_o < t_1 < \dots < t_N \leqslant 1\}$ be a given partition of I, $t_o \leqslant t_{k_o} < t_{k_1} < \dots < t_{k_n} \leqslant t_N$. Then there exist numbers α_j, $0 < \alpha_j < 1$ such that $\sum_{j=k_o}^{k_n-n} \alpha_j = 1$ and for any function f defined on I

$$(4) \qquad [t_{k_o},\dots,t_{k_n};f]_L = \sum_{j=k_o}^{k_n-n} \alpha_j [t_j,\dots,t_{j+n};f]_L .$$

To count the coefficients α_j we need the following

Lemma 1. Let φ_i be a polynomial w.r.t. the system U equal to zero at the points t_j, $j = i+1,\dots,i+n-1$ of the form $\varphi_i(t) = u_{n-1}(t) + \sum_{j=o}^{n-2} b_j u_j(t)$, $u_j \in U$ and let

$$\varphi_i^+(t) = \begin{cases} \varphi_i(t) & \text{for } t \geqslant t_{i+n-1} \\ 0 & \text{for } t < t_{i+n-1} \end{cases} .$$

Then

$$(5) \qquad [t_j,\dots,t_{j+n};\varphi_i^+]_L = \frac{c_i \delta_{ij}}{t_{i+n} - t_i} , \quad i,j = 0,\dots,N-n,$$

where the constants c_i satisfy the inequality $C_U^{-1} \leqslant c_i \leqslant C_U$, and the constant C_U depends only on the system U.

Proof. In the case $j < i$, $t_{j+n} \leqslant t_{i+n-1}$, whence $\varphi_i^+(t_k) = 0$ for $k = j,\dots,j+n$ and the left side of (5) is equal to zero. If now $j > i$, then $\varphi_i^+(t_k) = \varphi_i(t_k)$ for $k = j,\dots,j+n$ and we have the left side of (5) equal to zero again. Let now $i = j$. We can write the polynomial φ_i in the following form

$$\Psi_i(t) = \frac{D\begin{pmatrix} u_{n-1}, u_o & ,\dots, u_{n-2} \\ t & , t_{i+1}, \dots, t_{i+n-1} \end{pmatrix}}{D\begin{pmatrix} u_o, \dots, u_{n-2} \\ t_{i+1}, \dots, t_{i+n-1} \end{pmatrix}} .$$

Hence by (3)

$$[t_i, \dots, t_{i+n}; \Psi_i^+]_L = \frac{D\begin{pmatrix} u_{n-1}, u_o & ,\dots, u_{n-2} \\ t_{i+n}, t_{i+1}, \dots, t_{i+n-1} \end{pmatrix} \cdot D\begin{pmatrix} u_o, \dots, u_{n-1} \\ t_i, \dots, t_{i+n-1} \end{pmatrix}}{D\begin{pmatrix} u_o, \dots, u_{n-2} \\ t_{i+1}, \dots, t_{i+n-1} \end{pmatrix} \cdot D\begin{pmatrix} u_n & , u_o, \dots, u_{n-1} \\ t_{i+n}, t_i, \dots, t_{i+n-1} \end{pmatrix}}$$

$= \frac{A \cdot B}{C \cdot D}$, where A, B, C and D are the determinants from the numerator and the denominator, respectively. $B = \det[u_j(t_k),\ j = 0,\dots,n-1,\ k = i,\dots,i+n-1]$. We may assume that $u_o = 1$. Further

$$u_i(t_k) = \int_0^{t_k} w_1(\tau_1) \int_0^{\tau_1} w_2(\tau_2) \dots \int_0^{\tau_{i-1}} w_i(\tau_i)\, d\tau_i \dots d\tau_1 .$$

Substracting the kth column from its successor and factoring out the integrals from the function w_1, afteward expanding the determinant with respect to the first row and applying properties of determinants we obtain

$$B = \int_{t_i}^{t_{i+1}} w_1(y_1) \dots \int_{t_{i+n-2}}^{t_{i+n-1}} w_1(y_{n-1}) \det[a_{ij}]_{i,j=o}^{n-2} dy_1 \dots dy_{n-1} ,$$

where $a_{1j} = 1$, $a_{ij} = \int_0^{y_j} w_2(\tau_2) \int_0^{\tau_2} w_3(\tau_3) \dots \int_0^{\tau_{i-1}} w_i(\tau_i)\, d\tau_i \dots d\tau_2$

for $i = 2,\dots,n-1$, $j = 1,\dots,n-1$. Let $\tilde{B} = \det[t_j^k,\ j = 0,\dots,n-1,\ k = i,\dots,i+n-1]$. For the system $\{t^i\}_{i=o}^{n-1}$, $\tilde{w}_o = 1$, $\tilde{w}_i = i$ for $i \geqslant 1$. Since $w_i \in C(I)$ and $w_i > 0$, there exist positive constants c_i and d_i such that $c_i\tilde{w}_i \leqslant w_i \leqslant d_i\tilde{w}_i$. Applying this inequality we prove by induction that $c_B\tilde{B} \leqslant B \leqslant d_B\tilde{B}$, where $c_B = \prod_{j=1}^{n-1} c_j^{n-j}$, $d_B = \prod_{j=1}^{n-1} d^{n-j}$. Estimating the determinants A, C and D in the same way we obtain

$$c[t_i, \dots, t_{i+n}; \tilde{\Psi}_i^+] \leqslant [t_i, \dots, t_{i+n}; \Psi_i^+]_L \leqslant d[t_i, \dots, t_{i+n}; \tilde{\Psi}_i^+] ,$$

where the constants c and d depend only on the system U (operator L) and $\tilde{\varphi}_i$ is an algebraic polynomial equal to zero at the points t_j, $j = i+1,\dots,i+n-1$ of the form $\tilde{\varphi}_i(t) = t^{n-1} + \sum_{j=o}^{n-2} b_j t^j$ and we define $\tilde{\varphi}_i^+$ in the same way as φ_i^+. Hence by the equality $[t_i,\dots,t_{i+n};\tilde{\varphi}_i^+] = \frac{1}{t_{i+n}-t_i}$ we obtain the lemma.

Applying the above lemma and (4) to the function φ_i^+ we obtain

Theorem 3. Under the above assumption we have

$$\alpha_j = c_j^{-1}(t_{j+n} - t_j)[t_{k_o},\dots,t_{k_n};\varphi_j^+]_L .$$

In the algebraic case Theorem 2 was proved by T.Popoviciu in [6] (see also [1,2]) and the formula for the coefficient α_j was obtained by C.de Boor in [1] and in the complex case by P.M.Tamrazov in [10].*)

3. Basic properties of generalized moduli of smoothness. Let $f \in C(I)$ and let U and the operator L be defined as in the point 2. Put $\Delta_h^L f(t) = (n-1)!h^n[t,t+h,\dots,t+nh;f]_L$. Let q be a polynomial w.r.t. the system U interpolating the function f at the points $t+jh$, $j = 1,\dots,n$. In [13] we have proved the existence of constants α and β depending only on the system U such that

$$\alpha|\Delta_h^L f(t)| \leqslant |f(t) - q(t)| \leqslant \beta|\Delta_h^L f(t)| .$$

We define the modulus of smoothness of the function f w.r.t. the system U (operator L) by the formula

$$\omega_L(f,\delta) = \sup\{|\Delta_h^L f(t)|,\ 0 < h \leqslant \delta,\ t,\ t+nh \in I\}.$$

If $f \in L_p(I)$ for $1 \leqslant p < \infty$, we put

$$\omega_L^{(p)}(f,\delta) = \sup_{0<h\leqslant\delta}\left(\int_0^{1-nh}|\Delta_h^L f(t)|^p dt\right)^{\frac{1}{p}} .$$

For the operator $L = D^n$ we obtain the nth order modulus of smoothness.

The following properties of generalized moduli of smoothness have been proved in [13] $\left(\omega_L^{(\infty)}(f,\delta) = \omega_L(f,\delta)\right)$:

---

*) See Added in proof.

(P.1) $0 \leqslant \omega_L^{(p)}(f,\delta) \leqslant \omega_L^{(p)}(f,\delta')$ for $\delta \leq \delta'$.

(P.2) $\omega_L^{(p)}(f,\delta) \leqslant C\|f\|_p$, where the constant C depends only on the system U.

(P.3) $\omega_L^{(p)}(f+g,\delta) \leqslant \omega_L^{(p)}(f,\delta) + \omega_L^{(p)}(g,\delta)$.

(P.4) $\omega_L^{(p)}(f,m\delta) \leqslant m^n \omega_L^{(p)}(f,\delta)$, m positive integer.

(P.5) $\omega_L^{(p)}(f,\lambda\delta) \leqslant (1+\lambda)^n \omega_L^{(p)}(f,\delta)$, λ positive real number.

(P.6) $\dfrac{\omega_L^{(p)}(f,\delta_1)}{\delta_1^n} \leqslant 2^n \dfrac{\omega_L^{(p)}(f,\delta)}{\delta^n}$ for $0 < \delta \leqslant \delta_1$.

(P.7) If $f \in L_p(I)$ and $\omega_L^{(p)}(f,\delta) = o(\delta^n)$ by $\delta \to 0+$, then f is a polynomial w.r.t. the system U a.e.

(P.8) $\lim_{\delta\to 0} \omega_L^{(p)}(f,\delta) = 0$ for $f \in L_p(I)$.

Further we need the following

Lemma 2. There exist positive constants A and B depending only on the system U such that

$$A(t_n - t_o) \leqslant \begin{bmatrix} u_o,\dots,u_{n-1} \\ t_1,\dots,t_n \end{bmatrix} u_n \Big] - \begin{bmatrix} u_o,\dots,u_{n-1} \\ t_o,\dots,t_{n-1} \end{bmatrix} u_n \Big] \leqslant B(t_n - t_o)$$

for $0 \leqslant t_o < t_1 < \dots < t_n \leqslant 1$.

Proof. Applying Theorem 1 we obtain

$$\left[\begin{matrix} u_o,\dots,u_{n-1} \\ t_1,\dots,t_n \end{matrix}\middle| u_n\right] - \left[\begin{matrix} u_o,\dots,u_{n-1} \\ t_o,\dots,t_{n-1} \end{matrix}\middle| u_n\right] = \frac{\left[\begin{matrix} u_o,\dots,u_{n-1} \\ t_1,\dots,t_n \end{matrix}\middle| f\right] - \left[\begin{matrix} u_o,\dots,u_{n-1} \\ t_o,\dots,t_{n-1} \end{matrix}\middle| f\right]}{\left[\begin{matrix} u_o,\dots,u_n \\ t_o,\dots,t_n \end{matrix}\middle| f\right]}$$

for any function f such that the denominator is different from zero. Let us assume that $f(t_j) = 0$ for $j = 0,\dots,n-1$ and $f(t_n) = 1$. Then

$$\left[\begin{matrix} u_o,\dots,u_{n-1} \\ t_1,\dots,t_n \end{matrix}\middle| u_n\right] - \left[\begin{matrix} u_o,\dots,u_{n-1} \\ t_o,\dots,t_{n-1} \end{matrix}\middle| u_n\right] = \left[\frac{D\begin{pmatrix} u_o,\dots,u_{n-1} \\ t_o,\dots,t_{n-1} \end{pmatrix} \cdot D\begin{pmatrix} u_o,\dots,u_{n-1} \\ t_1,\dots,t_n \end{pmatrix}}{D\begin{pmatrix} u_o,\dots,u_n \\ t_o,\dots,t_n \end{pmatrix} \cdot D\begin{pmatrix} u_o,\dots,u_{n-2} \\ t_1,\dots,t_{n-1} \end{pmatrix}}\right]^{-1}.$$

In the algebraic case $(u_j = t^j)$ the last expression is equal to $t_n - t_o$.

Estimating the above determinants as in the proof of Lemma 1 we obtain Lemma 2.

Putting $t_j = t+jh$, $j = 0,\dots,n-1$, and $L_n = D_{n-1}\dots D_0$ we obtain

$$|\Delta_h^{L_n} f(t)| \leqslant C_U \left(|\Delta_h^{L_{n-1}} f(t+h)| + |\Delta_h^{L_{n-1}} f(t)| \right),$$

where the constant C_U depends only on the system U.

Applying this inequality we obtain

(P.9) There exists a constant C_{mk} depending only on the system U, m and k $(0 < k < m+k \leqslant n)$ such that

$$\omega^{(p)}_{L_{m+k}}(f,\delta) \leqslant C_{mk}\, \omega^{(p)}_{L_m}(f,\delta), \quad 1 \leqslant p \leqslant \infty.$$

4. <u>The Marchaud inequality.</u> Under the above assumptions we shall prove the following

<u>Theorem 4.</u> There exists a constant $C = C(U,r,k)$ depending only on the system U, r and k $(0 < k < r+k \leqslant n)$ such that for $f \in L_p(I)$, $1 \leqslant p \leqslant \infty$

$$(6) \qquad \omega^{(p)}_{L_r}(f,\delta) \leqslant C\delta^r \left[\omega^{(p)}_{L_r}\left(f,\tfrac{1}{r+k}\right) + \int_\delta^{\frac{1}{r+k}} \frac{\omega^{(p)}_{L_{r+k}}(f,s)}{s^{r+1}}\, ds \right],$$

where $0 < \delta(r+k) \leqslant 1$.

<u>Proof.</u> The idea of the proof is the same as in the algebraic case (see[2,11]). We shall prove (6) for $1 \leqslant p < \infty$. The proof for $p = \infty$ is analogous. We shall prove (6) by induction. It suffices to check (6) for $k = 1$. Introduce the following notations: $A = \frac{1}{r}$, $\omega(h) = \omega^{(p)}_{L_r}(f,h)$ and $\omega_1(h) = \omega^{(p)}_{L_{r+1}}(f,h)$. Let $\frac{A}{2} < \delta \leqslant A$. We obtain the inequality (6) for such δ straight.

$$\omega(\delta) \leqslant \omega(A) \leqslant \left(\tfrac{2}{A}\right)^r \delta^r \omega(A) \leqslant \left(\tfrac{2}{A}\right)^r \delta^r \left[\omega(A) + \int_\delta^A \frac{\omega_1(s)}{s^{r+1}}\, ds\right].$$

Let now $0 < \delta \leqslant \frac{A}{2}$. There exists a positive integer m such that $A2^{-(m+1)} < \delta \leqslant A2^{-m}$. Let $0 < h \leqslant \delta$. Applying Theorem 1 and Lemma 2 we prove that there exists a function $c(h,t)$ satisfying the inequalities

$0 < a \leqslant c(h,t) \leqslant b$ such that

$$(7) \qquad \Delta_h^{L_{r+1}} f(t) = c(h,t)\left(\Delta_h^{L_r} f(t+h) - \Delta_h^{L_r} f(t)\right),$$

where the constants a and b depend only on the system U.

Applying Theorem 2 we obtain

$$\Delta_{2h}^{L_r} f(t) = (r-1)!(2h)^r [t, t+2h, \dots, t+2rh; f]_{L_r}$$

$$= (r-1)!2^r h^r \sum_{j=0}^{r} \alpha_j [t+jh, \dots, t+(j+r)h; f]_{L_r}$$

$$= 2^r \sum_{j=0}^{r} \alpha_j \Delta_h^{L_r} f(t+jh), \text{ where } \sum_{j=0}^{r} \alpha_j = 1 \text{ and } 0 < \alpha_j < 1.$$

Put $c_j = c(h, t+jh)$, $f_j = \Delta_h^{L_r} f(t+jh)$. Then by (7) $f_{j+1} - f_j =$ $= \frac{1}{c_j} \Delta_h^{L_{r+1}} f(t+jh)$. Further

$$\Delta_{2h}^{L_r} f(t) - 2^r \Delta_h^{L_r} f(t) = 2^r \sum_{j=0}^{r} \alpha_j (f_j - f_o) = 2^r \sum_{j=1}^{r} \alpha_j \sum_{i=1}^{j} (f_i - f_{i-1})$$

$$= 2^r \sum_{j=1}^{r} \alpha_j \sum_{i=1}^{j} \frac{1}{c_{i-1}} \Delta_h^{L_{r+1}} f(t+(i-1)h) \text{ whence}$$

$$\left|\Delta_{2h}^{L_r} f(t) - 2^r \Delta_h^{L_r} f(t)\right| \leqslant \frac{2^r}{a} \sum_{j=1}^{r} \alpha_j \sum_{i=1}^{j} \left|\Delta_h^{L_{r+1}} f(t+(i-1)h)\right|,$$

where a is a constant which bounds the function $c(h,t)$ from below.

Put

$$\|f\|_p(h) = \left(\int_0^{1-h} |f|^p\right)^{\frac{1}{p}}, \quad 0 \leqslant h < 1, \ 1 \leqslant p < \infty.$$

Applying the Minkowski inequality we obtain

$$\|\Delta_{2h}^{L_r} f - 2^r \Delta_h^{L_r} f\|_p(2rh) \leqslant \frac{2^r}{a} \sum_{j=1}^{r} \alpha_j \sum_{i=1}^{j} \|\Delta_h^{L_{r+1}} f(t+(i-1)h)\|_p(2rh)$$

$$\leqslant \frac{r2^r}{a} \|\Delta_h^{L_{r+1}} f\|_p((r+1)h), \text{ whence}$$

$$(8) \quad \|2^{-r} \Delta_{2h}^{L_r} f - \Delta_h^{L_r} f\|_p(2rh) \leqslant C \|\Delta_h^{L_{r+1}} f\|_p((r+1)h), \text{ where } C = \frac{2^r}{a}.$$

Applying (8) and the Minkowski inequality to the identity

$$2^{-mr} \Delta_{2^m h}^{L_r} f - \Delta_h^{L_r} f = \sum_{j=1}^{m} 2^{-r(j-1)} \left(2^{-r} \Delta_{2(2^{j-1}h)}^{L_r} f - \Delta_{2^{j-1}h}^{L_r} f\right),$$

where m is the integer defined above, we obtain

$$\| 2^{-mr} \Delta_{2^m h}^{L\,r} f - \Delta_h^{L\,r} f\|_p(2^m rh)$$

$$\leqslant \sum_{j=1}^{m} 2^{-r(j-1)} \| 2^{-r} \Delta_{2(2^{j-1}h)}^{L\,r} f - \Delta_{2^{j-1}h}^{L\,r} f\|_p(2^j rh)$$

$$\leqslant C \sum_{j=o}^{m-1} 2^{-rj} \|\Delta_{2^j h}^{L\,r+1} f\|_p(2^j h(r+1)) .$$

Hence

$$\|\Delta_h^{L\,r} f\|_p(2^m rh) \leqslant \| 2^{-mr} \Delta_{2^m h}^{L\,r} f\|_p(2^m rh) + \tag{9}$$
$$+ \; C \sum_{j=o}^{m-1} 2^{-rj} \|\Delta_{2^j h}^{L\,r+1} f\|_p(2^j h(r+1)) .$$

Let now $g(t) = f(1-t)$, $t \in I$. We have $\Delta_h^{L\,r} g(t) = (-1)^r \Delta_h^{L\,r} f(1-t-rh)$ and $2^m rh < \frac{1}{2}$. Hence

$$\|\Delta_h^{L\,r} f\|_p(rh) \leqslant \|\Delta_h^{L\,r} f\|_p(2^m rh) + \|\Delta_h^{L\,r} g\|_p(2^m rh). \tag{10}$$

Further

$$\|2^{-mr} \Delta_{2^m h}^{L\,r} f\|_p(2^m rh) = \|2^{-mr} \Delta_{2^m h}^{L\,r} g\|_p(2^m rh), \tag{11}$$

and

$$\|\Delta_{2^j h}^{L\,r+1} f\|_p(2^j (r+1)h) = \|\Delta_{2^j h}^{L\,r+1} g\|_p(2^j (r+1)h). \tag{12}$$

Applying (9) - (12) we obtain

$$\|\Delta_h^{L\,r} f\|_p(rh) \leqslant 2\|2^{-mr} \Delta_{2^m h}^{L\,r} f\|_p(2^m rh) + 2C\sum_{j=o}^{m-1} 2^{-rj} \|\Delta_{2^j h}^{L\,r+1} f\|_p(2^j (r+1)h) . \tag{13}$$

We estimate the right side by the modulus of smoothness.

$$2\|2^{-mr} \Delta_{2^m h}^{L\,r} f\|_p(2^m rh) \leqslant 2\cdot 2^{-mr} \omega(2^m \delta) \leqslant 2\left(\frac{2}{A}\right)^r \delta^r \omega(A)$$

and

$$2C \sum_{j=o}^{m-1} 2^{-rj} \|\Delta_{2^j h}^{L\,r+1} f\|_p(2^j (r+1)h) \leqslant 2C \sum_{j=o}^{m-1} 2^{-rj} \omega_1(2^j \delta)$$

$$\leqslant \frac{Cr2^{r+1}}{2^r - 1} \delta^r \sum_{j=o}^{m-1} \int_{2^j\delta}^{2^{j+1}\delta} \frac{\omega_1(s)}{s^{r+1}} \, ds \leqslant \frac{Cr2^{r+1}}{2^r - 1} \delta^r \int_{\delta}^{A} \frac{\omega_1(s)}{s^{r+1}} \, ds .$$

Hence by (13)

$$\| \Delta_h^{L_r} f \|_p (rh) \leqslant B \delta^r \left(\omega(A) + \int_\delta^A \frac{\omega_1(s)}{s^{r+1}} ds \right),$$

where $B = \max \left(\frac{Cr2^{r+1}}{2^r - 1}, 2(2r)^r \right)$, whence we obtain the theorem.

Added in proof. The author has recently learned that Theorem 2 was proved by T.Popoviciu in: Sur le reste dans certaines formules lineaires d'approximation de l'analyse, Mathematica (Cluj), 1(24)1 (1959), 95-142.

References

[1] C.de Boor, Splines as linear combination of B-splines, Approximation Theory II, edited by G.G.Lorentz, C.K.Chui and L.L.Schumaker: 1 - 47, Academic Press, New York 1976.

[2] Z.Ciesielski, Lectures on spline functions (in Polish), Gdańsk University, 1979.

[3] A.O.Gelfond, Calculus of finite differences (in Russian), Fizmatgiz, Moscow 1959.

[4] S.Karlin, W.J.Studden, Tchebysheff systems: with applications in analysis and statistics, Interscience, New York 1966.

[5] G.Mühlbach, A recurrence formula for generalized divided differences and some applications, J. Approx. Theory 9(1973), 165 - 172.

[6] T.Popoviciu, Sur quelques properties des fonctiones d'une ou deux variables reelles, Mathematica 8(1934), 1 - 85, Cluj.

[7] K.Scherer, L.L.Schumaker, A dual basis for L-splines and applications, J. Approx. Theory 29 (1980), 151 - 169.

[8] I.J.Schoenberg, On trigonometric spline interpolation, J. Math. Mech. 13(1964), 795 - 825.

[9] L.L.Schumaker, Spline functions: basic theory, Wiley and Sons, New York 1981.

[10] P.M.Tamrazov, Smoothness and polynomial approximation (in Russian), Kiev 1975.

[11] A.F.Timan, Theory of approximation of function of a real variable (in Russian) Moscow 1960.

[12] Z.Wronicz, On some properties of LB-splines, Ann. Polon. Math. 46 1985 , 381 - 390.

[13] —— , Moduli of smoothness associated with Chebyshev systems and approximation by L-splines, Constructive Theory of Functions'84, 906 - 916, Sofia 1984.

ANALYTIC PROPERTIES OF TWO-DIMENSIONAL CONTINUED P-FRACTION EXPANSIONS WITH PERIODICAL COEFFICIENTS AND THEIR SIMULTANEOUS PADE-HERMITE APPROXIMANTS

A.I. Aptekarev
Keldysh Institute of Applied Mathematics
Miusskaya Sq. 4, 125047 Moscow A-47, USSR

V.A. Kalyagin
Polytechnic Institute, Gorki, USSR

1. INTRODUCTION

1.1. The problem of the asymptotic behaviour of Padé-Hermite polynomials and the convergence of the simultaneous rational approximants to analytic vector valued functions, in particular to functions of Markov (Stieltjes) type, can be considered as unsolved in its general form.

The task to be treated in this problem is the following: obtain conclusions about the convergence of simultaneous rational approximants and about the asymptotic behaviour of the numerator and the denominator polynomials from the information on branch points and the boundary values of the analytic functions to be approximated. In the case of the usual Padé approximants to functions of Markov type the solution for the convergence problem is given by a classical theorem of Markov [4] and the asymptotic behaviour of the numerator and denominator polynomials follows from Szegö's theory of orthogonal polynomials [12].

1.2 Let us recall the definition of Padé-Hermite approximants in the diagonal case. Let

$$\vec{f}(z) = (f_1, \ldots, f_m) \qquad , \qquad \vec{f}(\infty) = \vec{0}$$

be a vector valued function with components regular in $z=\infty$, and let $n=(n_1,\ldots,n_m)$ be a fixed m-tuple of positive integers, then the polynomials

$$Q_n(z),\ P_n^{(i)}(z),\ \deg[Q_n, P_n^{(i)}] \leqslant |n| = \sum_{i=1}^{m} n_i, \qquad i=1,\ldots,m$$

are called Padé-Hermite polynomials and the following vector of rational functions

$$\vec{\Pi}_n(z) = \left[\frac{P_n^{(1)}(z)}{Q_n(z)}, \ldots, \frac{P_n^{(m)}(z)}{Q_n(z)} \right] = (\pi_{n,1}(z), \ldots, \pi_{n,m}(z))$$

is called the (diagonal) simultaneous Padé-Hermite approximant of type $n=(n_1,\ldots,n_m)$ if the relations

$$(f_j - \pi_{n,j})(z) = \frac{A_{n,j}}{z^{|n|+n_j+1}} + \ldots \qquad j = 1, 2, \ldots, m \tag{1.1}$$

are satisfied.

In the case that the given vector-function has components of Markov type, i.e.

$$\vec{\mu}(z) = \left[\int \frac{d\mu_1(x)}{z - x}, \ldots, \int \frac{d\mu_2(x)}{z - x} \right], \qquad \operatorname{supp} \mu_j(x) \subset \mathbb{R}, \quad j=1,2,\ldots,m \tag{1.2}$$

then the relations (1.1) are equivalent to the following orthogonality relations

$$\int Q_n(x)\, x^s\, d\mu_j(x) = 0, \qquad s=0,1,\ldots,n_j-1, \qquad j=1,2,\ldots,m$$

The polynomials $P_n^{(j)}(z)$ are given by

$$P_n^{(j)}(z) = \int \frac{Q_n(z) - Q_n(x)}{z - x}\, d\mu_j(x)$$

In this case the Padé-Hermite polynomials are also called simultaneously orthogonal polynomials (see [5], [6], [3], [2]).

1.3. To present some qualitatively new features, which appear with the transition from the usual Padé approximants to Padé-Hermite approximants, it is useful to investigate some problems that are inverse to the problem of constructing Padé-Hermite approximants to a given vector function. For example, we will investigate the analytic properties of functions formally defined by Padé-Hermite approximants. For the definition we take multidimensional infinite continued $P^{(m)}$-fractions.

An analogous approach has recently been proposed by J. Nuttal [7] allowing him to make some successful conjectures.

Concerning the asymptotics of Padé-Hermite polynomials for a system (1.2) of Markov type, we study analytic properties of functions defined by $P^{(2)}$-fractions with periodic coefficients.

2. SOME GENERAL PROPERTIES OF $P^{(m)}$-FRACTIONS

2.1. The Jacobi-Perron's algorithm of expansion of the vector-functions into a m-dimensional continued fraction

$$\vec{f}(z) = \vec{P}_0(z) + \frac{1|}{|\vec{P}_1(z)} + \frac{1|}{|\vec{P}_2(z)} + \dots$$

(in the neighborhoud of infinity) is the following: a vector-function $\vec{f}(z)$ with components given by the formal power series in $z = \infty$, i.e.

$$\vec{f}(z) = (\sum f_{k,1} z^{-k}, \sum f_{k,2} z^{-k}, \dots, \sum f_{k,m} z^{-k})$$

can be writen as

$$\vec{f}(z) = \vec{P}_0(z) + \vec{f}^*(z) \tag{2.1}$$

where $\vec{P}_0(z)$ are vector-polynomials and the expansions of the components of $\vec{f}^*(z)$ do not contain negative powers. If we use the notation

$$\frac{1}{\vec{c}} = \frac{1}{(c_1, c_2, \dots, c_m)} = \left(\frac{1}{c_m}, \frac{c_1}{c_m}, \dots, \frac{c_{m-1}}{c_m}\right) \tag{2.2}$$

we can obtain $\vec{f}^{**}$ such that

$$\vec{f}^*(z) = 1/\vec{f}^{**}(z) \tag{2.3}$$

Therefore

$$\vec{f}(z) = \vec{P}_0(z) + 1/\ \vec{f}^{**}(z)$$

Iterating the transformations (2.1) and (2.3) we can obtain the next step of the Jacobi-Perron algorithm. Continuation of this process yields to an expansion of a vector function into an m-dimensional continued fraction.

It is clear that a continued fraction with a finite number of chains is a vector function with components that are rational functions with a common denominator. This feature of $P^{(m)}$-fractions provides the possibility to connect this fraction with simultaneous Padé-Hermite approximants.

This algorithm was investigated by V.I. Parusnikov, who has presented several interesting results [8], [9]. An analogous algorithm that forms a $C^{(m)}$-fraction of a vector function was investigated by M.G. de Bruin [1].

2.2. The following modification of a $P^{(m)}$-fraction will be useful in section 4. Defining the operation of multiplication for m=2 of a constant vector $\vec{r} = (r_1, r_2)$ by a vector function $\vec{f}(z) = (f_1(z), f_2(z))$ as follows

$$\vec{r}\,\vec{f}(z) = (r_1 f_1(z), r_2 f_2(z))$$

we obtain, according to the Jacobi-Perron algorithm, an expansion of $\vec{f}(z)$ ($\vec{f}(\infty) = (0,0)$) into the fraction

$$(f_1(z), f_2(z)) = \frac{\vec{r}_1 \rfloor}{\vec{P}_1(z) +} \frac{\vec{r}_2 \rfloor}{\vec{P}_2(z) +} \cdots \frac{\vec{r}_k \rfloor}{\vec{P}_k(z) +} \cdots \tag{2.4}$$

where we set $\vec{r}_k = (r_k, 1)$. The second component of $\vec{P}_k(z)$: $p_{2,k}(z)$ is a *monic* polynomial ($\deg p_{1,k} < \deg p_{2,k}$).

Numerators $P_k^{(1)}(z)$, $P_k^{(2)}(z)$ and the denominator $Q_k(z)$ of the k-th convergent of the fraction (2.4) satisfy the recurrence formulae

$$Y_k = p_{2,k} Y_{k-1} + p_{1,k} Y_{k-2} + r_k Y_{k-3} \qquad k=1,2, \ldots \tag{2.5}$$

with the initial conditions

$$\begin{array}{lll} Q_{-2} = 0 & Q_{-1} = 0 & Q_0 = 1 \\ P_{-2}^{(1)} = 1 & P_{-1}^{(1)} = 0 & P_0^{(1)} = 0 \\ P_{-2}^{(2)} = 1 & P_{-1}^{(2)} = 0 & P_0^{(2)} = 0 \end{array} \tag{2.6}$$

2.3 Let us prove some general properties of vector functions with periodic expansions. We denote by S an algebraic compact Riemann surface with l(S) sheets. M(S) stands for the field of meromorphic functions defined on S.

Proposition 2.1 *Let the vector function*

$$\vec{f}(z) = (f_1(z), f_2(z), \ldots, f_m(z))$$

have an expansion into an infinite $P^{(m)}$*– continued fraction*

$$\vec{f}(z) = \vec{P}_1(z) + \frac{1|}{|\vec{P}_2(z)+} \frac{1|}{|\vec{P}_3(z)+} \cdots \frac{1|}{|\vec{P}_p(z)+} \frac{1}{\vec{f}(z)} \qquad (2.7)$$

with period p. Then

a) *there exists a Riemann surface* $S_{m,p}(\vec{f})$*, such that* $1(S_{m,p}(\vec{f}))=m+1$ *and*
$f_j(z) \in M\{ S_{m,p}(\vec{f}) \}$, $\quad j=1,2,\ldots,m$.

b) *The collection of functions*

$$\{ 1, f_1(z), \ldots, f_m(z) \} \qquad (2.8)$$

is a basis of a field $M\{ S_{m,p}(\vec{f})\}$ *over the field of rational functions. In other words, any function which is meromorphic on* $S_{m,p}(\vec{f})$ *can be expressed as a linear combination of functions* (2.8) *with coefficients that are rational functions.*

Proof. a) The properties of the $P^{(m)}$–fractions (see [8]) and the following relation

$$(a_1, \ldots, a_m) = (p_1, \ldots, p_m) + \frac{1}{(b_1, \ldots, b_m)}$$

give

$$(1, b_1, \ldots, b_m) \times \begin{vmatrix} 0 & 0 & \ldots\ldots & 0 & 1 \\ 1 & 0 & \ldots\ldots & 0 & p_1 \\ 0 & 1 & \ldots\ldots & 0 & p_2 \\ \ldots & & & & \ldots \\ 0 & 0 & \ldots\ldots & 1 & p_m \end{vmatrix} = \text{const} \times (1, a_1, \ldots, a_m)$$

Then, denote

$$M_p = \prod_{i=1}^{p} \begin{vmatrix} 0 & 0 & \ldots & 1 \\ 1 & 0 & \ldots & p_{1,p-i+1}(z) \\ 0 & 1 & \ldots & p_{2,p-i+1}(z) \\ \ldots & & & \ldots \\ 0 & 0 & \ldots & p_{m,p-i+1}(z) \end{vmatrix}$$

where the polynomial $p_{j,k}$ (j=1,...,m and k=1,...,p) is the jth component of the vector $\vec{P}_k(z)$ from (2.7); so we have

$$(1, f_1(z), \ldots, f_m(z)) \times M_p(z) = \lambda(z)\,(1, f_1(z), \ldots, f_m(z)) \qquad (2.9)$$

where the function $\lambda(z)$, in order to satisfy (2.9), is the root of the algebraic equation with polynomial coefficients

$$\mathcal{L}(\lambda,z) = \det\{ M_p - \lambda \mathbf{I} \} = 0 \qquad (2.10)$$

For more information on this algebraic equation we refer to [9]. As $S_{m,p}(\vec{f})$ one can take the Riemann surface on the algebraic functions $\lambda(z)$.

To prove (a), it is sufficient to show that all components of $\vec{f}$ can be expressed as rational functions. To this end, we compare the jth component of the left hand side of (2.9) with that of the right hand side (j=1,...,m) and express $f_{j-1}(z)$ as a linear combination of the functions $f_j(z)$, ... ,$f_m(z)$ with rational coefficients. Finally we compare the (m+1)th components in (2.9) and obtain a rational expression for $f_m(z)$. By means of linear combinations of $f_{j-1}(z)$ we can now let j decrease again and complete the proof of (a).
The proposition (b) follows from the independence of (2.8) over the field of rational functions which in turn follows from (2.7) and Kronecker's criterion for a $P^{(m)}$-fraction [8], together with a classical Riemann theorem [11], [7] which states that any set of (n+1) functions from M(S), l(S)=n is linearly dependent over the polynomials.

3. $P^{(2)}(z)$-FRACTIONS WITH CONSTANT COEFFICIENTS

3.1. Let us consider the pair of functions $(u_1(z),u_2(z))$ for which

$$(u_1(z),u_2(z)) = 1/[(c,3z) + (u_1,u_2)]$$

According to (2.2) we have

$$\left| \begin{array}{l} 1/u_1 = u_2 + 3z \\ u_2/u_1 = c + u_1 \end{array} \right.$$

and consequently (u_1, u_2) are the branches of the solutions of the following algebraic equations

$$u_1^3(z) + c\,u_1^2(z) + 3z\,u_1(z) - 1 = 0$$
$$u_2^3(z) + 6z\,u_2^2(z) + (9z^2 - c)\,u_2(z) - 3cz - 1 = 0 \qquad (3.1)$$

Investigating those functions makes it possible to formulate the next conclusion:

Proposition 3.1 *The functions* $(u_1(z),u_2(z))$ *expanded in a* $P^{(2)}(z)$ *-fraction with constant coefficients*

$$(u_1(z),u_2(z)) = \frac{1|}{|(c,3z)+} \frac{1|}{|(c,3z)+} \cdots \qquad c \in \mathbb{R} \qquad (3.2)$$

with $c \leqslant -3$ *are Markov (Stieltjes) functions resulting from absolutely continuous measures with common support; in other words*

$$u_1(z) = \int_\Delta \frac{\rho_1(x)\,dx}{z-x}, \qquad u_2(z) = \int_\Delta \frac{\rho_2(x)\,dx}{z-x}$$

where Δ *is the interval of the real axis where the discriminant* $D(z)$ *of the equation* (3.1)

$$D(z) = z^3 - \frac{c^2}{12} z^2 + \frac{c}{2} z + \frac{1}{4} - \frac{c^3}{27}$$

is positive and the weight functions ρ_i *are given by*

$$i=1,2: \qquad \rho_i(z) = \frac{\sqrt{3}}{2\pi}\left[\,\omega_i(z) - \frac{q_i(z)}{\omega_i(z)}\right]$$

where $\omega_i(z) = \sqrt[3]{r_i(z) + \sqrt{D(z)}}$

$$r_1 = zc/2 + 1/2 - c^3/27 \qquad q_1 = z - c^2/9$$
$$r_2 = z^3 - zc/2 + 1/2 \qquad q_2 = -z^2 + c/3$$

3.2 Let us consider Padé-Hermite approximants $\vec{\pi}_n \equiv \{\pi_n^{(i)}(z) = \frac{P_{n-1}^{(i)}(z)}{Q_n(z)}\}_{i=1}^2$ to the Markov functions described above. So $\vec{\pi}_m$ is a convergent of the $P^{(2)}(z)$-fraction (3.2). The numerator and denominator of these Padé-Hermite approximants satisfy the fourth-term recurrence relation

$$Y_{n+3}(z) = 3z\,Y_{n+2}(z) + c\,Y_{n+1}(z) + Y_n(z) \qquad (3.3)$$

with initial conditions

$$Q_0 = 1 \qquad Q_1 = 3z \qquad Q_2 = 9z^2 + c$$
$$P_0^{(1)} = 0 \qquad P_1^{(1)} = 1 \qquad P_2^{(1)} = 3z$$
$$P_0^{(2)} = 0 \qquad P_1^{(2)} = 1 \qquad P_2^{(2)} = 3zc + 1 \tag{3.4}$$

It is well known that the general solution of (3.3) is given by

$$Y_n(z) = d_1(z)\, y_1^{-n}(z) + d_2(z)\, y_2^{-n}(z) + d_3(z)\, y_3^{-n}(z)$$

where the functions $y_i(z)$ for $i=1,2,3$ are the roots of equation (3.1) and the functions $d_i(z)$ can be determined from the initial conditions (3.4).

From properties of the roots of (3.1) we can write down the following asymptotic expansion for the Padé-Hermite numerators and denominators.

Proposition 3.2 *The asymptotic expansion*

$$Y_n(z) \sim y_1^{-n}(z)\,(d_1(z) + \bar{\bar{o}}(1))$$

is uniformly valid in compact sets in the complement of Δ, *where the root* $y_1(z)$ *of* (3.1) *decreases with* z *going to infinity.*

3.3. It is important to emphasize that the following one parameter (α) family of Stieltjes functions with congruent supports on the real axis ($i=1,2$: $\operatorname{supp} \mu_i^{\alpha} = \Delta \subset \mathbb{R}$)

$$\left\{ \int_\Delta \frac{\mu_1^{(\alpha)}(x)\,dx}{z-x},\ \int_\Delta \frac{\mu_2^{(\alpha)}(x)\,dx}{z-x} \right\}$$

exists, but that the asymptotic behaviour of the Padé-Hermite polynomial $|Q_n^{(\alpha)}(z)|^{1/n}$ for different α is different. It is well known [2] that in the case $\operatorname{supp} \mu_1 \subset \Delta_1$ and $\operatorname{supp} \mu_2 \subset \Delta_2$ with $\Delta_1 \cap \Delta_2 = \emptyset$ (which is called Angelesco's case) the asymptotic behaviour of $|Q_n|^{1/n}$ only depends on the geometrical properties of $\operatorname{supp} \mu_1$ and $\operatorname{supp} \mu_2$ and not on the weight functions.

Therefore the "weak" asymptotic expansion of $|Q_n|^{1/n}$ for Markov (Stieltjes) functions does not only depend on the support of the measures (as in the case of orthogonal polynomials and Angelesco's case) but on the Riemann surface (as stated in a conjecture by J. Nuttall [7]) formed by analytic continuations. When studying the case of Markov functions with support Δ, this Riemann surface looks like

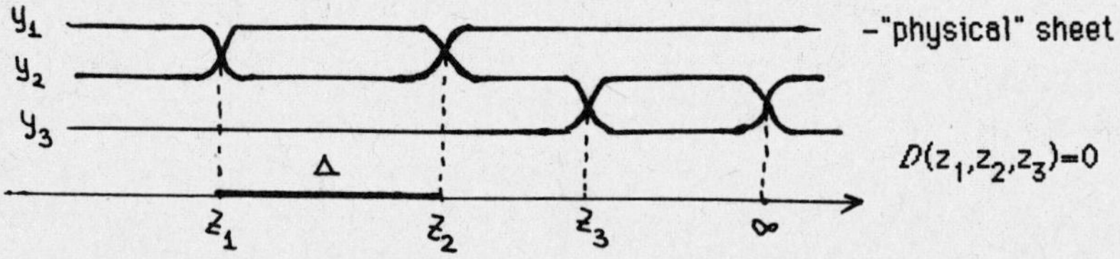

(3.5)

3.4. The asymptotic behaviour of the polynomials with increasing degree n is closely related to the existence of the weak limit of the probability measure $\frac{1}{n}\sum_{j=1}^{n}\delta(z-z_j)$ where the $\{z_j\}_{j=1}^{k}$ are the zeros of the polynomial of degree n. Designating the limit measure as σ and the correspondent logarithmic potential as $V_\sigma(z)$

$$V_\sigma(z) = \int \ln \frac{1}{|z-t|}\, d\sigma(t)$$

one can derive an asymptotic expansion for the nth root of the polynomials $Y_n(z)$

$$\lim_{n\to\infty} |Y_n(z)|^{1/n} = e^{-V_\sigma(z)}$$

(When the support of the measure contains infinity, the logarithmic potential can be defined as [10]

$$V_\sigma = \int \ln |H(z,t)|^{-1} d\sigma(t), \qquad H(z,t) = \begin{cases} z-t & |t| \leqslant 1 \\ (z-t)|t|^{-1} & |t| > 1 \end{cases}).$$

The limit measure σ related to the distribution of zeros of the Padé-Hermite polynomials is the equilibrium in some problem concerning a logarithmic potential. Let us now discuss what this problem is.

The unit charge $-|\sigma|$ is placed on the first plate of a condenser $\Delta=[z_1,z_2]$ and on the second plate $\Delta^*=[z_3,\infty]$ half of the unit charge $|\sigma^*|$ is placed. The transformation coefficient of a condenser is 1/2. The necessary conditions for the equilibrium distribution of the charge $\sigma(x)$ are

$$\begin{cases} V_\sigma(z) - \frac{1}{2} V_{\sigma^*}(z) = 0 & z \in \Delta \\ V_{\sigma^*}(z) - \frac{1}{2} V_\sigma(z) = 0 & z \in \Delta^* \end{cases}$$

The logarithmic branch of the algebraic function (3.1) constitutes the solution of this potential problem.

3.5 Let us consider the set **I** of points z where different branches of our algebraic function (3.1) are the same moduli

$$\mathbf{I} = \{\, z : |u^{(i)}/u^{(j)}| = 1, \quad i=1,2,3\,,\ i \neq j \,\}$$

Now we do not restrict ourselves to Markov type functions and consider the equation (3.1) for all real c. Let us emphasize that the set **I** and particularly its subset $\mathbf{I}_0$

$$\mathbf{I}_0 = \{ z : |u^{(i)}| = |u^{(j)}| < |u^{(k)}|, \quad i \neq j \neq k, \quad i,j,k=1,2,3\}$$

give very important information about the behaviour of Padé-Hermite approximants (see [9]). For example, there is uniform asymptotic convergence of Padé-Hermite approximants on a compact set Ω bounded by $\mathbf{I}_0$ and lying in the neighbourhood of infinity and their zeros are located in the set $\mathbb{C}\backslash\Omega$. In addition, the sets $\mathbf{I}_0$ and $\mathbf{I}_1=\mathbf{I}\backslash\mathbf{I}_0$ are the sets on which the harmonic potential obtains its equilibrium in problems with certain curved condensers. In that case, the equilibrium distribution of the charge on $\mathbf{I}_0$ is a function of the distribution of the zeros of the Padé-Hermite polynomials derived from $P^{(2)}$-continued fractions (3.2) with constant coefficients.

The set $\mathbf{I}$ is given by an algebraic curve. For the following algebraic function u (3.1)

$$u^3 + cu^2 + 3zc - 1 = 0$$

its equation is easily given. In this case the set $\mathbf{I}$ is the image of the equation's roots

$$\mathcal{J}(\nu,w) = (\nu + 2)\, w^3 - c^2w^2 + c(\nu^2 + 5\nu + 4)w + \nu^3 + 3\nu^2 + (3 - c^3)\nu - 2c^3 + 1 = 0$$

$$W = 3z$$

with $\nu \in [-2,2]$.

The curve $\mathbf{I}$ varies with c. In the figures below the curve $\mathbf{I}$ is given for c belonging to different intervals. The thick line on these figures indicates the subset $\mathbf{I}_0$.

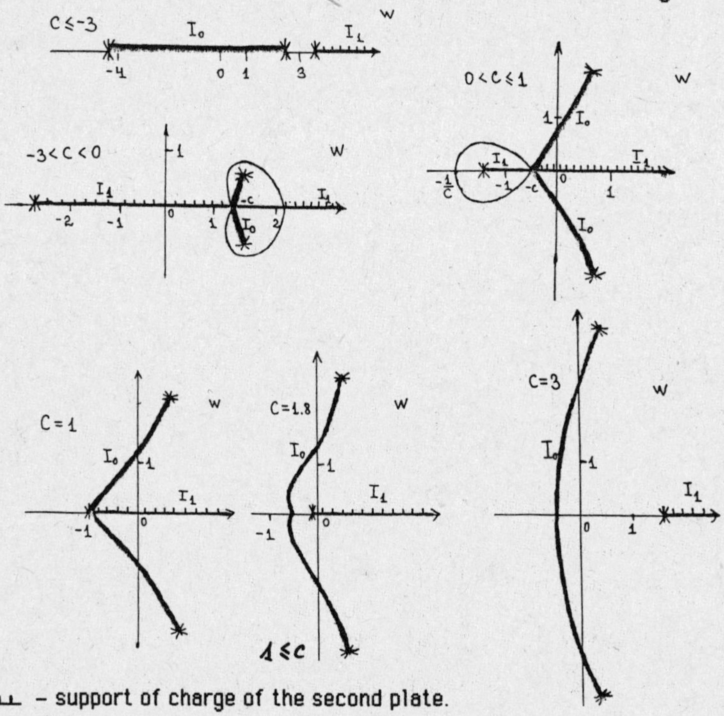

ıııııııı – support of charge of the second plate.

Analysing these figures we come to the conclusion that the distribution of the zeros of the Padé–Hermite polynomials defines the equilibrium of the potential on the condenser plates I_0 (first plate) and $I_1 \cap \mathbb{R}$ (second plate). This condenser probably has minimal capacity among the family of condenser plates which connect branch ponts of the function (3.1).

Let us emphasize that, when the plates intersect ($c \in [-3,-1]$), the equilibrium distribution of the charge on the second plate has a lacuna.

4. PERIODICAL $P^{(2)}$–FRACTIONS (WITH PERIOD 2) AND THEIR CONNECTION WITH THE SYSTEM OF ANGELESCO

4.1. Let us consider two Markov functions with non overlapping supports [-1,0] and [0,1] (Angelesco case). The formulae for the numerator and denominator of simultaneous Padé approximants for two logarithms

$$\vec{f}(z) = \left\{ \int_{-1}^{0} \frac{dx}{z-x}, \int_{0}^{1} \frac{dx}{z-x} \right\} \tag{4.1}$$

was obtained in [3]. The $P^{(2)}$–expansion of f into the form (2.4) is given by

$$\vec{f}(z) = \dots + \frac{(r_k, 1)\,|}{(e_k, z+d_k)} \dots \tag{4.2}$$

After the calculations, one can get the following formulae:

$$d_k = \begin{cases} -\dfrac{2n+1}{3n+1}\dfrac{B_n}{A_n} & k=2n+1 \\[2ex] \dfrac{2n+1}{3n+1}\dfrac{B_n}{A_n} & k=2n \end{cases}, \quad e_k = \begin{cases} \dfrac{(2n+1)^2}{(3n+1)^2}\dfrac{B_n^2}{A_n^2} - \dfrac{2}{3}\dfrac{(2n+1)^2}{(3n+1)(3n+2)} & k=2n+1 \\[2ex] -\dfrac{2n A_n}{3(3n-1)B_{n-1}} & k=2n \end{cases}$$

$$r_k = \begin{cases} \dfrac{2n(2n+1)B_n}{3(3n+1)(3n-1)B_{n-1}} & k=2n+1 \\[2ex] -\dfrac{2n(2n-1)A_n}{3(3n-1)(3n-2)A_{n-1}} & k=2n \end{cases}, \qquad n=0,1,2,\dots$$

where

$$A_n = \int_0^1 x^n(1-x^2)^n\,dx\,, \qquad B_n = \int_0^1 x^{n+1}(1-x^2)^n\,dx$$

We have thus established the following proposition.

Proposition 4.1. *The expansion of two logarithms* (4.1) *into a* $P^{(2)}$- *fraction* (4.2) *is limit-periodic with period* p=2. *Specifically*

$$\lim_{n\to\infty} d_{2n} = \alpha \qquad \lim_{n\to\infty} e_{2n} = -\alpha^2 \qquad \lim_{n\to\infty} r_{2n} = \alpha^3$$

$$\lim_{n\to\infty} d_{2n+1} = -\alpha \qquad \lim_{n\to\infty} e_{2n+1} = -\alpha^2 \qquad \lim_{n\to\infty} r_{2n+1} = -\alpha^3$$

where $\alpha = 2/(3\sqrt{3})$.

4.2 It is interesting to study the following periodic $P^{(2)}$-fraction

$$\frac{(-\alpha^3,1)}{(-\alpha^2,z-\alpha)+}\Big| \; \frac{(\alpha^3,1)}{(-\alpha^2,z+\alpha)+}\Big| \; \frac{(-\alpha^3,1)}{(-\alpha^2,z-\alpha)+}\Big| \; \cdots \tag{4.3}$$

with the period p=2. Let us construct the system of functions (φ_1, φ_2), which has the expansion (4.4) into a $P^{(2)}$-fraction. We have

$$(\varphi_1, \varphi_2) = \cfrac{(-\alpha^3, 1)}{(-\alpha^2, z-\alpha) + \cfrac{(\alpha^3, 1)}{(-\alpha^2, z+\alpha) + (\varphi_1, \varphi_2)}}$$

where

$$\varphi_1 = \frac{-\alpha^3(z+\alpha+\varphi_2)}{(z^2-2\alpha^2)+(z-\alpha)\varphi_2+\varphi_1}$$

$$\varphi_2 = \frac{-\alpha^2 z - \alpha^2\varphi_2}{(z-2\alpha^2)+(z-\alpha)\varphi_2+\varphi_1}.$$

After making the transformation, one can obtain an equation for the function $\varphi_2(z)$, namely

$$z\,\varphi_2(z)\,(z+\varphi_2(z))^2 + \alpha^2 z^2 = 0\,.$$

Making the further substitution $\varphi_2 = -zy/(y+1)$, we have

$$(y(z)+1)^3 - z^2\,y(z)/\alpha^2 = 0\,. \tag{4.5}$$

This equation was studied in [3] using the asymptotic expansion of Padé–Hermite polynomials for system (4.1). Analogously, we can obtain an expression for $\varphi_1(z)$. So the functions

$$\begin{cases} \varphi_1(z) = -\alpha \dfrac{z\, y_1(z)}{y_1(z)+1} - \alpha^2 y_1(z) & \qquad y_1(z)\Big|_{z=\infty} = \dfrac{1}{z^2} + \dots \\ \varphi_2(z) = -z \dfrac{y_1(z)}{y_1(z)+1} & \end{cases} \tag{4.6}$$

are meromorphic on the Riemann surface of the algebraic function (4.5). After the choise of the cut $[-1,1]$ in $\mathbb{C}$, the functions $\varphi_1(z)$ and $\varphi_2(z)$ become holomorphic. The character of their branch points are determined by the character of the branch points of the function (4.5). The three-sheeted Riemann surface of the function $y(z)$ given by (4.5) in the case $\alpha=2/(3\sqrt{3})$ is shown in the following figure

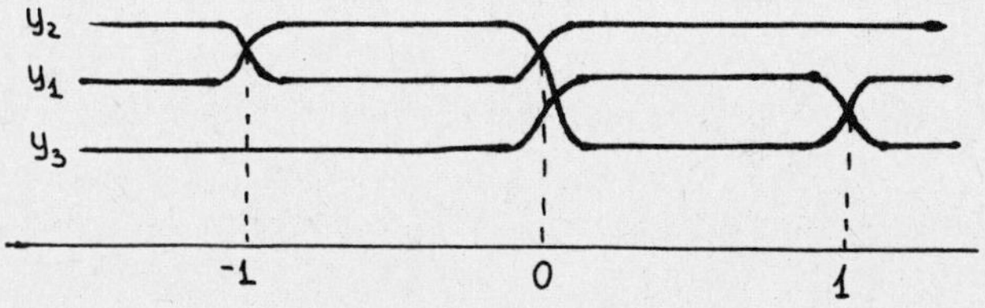

–"physical" sheet

(4.7)

$\alpha=2/(3\sqrt{3})$

The investigation of the boundary values φ_i^+ and φ_i^- of the functions $\varphi_1(z)$ and $\varphi_2(z)$, on the upper and lower edges of the cut, show, that $(\varphi_i^+ - \varphi_i^-)(z)$ has constant sign along the interval $[-1,1]$, and therefore the functions $\varphi_i(z)$ can be expressed as

$$\varphi_i(z) = -\int_{-1}^{1} \frac{\rho_i(x)}{z-x}\, dx, \qquad i=1,2, \qquad \alpha=2/(3\sqrt{3}) \tag{4.8}$$

where the weight functions $\rho_i(x)$ are analytic in $\mathbb{C}$ except at the points $z=-1,0,1$. Summarising, we have

Proposition 4.2 *The periodic ($p=2$)* $P^{(2)}$*– fraction* (4.3) *represents the collection of two Markov functions* (4.8) *with congruent supports. These functions are meromorphic on the Riemann surface of the algebraic function* (4.5).

4.3. By using the properties of periodicity, one can draw some conclusions about Padé–Hermite polynomials for the functions (4.6). From the recurrence relation (2.5), we can obtain the recurrence relation with constant coefficients for the even Padé-Hermite polynomials

$$Y_{2n}(z) = (z^2 - 3\alpha^2)\, Y_{2n-2}(z) - 3\alpha^4\, Y_{2n-4}(z) - \alpha^6\, Y_{2n-6}(z) \qquad (4.9)$$

with the initial conditions following from (2.6):

$$\begin{array}{lll} Q_0 = 1 & Q_2 = (z^2 - 2\alpha^2) & Q_4 = (z^2 - 3\alpha^2)(z^2 - 2\alpha^2) - 3\alpha^4 \\ P_0^{(1)} = 0 & P_2^{(1)} = -\alpha^3(z+\alpha) & P_4^{(1)} = -\alpha^3(z^2 - 3\alpha^2)(z+\alpha) - \alpha^6 \\ P_0^{(2)} = 0 & P_2^{(2)} = -\alpha^2 z & P_4^{(2)} = -\alpha^2 z\,(z^2 - 3\alpha^2) \end{array} \qquad (4.10)$$

The general solution of the relation (4.9) is

$$Y_{2n}(z) = c_1(z)\left(\frac{\alpha^2}{y_1(z)}\right)^n + c_2(z)\left(\frac{\alpha^2}{y_2(z)}\right)^n + c_3(z)\left(\frac{\alpha^2}{y_3(z)}\right)^n$$

where $y_i(z)$, $i=1,2,3$, are the roots of equation (4.5), and the functions $c_i(z)$ can be determined from the initial conditions (4.10).

Let us emphasize, that outside the interval $[-1,1]$ we have $|y_1| < |y_2|$ and $|y_1| < |y_3|$ (see [3]). Therefore we have

<u>Proposition 4.3.</u> *The asymptotic formulae for even Padé-Hermite polynomials of a collection of Markov function are* (4.6)

$$Y_{2n}(z) \sim c_1(z)\, \alpha^{2n}/y_1(z)^n \qquad z \in K \Subset \mathbb{C}\setminus[-1,1]$$

where $y_1(z)$ *is the root of the equation* (4.5) *decreasing at infinity and* $c_1(z)$ *can be determined from the initial conditions* (4.10).

4.4. In conclusion to this section, we would like to make two interesting remarks.

<u>Remark 1</u>. The Markov functions (4.1) are not meromorphic on the Riemann surface (4.7), but their "weakly" asymptotic Padé-Hermite polynomials coincide with "weakly" asymptotic Padé-Hermite polynomials of Markov functions (4.8), which are meromorphic on the Riemann surface (4.7). In other words, the same Riemann surface determines the asymptotic nature of the Padé-Hermite polynomials for essentially different collections of Markov functions, in accordance with the J. Nuttall conjecture [7].

<u>Remark 2</u>. We emphasize the equivalence of the Riemann surface (3.5) of the previous section and the Riemann surface (4.7) of the present section. Both surfaces have genus 0, but the $P^{(2)}$-fractions (3.2) and (4.3) represent collections of the functions on the

different sheets. It is interesting to observe that a $P^{(2)}$-fraction with the period p=2 (4.7) represents a meromorphic function on the Riemann surface of genus 0, like the $P^{(2)}$-fraction with the period p=1, just as it is known that a $P^{(1)}$-fraction with the period p=k would always represent a meromorphic function on a Riemann surface with genus k-1.

REFERENCES

[1] DE BRUIN M.G. , Convergence along steplines in a generalized Padé table, in *Padé and rational approximation*, ed. E.B. Saff, E.B. Varga (1977), 15-21.

[2] GONCHAR A. A. , RAKHMANOV E. A. , On the convergence of simultaneous Padé approximants for systems of functions of the Markov type, Proc. Steklov Math. Inst. **157** (1981), 31-48 (in Russian) (English transl. in Proc. Steklov Math. Inst. **3** (1983), 31-50)

[3] KALYAGIN V. A. , On a class of polynomials defined by two orthogonality relations, Mat. Sbornik **110** (152) (1979), 609-627 (in Russian), (English transl. in Math. USSR Sb. **38** (1981))

[4] MARKOV A. A. , Selected papers on continued fractions and the theory of functions deviating least from zero, ed. OGIZ, Moscow (1948) (in Russian)

[5] NIKISHIN E. M. , A system of Markov functions, Vestnik Moscow. Univ., Ser. I Mat. Mekh. No.4 (1979), 60-63 (in Russian), (English transl. in Moscow Univ. Math. Bull. **34** (1979)

[6] NIKISHIN E. M. , On simultaneous Padé approximants, Mat. Sbornik **113** (155) (1980), 499-519 (in Russian), (English transl. in Math. USSR Sb. **41** (1982)).

[7] NUTTALL J. , Asymptotics of diagonal Hermite-Padé polynomials, J. Approx. Theory **42**, No.4 (1984), 299-386.

[8] PARUSNIKOV V.I., The Jacobi-Perron algorithm and joint approximation of functions, Mat. Sbornik **114** (156) (1981), 322-333 (English transl. in Math. USSR Sb. 42 (1982)).

[9] PARUSNIKOV V.I., Limit periodic multidimensional continued fraction, Inst. Appl. Math. USSR Acad. of Sciences, No.62 (1983), 1-24 (preprint) (in Russian)

[10] RAKHMANOV E. A. , On asymptotic properties of polynomials which are orthogonal on the real axis, Mat. Sbornik **119** (1982), 163-202 (in Russian).

[11] RIEMANN B. , Oeuvres Mathématiques, ed. Albert Blanchard, Paris (1968), 353-363

[12] SZEGÖ G. , Orthogonal Polynomials, Providence (1939).

figures

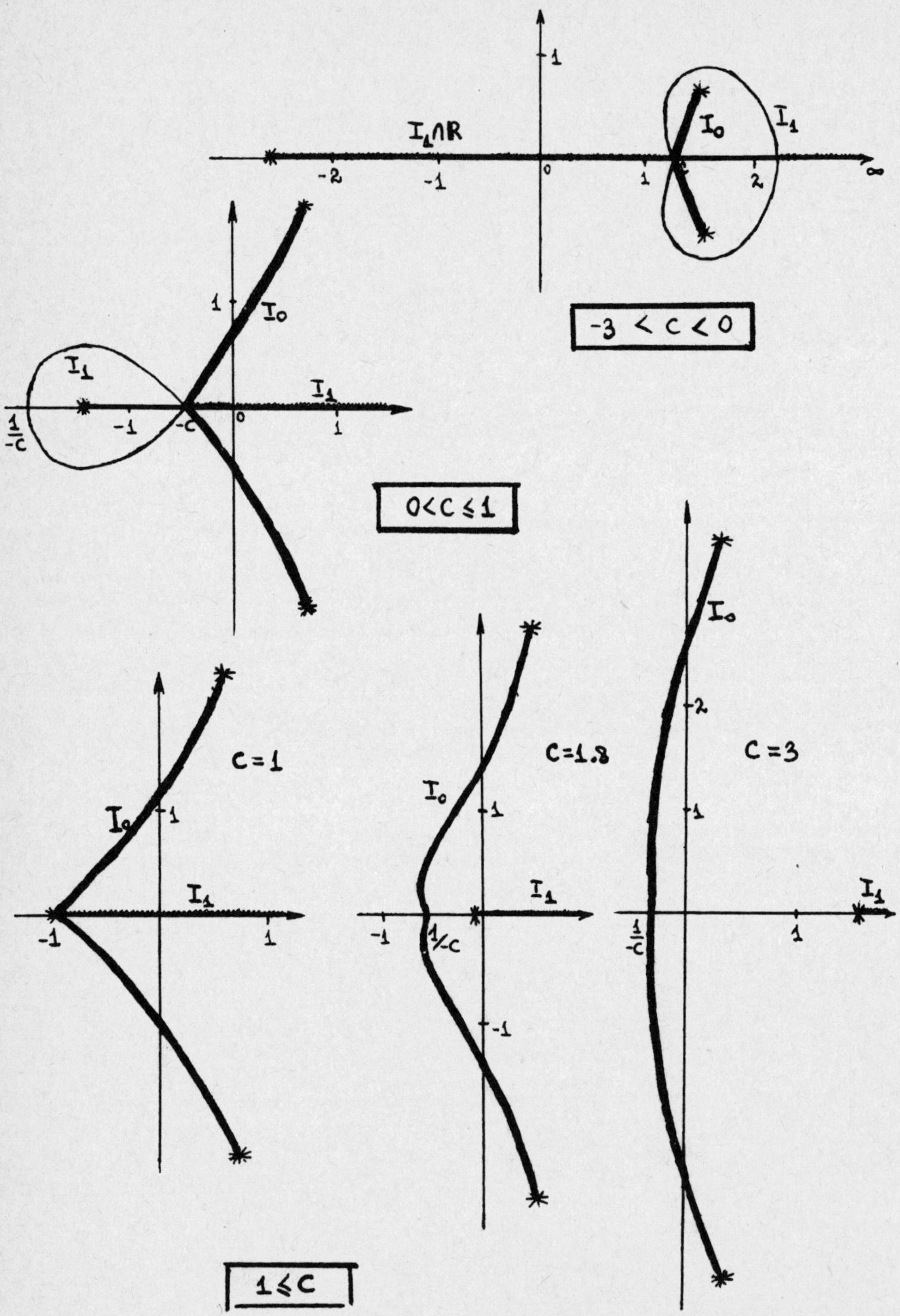

MODIFICATION OF GENERALISED CONTINUED FRACTIONS I

Definition and application to the limit-periodic case

Marcel G. de Bruin
Department of Mathematics
University of Amsterdam
Roetersstraat 15
1018 WB Amsterdam
The Netherlands

Lisa Jacobsen
Institutt for Matematikk og Statistikk
Universitetet i Trondheim, AVH
7055 Dragvoll
Norway

Summary

The aim of this paper is to generalise the concept of "modification", which has been so successfully exploited in the field of ordinary continued fractions, to generalised continued fractions.

The main result in this paper is the proof of convergence acceleration for a suitable modification in the case of limit-1-periodic n-fractions for which the auxiliary equation (connected with the underlying difference equation for numerators and denominators) has only simple roots with differing absolute values.

INTRODUCTION AND NOTATION

In the field of the analytic theory of continued fractions the concept of modification has proved to be a fruitful tool in many respects. Without going into detail, the machinery will be explained for ordinary continued fractions (for details the reader is referred to the papers [4], [8],[9],[10] and to the proceedings [11], containing a.o. the excellent survey by W.J. Thron and H. Waadeland, pages 38-66). Consider a continued fraction

$$\mathop{K}_{k=1}^{\infty} \frac{a_k}{b_k} = \frac{a_1}{b_1} + \frac{a_2}{b_2} + \cdots \qquad (a_k, b_k \in \mathbb{C};\ a_k \neq 0 \text{ for } k \geq 1) \tag{1}$$

and, connected with it, the sequence of *approximants* $\{f_k\}_{k=0}^{\infty}$

$$f_k = \mathop{K}_{v=1}^{k} \frac{a_v}{b_v} \quad (k \geq 1),\ f_0 = 0 \tag{2}$$

which exists as a sequence in the extended complex plane $\hat{\mathbb{C}}$. In case of convergence, the limit

$$f = \lim_{k\to\infty} f_k \tag{3}$$

is called the *value* of the continued fraction ($f \in \hat{\mathbb{C}}$).

Looking closer at (2),(3) it appears that the limit is calculated using expressions that arise from (1) by *equating to zero* the so-called *tails* $f^{(m)}$ given by

$$f^{(m)} = \underset{k=m+1}{\overset{\infty}{K}} \frac{a_k}{b_k} \quad (m \geq 1) \tag{4}$$

It is a well known fact that the continued fraction (1) converges if and only if its mth tail (4) converges $(m \geq 1)$.

The reader is possibly aware of the close connection between continued fractions, approximants and Möbius-transformations (sometimes called linear fractional transformations) :

$$f_k = \frac{A_k}{B_k} = S_k(0) \tag{5}$$

where the approximants can be calculated using

$$\begin{cases} A_{-1}=1, A_0=0 \; ; \; A_k=b_k A_{k-1}+a_k A_{k-2} & (k \geq 1) \\ B_{-1}=0, B_0=1 \; ; \; B_k=b_k B_{k-1}+a_k B_{k-2} & (k \geq 1) \end{cases} \tag{6}$$

and $\{S_k\}$ can be generated by

$$s_k(w) = \frac{a_k}{b_k+w} \quad (k \geq 1); \; S_1(w)=s_1(w), \; S_k(w)=S_{k-1}(s_k(w)) \text{ for } k \geq 2 \tag{7}$$

The connection between (6) and (7) is the well-known formula

$$S_k(w) = \frac{A_k+A_{k-1}w}{B_k+B_{k-1}w} \quad (k \geq 1) \tag{8}$$

With this notation the value of a convergent continued fraction is given by $f = S_k(f^{(k)})$ $(k \geq 1)$, while formula (3) implies, using (5), $f = \lim_{k\to\infty} S_k(0)$. However, since $\{f^{(k)}\}_{k=1}^{\infty}$ usually does not converge to 0, it looks as if one might be better off by considering *modified approximants* $\{S_k(w_k)\}$, $w_k \in \hat{\mathbb{C}}$. Obviously the optimal choice is $w_k=f^{(k)}$ but unfortunately the explicit values of the tails are rarely known. Still it is often possible to find a sequence $\{w_k\}$ better suited for convergence purposes (and others) than the constant sequence $\{0\}$.

Of course, this is only part of the story: there are other reasons why the use of modified approximants might help considerably in solving problems in the analytic theory of continued fractions; cf. the references quoted above.

Now it is time to turn to the generalisation mentioned in the title of the paper. The notation used will almost entirely be that from [1], apart from a slight difference in the definition of the tails.

Consider a sequence of complex (n+1)-tuples $(b_k, a_k^{(1)}, \ldots, a_k^{(n)})$, where $n \in \mathbb{N}$ and the choice $n=1$ nearly always leads to formulae for ordinary continued fractions, and write this sequence, for notational convenience, in the following compact way

$$\begin{pmatrix} a_1^{(1)} & a_2^{(1)} & \dots & a_k^{(1)} & \dots \\ \cdot & \cdot & & \cdot & \\ \cdot & \cdot & & \cdot & \\ \cdot & \cdot & & \cdot & \\ a_1^{(n)} & a_2^{(n)} & \dots & a_k^{(n)} & \dots \\ b_1 & b_2 & \dots & b_k & \dots \end{pmatrix} \quad \text{or} \quad \overset{\infty}{\underset{k=1}{K}} \begin{pmatrix} a_k^{(1)} \\ \cdot \\ \cdot \\ \cdot \\ a_k^{(n)} \\ b_k \end{pmatrix},$$

$$\text{with } a_k^{(1)} \neq 0 \quad (k \geq 1) \tag{9}$$

The n-fraction (9) is then given by its sequence of *approximant n-tuples* $\{(A_k^{(1)}/B_k,\dots,A_k^{(n)}/B_k)\}_{k=1}^{\infty}$, where numerators and denominators all satisfy the same recurrence relation

$$X_k = b_k X_{k-1} + a_k^{(n)} X_{k-2} + a_k^{(n-1)} X_{k-3} + \dots + a_k^{(1)} X_{k-n-1} \quad (k \geq 1) \tag{10}$$

and have initial values given by

$$\left.\begin{aligned} A_{-j}^{(i)} &= \delta_{i+j,n+1}, \quad B_{-j} = 0 \quad (j=1,2,\dots,n) \\ A_0^{(i)} &= 0 \qquad\quad\;\;, \quad B_0 = 1 \end{aligned}\right\} (i=1,2,\dots,n) \tag{11}$$

and *where we tacitly assumed that the n-tuples are well-defined for* $k \geq k_0$ (i.e. $A_k^{(i)}/B_k$ is not of the form 0/0 for $k \geq k_0$ and all $i=1,2,\dots,n$).

Introduce for $k \geq 1$ Möbius-transformations by

$$s_k^{(1)}(w^{(1)},\dots,w^{(n)}) = \frac{a_k^{(1)}}{b_k+w^{(n)}}, \quad s_k^{(i)}(w^{(1)},\dots,w^{(n)}) = \frac{a_k^{(i)}+w^{(i-1)}}{b_k+w^{(n)}}$$

$$(i=2,\dots,n) \tag{12}$$

and

$$\begin{cases} S_1^{(i)}(w^{(1)},\dots,w^{(n)}) = s_1^{(i)}(w^{(1)},\dots,w^{(n)}), \\ S_k^{(i)}(w^{(1)},\dots,w^{(n)}) = S_{k-1}^{(i)}(s_k^{(1)},\dots,s_k^{(n)}) \quad (k \geq 2) \end{cases} \quad (i=1,\dots,n) \tag{13}$$

It is very important, to realise that these Möbius-transformations $s_k^{(i)}$, and $S_k^{(i)}$ do not need to be well-defined for all values of the variables $w^{(1)},\dots,w^{(n)} \in \hat{\mathbb{C}}$. Because of the assumption that the approximant n-tuples are well-defined and the connection between (10),(11) and (12), (13) to be given below in (14), we see that the origin in $\hat{\mathbb{C}}^n$ (i.e. $w^{(1)}=\ldots=w^{(n)}=0$) must be regular point. This fact of the origin playing a special role already strongly suggests that the concept of convergence sometimes might be better introduced through the intermediate step of considering $\{S_k^{(i)}(w_k^{(1)},\ldots,w_k^{(n)})\}_{i=1}^{n}{}_{k=1}^{\infty}$ at points differing from the origin.

The connection formula in the case of generalised continued fractions (cf. [1]) is a quite simple extension of (8):

$$S_k^{(i)}(w^{(1)},\ldots,w^{(n)}) = \frac{A_k^{(i)}+A_{k-1}^{(i)}w^{(n)}+A_{k-2}^{(i)}w^{(n-1)}+\ldots+A_{k-n}^{(i)}w^{(1)}}{B_k+B_{k-1}w^{(n)}+B_{k-2}w^{(n-1)}+\ldots+B_{k-n}w^{(1)}}$$

$$(i=1,\ldots,n;k\geq 1) \tag{14}$$

As before we find that the approximants are the images of the origin

$$S_k^{(i)}(0,\ldots,0) = A_k^{(i)}/B_k \quad (i=1,\ldots,n;k\geq 1) \tag{15}$$

It is clear now, how to define the concept of modification:

DEFINITION 1

Given a sequence of n-tuples $\{(w_k^{(1)},\ldots,w_k^{(n)})\}_{k=1}^{\infty}$ of numbers from $\hat{\mathbb{C}}$, a *modification* of (9) is given by the sequence of n-tuples

$$\{S_k^{(i)}(w_k^{(1)},\ldots,w_k^{(n)})\}_{i=1}^{n}{}_{k=1}^{\infty} \tag{16}$$

(assuming that this sequence is well defined from some k on)

Before we can turn towards one of the goals to achieve by modification, i.e. convergence acceleration, it is necessary to introduce what is meant by a *convergent n-fraction*

DEFINITION 2

The n-fraction (9) is called *convergent* in $\hat{\mathbb{C}}^n$ if the following limit exists in $\hat{\mathbb{C}}^n$

$$(\xi_0^{(1)},\ldots,\xi_0^{(n)}) = \lim_{k\to\infty}(A_k^{(1)}/B_k,\ldots,A_k^{(n)}/B_k) \tag{17}$$

Remark: of course one could use $\lim_{k\to\infty} S_k^{(i)}(0,\ldots,0)$ also.

There does exist some literature concerning convergence of generalised continued fractions (a.o. [1],[5],[6]), but always convergence is investigated from the viewpoint of the sequence $\{S_k^{(i)}(0,\ldots,0)\}_{i=1}^{n}{}_{k=1}^{\infty}$: i.e. the *tails*

$$\begin{bmatrix} \xi_m^{(1)} \\ \cdot \\ \cdot \\ \cdot \\ \xi_m^{(n)} \end{bmatrix} = \overset{\infty}{\underset{k=m+1}{K}} \begin{bmatrix} a_k^{(1)} \\ \cdot \\ \cdot \\ \cdot \\ a_k^{(n)} \\ b_k \end{bmatrix} \qquad (m \geqq 1) \tag{18}$$

are replaced by zero again although $\{\xi_m^{(i)}\}_{m=1}^{\infty}$ *converges to zero in exceptional cases only.*

A simple example is the case of the 1-periodic n-fraction with all entries equal to 1:

$$\overset{\infty}{\underset{k=1}{K}} \begin{pmatrix} 1 \\ \cdot \\ \cdot \\ \cdot \\ 1 \\ 1 \end{pmatrix} \quad \text{or recurrence } X_k = X_{k-1} + X_{k-2} + \ldots + X_{k-n-1} \quad (k \geqq 1) \tag{19}$$

The limits of this n-fraction (cf. [5]), a sort of Multi-nacci type situation, are connected with the single positive real root of $r^{n+1} - r^n - r^{n-1} - \ldots - r - 1 = 0$, none of them is zero.

It is obvious that, as was the case for the ordinary continued fraction, extreme convergence acceleration as well as extremely bad behaviour can occur here

- if all $\xi_k^{(i)}$ exist $(i=1,\ldots,n; k \geqq 0)$, the choice $w_k^{(i)} = \xi_k^{(i)}$ $(i=1,\ldots,n; k \geqq 1)$ leads to stationary sequences

$$S_k^{(i)}(\xi_k^{(1)},\ldots,\xi_k^{(n)}) = \xi_0^{(i)} \qquad (i=1,\ldots,n; k \geqq 1)$$

- consider the situation that $(\eta^{(1)},\ldots,\eta^{(n)}) \neq (\xi_0^{(1)},\ldots,\xi_0^{(n)})$ is an arbitrary point from $\mathbb{C}^n$; then

$$\begin{bmatrix} A_k^{(1)} & A_{k-1}^{(1)} & \ldots & A_{k-n-1}^{(1)} \\ A_k^{(2)} & A_{k-1}^{(2)} & \ldots & A_{k-n-1}^{(2)} \\ \cdot \\ \cdot \\ \cdot \\ A_k^{(n)} & A_{k-1}^{(n)} & \ldots & A_{k-n-1}^{(n)} \\ B_k & B_{k-1} & \ldots & B_{k-n-1} \end{bmatrix} \begin{bmatrix} 1 \\ w_k^{(n)} \\ \cdot \\ \cdot \\ \cdot \\ w_k^{(1)} \end{bmatrix} = \begin{bmatrix} \eta^{(1)} \\ \cdot \\ \cdot \\ \cdot \\ \eta^{(n)} \\ 1 \end{bmatrix}$$

has a solution $(w_k^{(1)},\ldots,w_k^{(n)}) \in \mathbb{C}^n$ for all k (the determinant of the matrix on the left hand side is in absolute value $\prod_{v=1}^{k} |a_v^{(1)}|$ i.e.

different from zero!): the modification using the $w_k^{(i)}$ leads to a stationary sequence, having "the wrong value".

For the sequel we will restrict ourselves to the case that the n-fraction converges in $\mathbb{C}^n$. This is done, since the characterisation of convergence acceleration is chosen to be

$$\lim_{k\to\infty} \frac{\xi_0^{(i)} - S_k^{(i)}(w_k^{(1)},\dots,w_k^{(n)})}{\xi_0^{(i)} - S_k^{(i)}(0,\dots,0)} = 0 \quad (i=1,\dots,n).$$

Now this type of n-fractions having finite limits has been studied before. A famous theorem for continued fractions due to S. PINCHERLE [7] has been generalised to n-fractions in the work by P. van der CRUYSSEN [3]:

THEOREM 1

The following statements are equivalent:

(a) the n-fraction (10),(11) converges to finite limits

(b) there exists a solution $\{X_k^{(1)}\}_{k=1}^{\infty}$ of (10), that dominates an n-dimensional subspace of the linear space of all solutions of (10), which is spanned by $\{X_k^{(2)}\},\dots,\{X_k^{(n+1)}\}$ satisfying

$$\begin{vmatrix} X_{-n}^{(2)} & X_{-n+1}^{(2)} & \dots & X_{-1}^{(2)} \\ X_{-n}^{(3)} & X_{-n+1}^{(3)} & \dots & X_{-1}^{(3)} \\ \vdots & \vdots & & \vdots \\ X_{-n}^{(n+1)} & X_{-n+1}^{(n+1)} & \dots & X_{-1}^{(n+1)} \end{vmatrix} \neq 0.$$

($\{X_k^{(1)}\}$ dominates $\{Y_k\}$ if $\lim_{k\to\infty} Y_k/X_k^{(1)}=0$) □

Here we get a first glimpse of the important concept of *dominance in the setting of the linear space of solutions of a linear recurrence relation.* Although this concept can be defined without connections to linear recurrence relations (cf.[2]) it is a tool used almost entirely within the context of relations like (10); the importance of the concept of dominance for the main result of this paper begins to show in the famous POINCARE-PINCHERLE-PERRON theorem (cf.[2]for the history) that will be a tool in the sequel:

THEOREM 2

Let the coefficients of the n-fraction (9) satisfy

$$\lim_{k\to\infty} a_k^{(i)} = a^{(i)} \in \mathbb{C} \quad (i=1,\ldots,n), \quad \lim_{k\to\infty} b_k = b \in \mathbb{C} \tag{20}$$

and let the zeroes $r_1, r_2, \ldots, r_{n+1}$ of the auxiliary equation

$$r^{n+1} = br^n + a^{(n)} r^{n-1} + a^{(n-1)} r^{n-2} + \ldots + a^{(2)} r + a^{(1)} \tag{21}$$

satisfy the inequalities

$$|r_1| > |r_2| > \ldots > |r_n| > |r_{n+1}| \tag{22}$$

Then there exists a basis $\{D_k^{(i)}\}_{k=-n}^{\infty}$ $(i=1,\ldots,n+1)$ for the linear space V of all solutions of (10), that satisfies

$$\lim_{k\to\infty} D_{k+1}^{(i)} / D_k^{(i)} = r_i \quad (i=1,\ldots,n+1) \tag{23}$$

Remark: Because of (22) we see that this basis is *ordered by domination:*

$$\lim_{k\to\infty} D_k^{(j)} / D_k^{(i)} = 0 \quad (i+1 \leq j \leq n\ ;\ i=1,\ldots,n-1)$$

This strongly suggests that the n-fraction converges under these assumptions. And indeed, in [2] the following theorem has been proved

THEOREM 3

Under the conditions of Theorem 2, the n-fraction (9) converges in $\hat{\mathbb{C}}^n$.

It is within the context of these so-called limit-1-periodic n-fractions in Theorem 2,3,that the main result of this paper (concerned with convergence acceleration) will be given in the next section.

MAIN RESULT

For the reader familiar with introductory matters, who has skipped the first part of the paper, the theorem will recall formulae given in the introduction that are needed; for sake of convenience the formulae will have the same numbers as before.

THEOREM 4

Given the limit-1-periodic n-fraction

$$\mathop{K}_{k=1}^{\infty} \begin{pmatrix} a_k^{(1)} \\ \cdot \\ \cdot \\ \cdot \\ a_k^{(n)} \\ b_k \end{pmatrix} \tag{9}$$

with

$$\lim_{k\to\infty} a_k^{(i)} = a^{(i)} \in \mathbb{C} \quad (i=1,\dots,n), \quad \lim_{k\to\infty} b_k = b \in \mathbb{C} \tag{20}$$

such that the zeroes $r_1, r_2, \dots, r_{n+1}$ of

$$r^{n+1} = br^n + a^{(n)}r^{n-1} + a^{(n-1)}r^{n-2} + \dots + a^{(2)}r + a^{(1)} \tag{21}$$

satisfy

$$|r_1| > |r_2| > \dots > |r_n| > |r_{n+1}| > 0 \tag{22'}$$

(i.e. $a^{(1)} \neq 0$!).

Then the values $(\xi_m^{(1)},\dots,\xi_m^{(n)})$ of its tails satisfy

$$\lim_{m\to\infty} \xi_m^{(i)} = w^{(i)} \quad (i = 1,\dots,n) \tag{24}$$

where

$$w^{(1)} = \frac{a^{(1)}}{r_1}, \quad w^{(i)} = \frac{a^{(i)} + w^{(i-1)}}{r_1} \quad (i=2,\dots,n) \tag{25}$$

This theorem combined with (20) suggests that $w_k^{(i)} = w^{(i)}$ $(i=1,\dots,n;k\geq 1)$ is a sensible choice of modifying factors. Just this is

THEOREM 5

Under the assumptions of Theorem 4, we have

$$\lim_{k\to\infty} \frac{\xi_0^{(i)} - S_k^{(i)}(w^{(1)},\dots,w^{(n)})}{\xi_0^{(i)} - S_k^{(i)}(0,\dots,0)} = 0 \quad (i=1,\dots,n) \tag{26}$$

if $\xi_0^{(i)} \neq \infty$ for all $i \in \{1,\dots,n\}$.

Remarks

(a) Thus the modification induced convergence acceleration.

(b) From (25) and (21) we find: $w^{(n)} = \frac{a^{(n)}}{r_1} + \frac{a^{(n-1)}}{r_1^2} + \dots + \frac{a^{(1)}}{r_1^n} = r_1 - b.$

Before proceeding to the section containing the proof of this theorem, a numerical example will be treated.

Example

Consider an n-fraction (9) with n=2 of limit-1-periodic type

$$\begin{cases} a_k^{(1)} = (0.4 + \frac{1}{50k})(0.6 + \frac{1}{50k}) \\ a_k^{(2)} = -(1.0 + \frac{1}{50k} + \frac{1}{50(k+1)} + a_k^{(1)}) \\ b_k = 2.0 + \frac{1}{50k} + \frac{1}{50(k+1)} \end{cases}$$

Then

$$a^{(1)} = \lim_{k\to\infty} a_k^{(1)} = 0.24 \ , \ a^{(2)} = \lim_{k\to\infty} a_k^{(2)} = -1.24 \ , \quad b = \lim_{k\to\infty} b_k = 2.0$$

and

$$r^3 - br^2 - a^{(2)}r - a^{(1)} = (r-1.0)(r-0.6)(r-0.4) \ ,$$

which leads to the constant sequence of modifying factors

$$\{w^{(1)}\}_{k=1}^{\infty} \text{ with } w^{(1)} = 0.24 \ , \ \{w^{(2)}\}_{k=1}^{\infty} \text{ with } w^{(2)} = 1.0$$

The calculations were performed on a CYBER 750 using FTN5.1 + 564 and double precision (results correct up to 29 digits) this led to (all digits correct):

$$\xi_0^{(1)} = .25640\ 85890\ 03351\ 53012 \ ; \ \xi_0^{(2)} = -1.02592\ 79542\ 28502\ 35227$$

In the sequel in table 1 values of $S_k^{(i)}(0,0)$ (=non-modified) and $S_k^{(i)}(w^{(1)},w^{(2)})$ (=modified) will be given and in table 2 the *accelera-tion coefficients*

$$\mathbf{ac}_k^{(i)} := |\xi_0^{(i)} - S_k^{(i)}(w^{(1)},w^{(2)})| \ / \ |\xi_0^{(i)} - S_k^{(i)}(0,0)| \ .$$

ν	$S_\nu^{(1)}(0,0)$	$S_\nu^{(1)}(w^{(1)},w^{(2)})$	$S_\nu^{(2)}(0,0)$	$S_\nu^{(2)}(w^{(1)},w^{(2)})$
1	.12827 58620	.25281 55339	-0.63566 50246	-1.01980 58252
2	.18575 43743	.25527 77063	-0.83203 08446	-1.02385 57469
3	.21564 25138	.25598 17275	-0.91972 49920	-1.02509 11312
4	.23234 91463	.25622 67439	-0.96505 82558	-1.02555 03057
5	.24204 53280	.25632 08674	-0.99022 34061	-1.02574 43307
6	.24778 53536	.25636 67858	-1.00472 64382	-1.02583 40403
7	.25121 82784	.25638 69730	-1.01325 61708	-1.02587 82460
8	.25328 16462	.25639 70610	-1.01832 84845	-1.02590 10213
9	.25452 45081	.25640 23078	-1.02136 27034	-1.02591 31239
10	.25527 36488	.25640 51149	-1.02318 33361	-1.02591 96946
15	.25631 90601	.25640 83874	-1.02571 21794	-1.02592 74695
20	.25640 15663	.25640 85763	-1.02591 10363	-1.02592 79236
25	.25640 80393	.25640 85881	-1.02592 66302	-1.02592 79522
30	.25640 85460	.25640 85889	-1.02592 78507	-1.02592 79540
40	.25640 85887	.25640 85890	-1.02592 79535	-1.02592 79542
50	.25640 85890	.25640 85890	-1.02592 79542	-1.02592 79542

Table 1

ν	$ac_{\nu}^{(1)}$	$ac_{\nu}^{(2)}$
1	.02804 16651	.01568 71906
2	.01600 58767	.01068 71487
3	.01047 09964	.00787 94694
4	.00755 81565	.00620 42113
5	.00588 32222	.00514 28603
6	.00484 77351	.00442 95828
7	.00416 46818	.00392 27489
8	.00368 66374	.00354 40410
9	.00333 37915	.00324 85103
10	.00306 09878	.00309 38665
15	.00225 07966	.00224 61089
20	.00180 82140	.00180 77399
25	.00151 38357	.00151 37850
30	.00130 18859	.00130 18803
40	.00101 67504	.00101 67503
50	.00083 38962	.00083 38962

Table 2

REMARK

It has to be pointed out here, that the calculation of the modified approximants $S_k^{(i)}(w^{(1)},\ldots,w^{(n)})$ requires no extra operations compared to the calculation of $S_k^{(i)}(0,\ldots,0)$, except for the calculation of $w^{(1)},\ldots,w^{(n)}$ which can be done once. This means that the convergence acceleration is practically for free using this method!

One could even go one step further and use modified approximants to define convergence and then use other methods of convergence acceleration on this sequence of modified approximants in stead of applying that other method to the sequence of non-modified approximants!

PROOF OF THE MAIN RESULT

Proof of Theorem 4

Obviously the tails satisfy for $k \geq 1$ (cf. [1])

$$\xi_k^{(1)} = \frac{a_{k+1}^{(1)}}{b_{k+1} + \xi_{k+1}^{(n)}} \;,\; \xi_k^{(i)} = \frac{a_{k+1}^{(i)} + \xi_{k+1}^{(i-1)}}{b_{k+1} + \xi_{k+1}^{(n)}} \quad (i=2,\ldots,n) \tag{27}$$

It is now sufficient to prove $\lim_{k\to\infty} \xi_k^{(n)} = w^{(n)}$, because then the fact that $\lim_{k\to\infty} (b_k + \xi_k^{(n)}) = b + w^{(n)} = b + r_1 - b = r_1 \neq 0$ and the recursive definition of the w's in (25) automatically implies the correctness of the other limits.

Consider a basis $\{D_k^{(1)}\},\{D_k^{(2)}\},\dots,\{D_k^{(n+1)}\}$ for the space of all solutions of the recurrence relation (10), which is ordered by domination (cf. Theorem 2 in the Introduction). Furthermore consider the *shifted* recurrence relation

$$X_m = b_{m+k}X_{m-1} + a_{m+k}^{(n)}X_{m-2} + \dots + a_{m+k}^{(1)}X_{m-n-1} \qquad (m \geq 1) \tag{10'}$$

The tails $\xi_k^{(i)}$ satisfy: $\xi_k^{(i)} = \lim\limits_{k\to\infty} A_{k,m}^{(i)} / B_{k,m}$, where the $A_{k,m}^{(i)}$ and $B_{k,m}$ follow from (10') with the initial values (11) as *for the non-shifted case*. Because the D's form a basis for (10), the *shifted* D's form a basis for (10') i.e. (concentrating on $\xi_k^{(n)}$) :

$$\left.\begin{aligned} A_{k,m}^{(n)} &= \alpha_1^{(k)}D_{m+k}^{(1)} + \alpha_2^{(k)}D_{m+k}^{(2)} + \dots + \alpha_{n+1}^{(k)}D_{m+k}^{(n+1)} \\ B_{k,m} &= \beta_1^{(k)}D_{m+k}^{(1)} + \beta_2^{(k)}D_{m+k}^{(2)} + \dots + \beta_{n+1}^{(k)}D_{m+k}^{(n+1)} \end{aligned}\right\} \tag{28}$$

While the $\{D_k^{(i)}\}$ are ordered by domination, we are interested in $\alpha_1^{(k)}$ and $\beta_1^{(k)}$ only (division by $D_{m+k}^{(1)}$ in numerator and denominator of $A_{k,m}^{(n)} / B_{k,m}$ "kills" the other D's). Define

$$\mathfrak{D}_k = \begin{pmatrix} D_{k-n}^{(1)} & D_{k-n}^{(2)} & \dots & D_{k-n}^{(n+1)} \\ D_{k-n+1}^{(1)} & D_{k-n+1}^{(2)} & \dots & D_{k-n+1}^{(n+1)} \\ \vdots & \vdots & & \vdots \\ D_k^{(1)} & D_k^{(2)} & \dots & D_k^{(n+1)} \end{pmatrix} \tag{29}$$

then $\alpha_1^{(k)}$ and $\beta_1^{(k)}$ follow from

$$\begin{aligned} &\mathfrak{D}_k\,(\alpha_1^{(k)},\dots,\alpha_{n+1}^{(k)})^T = (0,\dots,0,1,0)^T, \\ &\mathfrak{D}_k\,(\beta_1^{(k)},\dots,\beta_{n+1}^{(k)})^T = (0,\dots,0,1)^T \end{aligned} \tag{30}$$

The $\{D_k^{(i)}\}$ form a basis, thus $\det\mathfrak{D}_k \neq 0$ and Cramer's rule leads to

$$\alpha_1^{(k)} = (-1)^{n+1}\frac{\det\mathfrak{D}_k^{(n,1)}}{\det\mathfrak{D}_k}, \quad \beta_1^{(k)} = (-1)^n\frac{\det\mathfrak{D}_k^{(n+1,1)}}{\det\mathfrak{D}_k} \tag{31}$$

where $\mathfrak{D}_k^{(p,1)}$ is the matrix arising from $\mathfrak{D}_k$ by removing the p-th row and 1st column.

As $\lim_{k\to\infty} D_k^{(i)} / D_{k-1}^{(i)} = r_i \quad (i=1,\ldots n+1), \quad \prod_{i=2}^{n+1} D_{k-n}^{(i)} \neq 0$ for $k \geq k_0$.

Taking out these factors from det $\mathfrak{D}_k^{(n+1,1)}$, we find

$$\lim_{k\to\infty} \frac{\det \mathfrak{D}_k^{(n+1,1)}}{\prod_{i=2}^{n+1} D_{k-n}^{(i)}} = \lim_{k\to\infty} \begin{vmatrix} 1 & \ldots & 1 \\ D_{k-n+1}^{(2)} / D_{k-n}^{(2)} & \ldots & D_{k-n+1}^{(n+1)} / D_{k-n}^{(n+1)} \\ \cdot & & \cdot \\ \cdot & & \cdot \\ \cdot & & \cdot \\ D_{k-1}^{(2)} / D_{k-n}^{(2)} & \ldots & D_{k-1}^{(n+1)} / D_{k-n}^{(n+1)} \end{vmatrix} =$$

$= V(r_2, r_3, \ldots, r_{n+1}) \neq 0$,

where $V(r_2,\ldots,r_{n+1})$ is the well-known Vandermonde determinant with $r_i \neq r_j$!

Applying the same method for $\mathfrak{D}_k^{(n,1)}$, we find from (31) :

$$\lim_{k\to\infty} \frac{\alpha_1^{(k)}}{\beta_1^{(k)}} = - \begin{vmatrix} 1 & & 1 \\ r_2 & \ldots & r_{n+1} \\ \cdot & & \cdot \\ \cdot & & \cdot \\ \cdot & & \cdot \\ r_2^{n-2} & \ldots & r_{n+1}^{n-2} \end{vmatrix} / V(r_2, r_3, \ldots, r_{n+1}) \qquad (32)$$

Obviously a finite number.

It is now a matter of simple analysis to prove that the quotient of the two determinants in (32) leads to the simple expression

$$-(r_2 + r_3 + \ldots + r_{n+1})$$

(for instance by viewing the numerator as a polynomial in r_2 of degree n with coefficient of r_2^{n-1} equal to zero etc; cf. problem 10, page 30 in G.E. Shilov, Linear Algebra, Dover (1977), New York).

This implies at once

$$\lim_{m\to\infty} A_{k,m}^{(n)} / B_{k,m} = \lim_{m\to\infty} \frac{\alpha_1^{(k)} + \Sigma_{j=2}^{n+1} \alpha_j^{(k)} D_{m+k}^{(j)} / D_{m+k}^{(1)}}{\beta_1^{(k)} + \Sigma_{j=2}^{n+1} \beta_j^{(k)} D_{m+k}^{(j)} D_{m+k}^{(1)}} = \frac{\alpha_1^{(k)}}{\beta_1^{(k)}}$$

$$\lim_{k\to\infty} \xi_k^{(n)} = \lim_{k\to\infty} \alpha_1^{(k)}/\beta_1^{(k)} = -(r_2+\ldots+r_{n+1}) = r_1-(r_1+r_2+\ldots+r_{n+1}) =$$

$$= r_1 - b = w^{(n)}.$$ □

Lemma 1

$$g(r) := r^n + w^{(n)} r^{n-1} + w^{(n-1)} r^{n-2} + \ldots + w^{(2)} r + w^{(1)} = \prod_{i=2}^{n+1} (r-r_i).$$

Proof of Theorem 5

Multiply by $r-r_1$, use the definition (25) of the w's and (21). □

Corollary

$$g(r_j) = 0 \quad (j=2,3,\ldots,n+1) \ , \ g(r_1) \neq 0.$$

Proof of

The rest of the proof consists of three steps (i is kept fixed, $i \in \{1,\ldots,n\}$):

(a) $S_{p+k}^{(i)}(0,\ldots,0) - S_k^{(i)}(0,\ldots,0) \neq 0$ for $k \geq k_0$, $p \geq p_0$.

(b) $$\lim_{p\to\infty} \frac{\{S_{p+k}^{(i)}(0,\ldots,0) - S_k^{(i)}(w^{(1)},\ldots,w^{(n)})\}}{\{S_{p+k}^{(i)}(0,\ldots,0) - S_k^{(i)}(0,\ldots,0)\}} \in \mathbb{C}$$

(c) the proof of (26).

Because the sequences $\{A_k^{(i)}\}$ $(i=1,\ldots,n)$, $\{B_k\}$ form a basis for the solution space of the linear recurrence relation (10), just like the $\{D_k^{(i)}\}$ $(i=1,\ldots,n+1)$, and the fact that all $\xi_0^{(i)}$ $(i=1,\ldots,n)$ are finite, we have

$$A_k^{(i)} = \Sigma_{q=1}^{n+1} \gamma_q^{(i)} D_k^{(q)} \quad (i=1,\ldots,n)$$

$$B_k = \Sigma_{q=1}^{n+1} \gamma_q D_k^{(q)} \ , \ \gamma_1 \neq 0 \qquad (k \geq 1) \tag{34}$$

Indeed, if $\gamma_1 = 0$ (i.e. $D_k^{(1)}$ is not present in B_k), the fact that all limits are finite and the fact that the basis $\{D_k^{(i)}\}$ is ordered by domination, together imply $\gamma_1^{(i)}=0$ *for all* $i \in \{1,\ldots,n\}$: this is in contradiction with the fact that $\{A_k^{(i)}\}$ $(i=1,\ldots,n)$, $\{B_k\}$ is a basis for the space of all solutions of (10)!

Thus

$$\lim_{k\to\infty} \frac{B_k}{B_{k-1}} = r_1 \ , \ \lim_{k\to\infty} \frac{D_k^{(i)}}{B_k} = 0 \quad (i=2,\ldots,n) \tag{35}$$

and therefore *we may replace the sequence* $\{D_k^{(i)}\}$ *by the sequence* $\{B_k\}$ *and we still have a basis ordered by domination for the solution space:* $\{B_k\},\{D_k^{(i)}\}$ $(i=2,\dots,n+1)$.
This implies that (34) can be replaced by

$$A_k^{(i)} = \delta_1^{(i)} B_k + \Sigma_{q=2}^{n+1} \delta_q^{(i)} D_k^{(q)} \qquad (i=1,\dots,n;\ k \geqq 1) \tag{34'}$$

where, because $\{A_k^{(i)}\}$ $(i=1,\dots,n)$, $\{B_k\}$ is a basis too, we have

$$\forall\ i \in \{1,\dots,n\}\ \exists\ t \in \{2,\dots,n+1\}\ (\delta_q^{(i)} = 0\,(q=2,\dots,t-1) \wedge \delta_t^{(i)} \neq 0) \tag{34"}$$

We are now ready to take the three steps needed for the proof.

(a) Using (34') and (34") we have

$$| S_{k+p}^{(i)}(0,\dots,0) - S_k^{(i)}(0,\dots,0) | = | A_{k+p}^{(i)}/B_{k+p} - A_k^{(i)}/B_k |$$

$$= | \sum_{q=t}^{n+1} \delta_q^{(i)} \left(\frac{D_{k+p}^{(q)}}{B_{k+p}} - \frac{D_k^{(q)}}{B_k} \right) | = | \sum_{q=t}^{n+1} \delta_q^{(i)} \frac{D_k^{(q)}}{B_k} \left(\frac{D_{k+p}^{(q)} B_k}{D_k^{(q)} B_{k+p}} - 1 \right) |$$

$$\geq \frac{1}{2} | \delta_t^{(i)} \frac{D_k^{(t)}}{B_k} \left(\frac{D_{k+p}^{(t)} B_k}{D_k^{(t)} B_{k+p}} - 1 \right) | \geq \frac{1}{4} | \delta_t^{(i)} D_k^{(t)} / B_k | > 0$$

for k,p sufficiently large.

(b) Because of (a), the quotient is well-defined for $k \geqq k_0$, $p \geqq p_0$ and we rewrite the expression using (14):

$$\frac{S_{p+k}^{(i)}(0,\dots,0) - S_k^{(i)}(w^{(1)},\dots,w^{(n)})}{S_{p+k}^{(i)}(0,\dots,0) - S_k^{(i)}(0,\dots,0)} = \frac{\dfrac{A_{p+k}^{(i)}}{B_{p+k}} - \dfrac{A_k^{(i)} + \Sigma_{j=0}^{n-1} w^{(n-j)} A_{k-j-1}^{(i)}}{B_k + \Sigma_{j=0}^{n-1} w^{(n-j)} B_{k-j-1}}}{\dfrac{A_{p+k}^{(i)}}{B_{p+k}} - \dfrac{A_k^{(i)}}{B_k}}$$

$$= \left(1 + \Sigma_{j=0}^{n-1} w^{(n-j)} \frac{B_{k-j-1}}{B_k} \right)^{-1} \Sigma_{j=-1}^{n-1} w^{(n-j)} \frac{A_{p+k}^{(i)} B_{k-j-1} - A_{k-j-1}^{(i)} B_{p+k}}{A_{p+k}^{(i)} B_k - A_k^{(i)} B_{p+k}}$$

where $w^{(n+1)} := 1$.
The first factor satisfies

$$\lim_{k\to\infty} \left(1 + \Sigma_{j=0}^{n-1} w^{(n-j)} \frac{B_{k-j-1}}{B_k} \right) = 1 + \Sigma_{j=0}^{n-1} w^{(n-j)} r_1^{-j-1} = \frac{g(r_1)}{r_1^{\,n}} \neq 0 ,$$

according to the corollary to Lemma 1.

A typical term in the sum, denote it by $w^{(n-j)}\, q^{(i)}_{p,k,j}$, can be written as follows, using (34'):

$$q^{(i)}_{p,k,j} = \frac{A^{(i)}_{p+k}\, B_{k-j-1} - A^{(i)}_{k-j-1}\, B_{p+k}}{A^{(i)}_{p+k}\, B_k - A^{(i)}_k\, B_{p+k}} =$$

$$= \frac{\sum_{q=t}^{n+1} \delta^{(i)}_q \,(D^{(q)}_{k+p} B_{k-j-1} - D^{(q)}_{k-j-1} B_{k+p})}{\sum_{q=t}^{n+1} \delta^{(i)}_q \,(D^{(i)}_{k+p} B_k - D^{(q)}_k B_{k+p})} =$$

$$= \sum_{q=t}^{n+1} \delta^{(i)}_q \Big(\frac{D^{(q)}_{k+p} B_{k-j-1}}{B_{k+p} D^{(t)}_k} - \frac{D^{(q)}_{k-j-1}}{D^{(t)}_k} \Big) \Big/ \sum_{q=t}^{n+1} \delta^{(i)}_q \Big(\frac{D^{(q)}_{k+p} B_k}{B_{k+p} D^{(t)}_k} - \frac{D^{(q)}_k}{D^{(t)}_k} \Big).$$

Since $\lim D^{(q)}_{k+p} / B_{k+p} = 0$ for $p \to \infty$ for all $q \geq t \geq 2$, we get

$$\lim_{p \to \infty} q_{p,k,j} = \sum_{q=t}^{n+1} \delta^{(i)}_q D^{(q)}_{k-j-1} / D^{(t)}_k \Big/ \sum_{q=t}^{n+1} \delta^{(i)}_q D^{(q)}_k / D^{(t)}_k$$

where the numerator is finite and the denominator satisfies

$$\Big| \sum_{q=t}^{n+1} \delta^{(i)}_q D^{(q)}_k / D^{(t)}_k \Big| \geq \frac{1}{2} \big| \delta^{(t)}_q \big| > 0$$

for sufficiently large k, since $D^{(q)}_k / D^{(t)}_k \to 0$ as $k \to \infty$ for all $q > t$.

As the expression we started with is constructed from a finite sum of these numbers from $\mathbb{C}$, multiplied by something that has a finite limit, part (b) has been proved.

(c) The first observation to be made, is

$$\lim_{p\to\infty} S^{(i)}_{p+k}(0,\ldots,0) = \xi^{(i)}_0 \qquad (k \geq 1)$$

The acceleration coefficients then satisfy (use $\delta^{(i)}_t \neq 0$) :

$$\lim_{k\to\infty} \frac{\xi^{(i)}_0 - S^{(i)}_k(w^{(1)},\ldots,w^{(n)})}{\xi^{(i)}_0 - S^{(i)}_k(0,\ldots,0)} = \lim_{k\to\infty}\lim_{p\to\infty} \frac{S^{(i)}_{k+p}(0,\ldots,0) - S^{(i)}_k(w^{(1)},\ldots,w^{(n)})}{S^{(i)}_{k+p}(0,\ldots,0) - S^{(i)}_k(0,\ldots,0)}$$

$$= \lim_{k\to\infty}\lim_{p\to\infty} \Big(1 + \Sigma_{j=0}^{n-1} w^{(n-j)} \frac{B_{k-j-1}}{B_k}\Big)^{-1} \Sigma_{j=-1}^{n-1} w^{(n-j)}\, q^{(i)}_{p,k,j}$$

$$= \lim_{k\to\infty} \left(1 + \Sigma_{j=0}^{n-1} w^{(n-j)} \frac{B_{k-j-1}}{B_k}\right)^{-1} \Sigma_{j=-1}^{n-1} w^{(n-j)} \lim_{p\to\infty} q_{p,k,j}^{(i)}$$

$$= \lim_{k\to\infty} \left(1 + \Sigma_{j=0}^{n-1} w^{(n-j)} \frac{B_{k-j-1}}{B_k}\right)^{-1} \Sigma_{j=-1}^{n-1} w^{(n-j)} \frac{\Sigma_{q=t}^{n+1} \delta_q^{(i)} D_{k-j-1}^{(q)}/D_{k-j-1}^{(t)}}{\Sigma_{q=t}^{n+1} \delta_q^{(i)} D_k^{(q)}/D_{k-j-1}^{(t)}}$$

$$= \frac{r_1^{\,n}}{g(r_1)} \Sigma_{j=-1}^{n-1} w^{(n-j)} \frac{\delta_t^{(i)}}{\delta_t^{(i)} r_t^{\,j+1}} = \frac{r_1^{\,n}}{g(r_1)} \cdot \frac{g(r_t)}{r_t^{\,n}} = 0$$

(again Lemma 1). □

REFERENCES

1. BRUIN, Marcel G. de: Convergence of generalized C-fractions, J.Approximation Theory 24, 177-207 (1978).
2. BRUIN, Marcel G. de and Lisa JACOBSEN: The dominance concept for linear recurrence relations with applications to continued fractions, Nw. Arch. v. Wiskunde (4),3, 253-266 (1985).
3. CRUYSSEN, P. van der: Linear difference equations and generalized continued fractions, Computing 22, 269-278 (1979).
4. JACOBSEN, Lisa: Modified approximants for continued fractions, construction and applications, Skr., K.Nor. Vidensk.Selsk., no.3, 1983.
5. PERRON, Oskar: Grundlagen für eine Theorie des Jacobischen Kettenbruch-algorithmus, Math.Ann. 64, 1-76 (1907).
6. PERRON, Oskar: Über die Konvergenz der Jacobi-Kettenalgorithmen mit komplexen Elementen, Sitzungsber. Bayer.Akad.Wiss.Math.-Naturwiss.kl. 37, 401-482 (1907).
7. PINCHERLE, S.: Delle funzioni ipergeometriche e di varie questioni ad esse attinenti, Giorn.Math.Battaglini 32, 209-291 (1894).
8. THRON, Wolfgang J. and Haakon WAADELAND: Accelerating convergence of limit-periodic continued fractions $K(a_n/1)$, Numer.Math. 34, 155-170 (1980).
9. THRON, Wolfgang J. and Haakon WAADELAND: Analytic continuation of functions defined by means of continued fractions, Math.Scand. 47, 72-90 (1980).
10. THRON, Wolfgang J. and Haakon WAADELAND: Convergence questions for limit-periodic continued fractions, Rocky Mt.J.Math. 11, 641-657 (1981).
11. Analytic Theory of Continued Fractions, proceedings Loen, Norway 1981, Springer-Verlag, Berlin, Lect. Notes Math. 932 (1982), (eds. W.B. Jones, W.J. Thron and H. Waadeland).

CONVERGENCE ACCELERATION FOR CONTINUED FRACTIONS $K(a_n/1)$, WHERE $a_n \to \infty$

L. Jacobsen
Dept. of Math.
University of Trondheim
N-7055 Dragvoll
Norway

W. B. Jones
Dept. of Math.
University of Colorado
Boulder, Colorado 80302
U.S.A.

H. Waadeland
Dept. of Math.
University of Trondheim
N-7055 Dragvoll
Norway

Abstract.

We introduce a method of convergence acceleration for a class of continued fractions $K(a_n/1)$ where $a_n \to \infty$. By using the modifying factors $w_n = \sqrt{a_{n+1} + 1/4} - 1/2$, we obtain an improvement roughly like $|f-S_n(w_n)|/|f-S_n(0)| \leq C|a_{n+1}|^{-1}$.

1. Introduction

Continued fraction expansions

$$\mathop{K}_{n=1}^{\infty} \frac{a_n(z)}{b_n(z)} = \frac{a_1(z)}{b_1(z)} + \frac{a_2(z)}{b_2(z)} + \ldots = \cfrac{a_1(z)}{b_1(z) + \cfrac{a_2(z)}{b_2(z) + \ddots}}, \quad a_k(z) \not\equiv 0, \tag{1.1}$$

of complex functions $f(z)$ are useful tools in approximation theory. If $a_k(z)$, $b_k(z)$ are polynomials, and $K(a_n(z)/b_n(z))$ converges to $f(z)$ in some metric, (that is, its sequence of approximants,

$$f_n(z) = \frac{a_1(z)}{b_1(z)} + \frac{a_2(z)}{b_2(z)} + \ldots + \frac{a_n(z)}{b_n(z)}, \quad n = 1, 2, 3, \ldots \tag{1.2}$$

converges to $f(z)$) then $f_n(z)$ is a rational approximation to $f(z)$ in this metric. If in particular we use the correspondence metric λ, as described in [3,p. 149], and (1.1) has the form

$$a_k(z) = \beta_k z^{n_k} , \quad b_k(z) = 1, \ k = 1,2,3,.., \ \beta_k \neq 0 , \ n_k \in \mathbb{N} , \tag{1.3}$$

with some f(0) added to (1.1), then $f_n(z)$ are the Padé approximants of f(z). ((1.1) is then called a C-fraction. Also other forms of $a_k(z)$, $b_k(z)$ can give Padé approximants.)

A well known example is the regular C-fraction expansions (i.e. $n_k = 1$ for all k) of the hypergeometric functions ${}_2F_1(a,1;c;z)$ and ${}_1F_1(1;c;z)$, and of the ratios

$$\frac{{}_2F_1(a, b; c; z)}{{}_2F_1(a, b+1; c+1; z)} , \frac{{}_1F_1(b; c; z)}{{}_1F_1(b+1;c+1; z)} , \frac{{}_0F_1(c; z)}{{}_0F_1(c+1; z)} ,$$

$$\frac{{}_2F_0(a, b; z)}{{}_2F_0(a, b+1; z)} \tag{1.4}$$

of hypergeometric functions, [3, Chapter 6]. These expansions also have the following nice properties:

(i) They converge pointwise to the functions in the whole complex plane, except possibly at a ray. (They converge to ∞ at the poles of the functions.)

(ii) They are limit periodic, that is, $\beta_k \to \beta \in \hat{\mathbb{C}} = \mathbb{C} \cup \{\infty\}$. ($\beta = 0$, $-1/4$ or ∞.)

The advantage of property (i) is obvious. One of the advantages of property (ii) is that we have an easy and effective method of convergence acceleration, at least in some cases. It consists of using modified approximants

$$S_n(w_n,z) = 1 + \frac{\beta_1 z}{1} + \frac{\beta_2 z}{1} + \ldots + \frac{\beta_n z}{1 + w_n(z)} , \ n = 1, 2,.. \tag{1.5}$$

instead of $f_n(z)$. If $\beta = -1/4$, we have for some $C > 0$

$$\left| \frac{f(z) - S_n(w_n,z)}{f(z) - S_n(0,z)} \right| \leq C|\beta - \beta_n||z| \to 0 \ \text{ if } \ f \neq \infty , \tag{1.6}$$

at least from some n on, when

$$w_n(z) = w(z) = (\sqrt{1 - z} - 1)/2 , \ \mathrm{Re}\sqrt{1 - z} > 0 , \ [7]. \tag{1.7}$$

If $\beta = 0$, $|\sqrt{1 + 4\beta_n z} - 1| \leq 2/5$ and

$$\max \{|\sqrt{1 + 4\beta_m z} - \sqrt{1 + 4\beta_{m+1} z}|; \ m \geq n\} \leq \varepsilon_n |\sqrt{1 + 4\beta_{n+1} z} - 1| \tag{1.8}$$

for all $n \geq 1$, where $\mathrm{Re}\sqrt{\ldots} > 0$ and $0 \leq \varepsilon_n \leq 1$, then

$$\left|\frac{f(z)-S_n(w_n,z)}{f(z)-S_n(0,z)}\right| < \varepsilon_n \frac{\max\{|w_m(z)|;\ m\geq n-1\}}{(1-5\max\{|w_m(z)|;\ m\geq n-1\})^2} \to 0 \tag{1.9}$$

when

$$w_n(z) = (\sqrt{1+4\beta_{n+1}z}-1)/2\ ,\ n = 1, 2, 3, \ldots ,[1]. \tag{1.10}$$

This method of convergence acceleration is almost "free of charge". We do not need to compute the ordinary approximants (1.2) at all, and the computation of (1.5) takes the same number of operations as it would take to compute (1.2) except for the computation of $w_n(z)$. Other methods of convergence acceleration can then be applied to this "improved" sequence of approximants, if so wanted.

In this paper we shall introduce some modifying factors $w_n(z)$ which sometime give convergence acceleration for the case $\beta = \infty$. Or more generally (since we are considering pointwise convergence), we shall introduce a method of convergence acceleration for continued fractions

$$K\frac{a_n}{1}\ ,\ a_n \to \infty, \tag{1.11}$$

where a_n is eventually contained in some parabolic region, and the sequence $\{a_{n+1}-a_n\}$ is bounded and has its set of limit points contained in some disk.

We shall need the concept of the nth tail of a continued fraction $K(a_n/1)$; that is

$$\overset{\infty}{\underset{m=1}{K}}\frac{a_{n+m}}{1} = \frac{a_{n+1}}{1} + \frac{a_{n+2}}{1} + \ldots\ ,\quad n\in \mathbb{N}\cup\{0\}. \tag{1.12}$$

If $K(a_n/1)$ converges, then so does also (1.12), and we denote the value of (1.12) by $f^{(n)}$. Clearly then $f = S_n(f^{(n)})$.

2. The incomplete Γ-function

As an example of a function with a continued fraction expansion satisfying such conditions, we shall consider the incomplete Γ-function

$$\Gamma(a,z) = \int_z^\infty e^{-t}\, t^{a-1}\, dt,\ z\in S_\pi = \{v\in\mathbb{C};\ |\arg v| < \pi\},\ (\text{Princ.val.}) \tag{2.1}$$

where $a\in\mathbb{C}\setminus\mathbb{Z}$ and the path of integration is the horizontal ray $t = z + \tau$, $0 \leq \tau < \infty$. $\Gamma(a,z)$ has the continued fraction expansion

$$\frac{z^{a-1}e^{-z}}{1} + \frac{(1-a)\xi}{1} + \frac{1\cdot\xi}{1} + \frac{(2-a)\xi}{1} + \frac{2\cdot\xi}{1} + \frac{(3-a)\xi}{1} + \ldots \tag{2.2}$$

$$= z^a . e^{-z} K(a,z), \text{ [4]},$$

where $\xi = 1/z$. This continued fraction converges to $\Gamma(a, z)$ in S_π, and it clearly satisfies (1.11). If the limit points of $\{a_{n+1} - a_n\}$ i.e. $a\xi$ and $(1-a)\xi$, are not contained in a permissible disk, we can use the alternative continued fraction expansion

$$z^a e^{-z} \{-\sum_{k=1}^{r} \frac{(1-a)_{k-1}}{(-z)^k} + \frac{(1-a)_r}{(-z)^r} K(a-r, z)\}, \text{ [4]}, \tag{2.3}$$

for appropriate choice of $r \in \mathbb{N}$.

3. Main results

We shall prove the following:

Theorem 3.1. Let the continued fraction $K(a_n/1)$ satisfy $a_n \to \infty$,

$$a_n \in P_\alpha = \{v \in \mathbb{C};\ |v| - \mathrm{Re}(ve^{-2i\alpha}) \le \tfrac{1}{2}\cos^2\alpha\} \text{ from some } n \text{ on}, |\alpha| < \pi/2, \tag{3.1}$$

and let the limit points of $\{a_{n+1} - a_n\}$ be contained in a disk

$$\{v \in \mathbb{C};\ |v - 2\rho^2 e^{i2\alpha}| \le 2R\}, 0 < R < \rho \cos\alpha. \tag{3.2}$$

Then the following hold:

A. $K(a_n/1)$ converges to a value $f \in \hat{\mathbb{C}} = \mathbb{C} \cup \{\infty\}$.

B. Let

$$w_n = \sqrt{a_{n+1} + 1/4} - 1/2, \quad \mathrm{Re}\sqrt{\ldots} > 0. \tag{3.3}$$

If

$$f \neq \infty, \tag{3.4}$$

then

$$\left| \frac{f - S_n(w_n)}{f - S_n(0)} \right| \to 0. \tag{3.5}$$

Remarks 3.2

(i) S_n is as in the introduction.

(ii) P_α is a region whose boundary is a parabola with focus at 0 and axis along the ray arg $v = 2\alpha$, [6]. Clearly (3.1) holds if all a_n from some n on are contained in a strip between the two rays $\{v \in \mathbb{C};\ \arg(v \pm t) = 2\alpha\}$, $t>0$. This is for instance the case for (2.2).

(iii) The choice (3.3) for w_n is the same as Gill suggested to use in the case $a_n \to 0$, (1.10). It corresponds to using the modified approximant

$$S_n(w_n) = \frac{a_1}{1} + \frac{a_2}{1} + \ldots + \frac{a_n}{1} + \frac{a_{n+1}}{1} + \frac{a_{n+1}}{1} + \ldots \tag{3.6}$$

(iv) The "moral" of Theorem 3.1B is the following: Do not compute the ordinary approximants! Instead we

(α) compute $w_n = \sqrt{a_{n+1} + 1/4} - 1/2$,

(β) compute the modified approximant $S_n(w_n)$ by for instance the backward recurrence algorithm.

For truncation error bounds for $S_n(w_n)$ we refer to Theorem 3.5.
To prove Theorem 3.1, we shall use the following two lemmas.

Lemma 3.3. Let $\{C_n\}_{n=o}^{\infty}$ be a given sequence and ρ a positive number such that

$$C_n \in \mathbb{C},\ |C_{n-1}| \leq |1 + C_n|,\ 0<\rho<|1+C_n| \text{ for all } n \geq 1. \tag{3.7}$$

Then

$$E_n = \{v \in \mathbb{C};\ |v(1+\overline{C}_n) - C_{n-1}(|1+C_n|^2-\rho^2)| + \rho|v| < \rho(|1+C_n|^2-\rho^2)\},\ n = 1, 2, 3, \ldots \tag{3.8}$$

and

$$V_n = \{v \in \mathbb{C};\ |v-C_n| < \rho\},\ n = 0, 1, 2, \ldots \tag{3.9}$$

are corresponding sequences of element and pre value regions, [2].

This means that

$$E_n/(1+V_n) \subseteq V_{n-1} \text{ for all } n \in \mathbb{N}. \tag{3.10}$$

The proof is a straightforward verification. We have $E_n \neq \emptyset$ since $|C_{n-1}| \leq |1+C_n|$. (Equivalence.) The important consequence of this is that if $a_n \in E_n$ and $w_n \in V_n$ for all $n > n_o$, then

$$S_m^{(N)}(w_{N+m}) = \frac{a_{N+1}}{1} + \frac{a_{N+2}}{1} + \ldots + \frac{a_{N+m}}{1+w_{N+m}} \in V_N,\ N > n_o. \tag{3.11}$$

If moreover $\{S_n(w_n)\}$ and $\{S_n(0)\}$ converge to the same value, then the value $f^{(N)}$ of the Nth tail of $K(a_n/1)$ satisfies

$$f^{(N)} = \lim_{m\to\infty} S_m^{(N)}(w_{N+m}) \in V_N,\ N > n_o. \tag{3.12}$$

The regions E_n are convex sets, symmetric about the line

$$z = kC_{n-1}(1+C_n),\ -\infty < k < \infty. \tag{3.13}$$

They are known as Cartesian ovals, [2].

<u>Lemma</u> 3.4. <u>If</u> $C_n = w_n$ <u>from some</u> n <u>on, then</u> $a_n \in E_n$ <u>from some</u> n <u>on.</u>

<u>Proof</u>: We need to prove that

$$|a_n(1+\bar{w}_n) - w_{n-1}(|1+w_n|^2-\rho^2)| + \rho|a_n| < \rho(|1+w_n|^2-\rho^2) \tag{3.14}$$

for n large enough. Let $a_n' = w_{n-1}(1+w_n)$. Then (3.14) can be written

$$|a_n - a_n' + \frac{w_{n-1}}{1+\bar{w}_n}\rho^2| < \rho\frac{|1+w_n|^2 - \rho^2 - |a_n|}{|1+w_n|}. \tag{3.15}$$

Moreover,

$$w_n - w_{n-1} = \frac{a_{n+1} - a_n}{\sqrt{a_{n+1} + 1/4} + \sqrt{a_n + 1/4}} = \varepsilon_n \to 0, \tag{3.16}$$

$$w_{n-1}/(1+\bar{w}_n) \to e^{i2\alpha},\ a_{n+1}/a_n \to 1, \tag{3.17}$$

$$\begin{aligned} a_n - a_n' &= a_n + 1/4 + \varepsilon_n/2 - \sqrt{(a_{n+1} + 1/4)(a_n + 1/4)} \\ &= \varepsilon_n/2 + \frac{(a_n + 1/4)^2 - (a_{n+1}+1/4)(a_n + 1/4)}{a_n + 1/4 + \sqrt{(a_{n+1}+1/4)(a_n+1/4)}} \\ &= \varepsilon_n/2 + \frac{a_n - a_{n+1} + (1-a_{n+1}/a_n)/4}{1 + \frac{1}{4a_n} + \sqrt{(\frac{a_{n+1}}{a_n} + \frac{1}{4a_n})(1 + \frac{1}{4a_n})}}, \end{aligned} \tag{3.18}$$

that is, $a_n - a_n' = (a_n - a_{n+1})/2 + \varepsilon_n'$, where $\varepsilon_n' \to 0$. Finally, by a geometrical argument it follows that

$$|1 + w_n| - |w_{n-1}| \to \cos \alpha. \tag{3.19}$$

This means that (3.15) can be written

$$\left| \frac{a_n - a_{n+1}}{2} + \tilde{\varepsilon}_n + \rho^2 e^{i2\alpha} \right| < \rho \cos \alpha + \hat{\varepsilon}_n, \tag{3.20}$$

where $\tilde{\varepsilon}_n \to 0$ and $\hat{\varepsilon}_n \to 0$. Since $\arg a_n \to 2\alpha$, it follows that (3.20) holds from some n on because of (3.2). Since $E_n \neq \emptyset$, we automatically have $|C_{n-1}| \leq |1 + C_n|$. □

We can now prove Theorem 3.1:

<u>Proof of Theorem</u> 3.1:

A. All a_n are eventually contained in P_α. Since $|a_n| = O(n)$ we also have

$$\sum_{n=1}^{\infty} 1/\sqrt{|a_n|} = \infty. \tag{3.21}$$

From [6] follows therefore that $K(a_n/1)$ converges.

B. To prove (3.5) we observe that

$$\left| \frac{f-S_n(w_n)}{f-S_n(0)} \right| = \left| \frac{h_n}{h_n+w_n} \right| \left| \frac{f^{(n)}-w_n}{f^{(n)}} \right| , \quad n = 1, 2, 3, \ldots , \quad [7](3.22)$$

where

$$h_n = -S_n^{-1}(\infty) = 1 + \frac{a_n}{1} + \frac{a_{n-1}}{1} + \cdots + \frac{a_2}{1}, \quad n = 1, 2, 3, \ldots . \tag{3.23}$$

From Lemma 3.4 it follows that $a_n \in E_n$ (with $C_n = w_n$) from some n on. In view of (3.11) it follows from Lemma 3.2 that $f^{(n)} \in V_n$ from some n on if

$$\lim S_n(w_n) = f. \tag{3.24}$$

Since $a_n \in P_\alpha$ and clearly w_n is contained in the half plane

$$V_\alpha = \{v \in \mathbb{C};\ \mathrm{Re}(ve^{-i\alpha}) \geq -\tfrac{1}{2} \cos \alpha \} \tag{3.25}$$

from some n on, (3.24) follows from the proof of the result in [6]. We therefore have for sufficiently large n

$$\left| \frac{f^{(n)}-w_n}{f^{(n)}} \right| < \frac{\rho}{|w_n|-\rho} \to 0. \qquad (3.26)$$

It remains to prove that limsup $|h_n/(h_n + w_n)| < \infty$. Since $S_n(-h_n) = \infty$ for all n and $f = \lim S_n(z_n) \neq \infty$ for all $z_n \in V_\alpha$, it follows that $-h_n \notin V_\alpha$ from some n on. Hence

$$|h_n + w_n| \geq \mathrm{Re}(w_n e^{-i\alpha} - (-h_n e^{-i\alpha})) > |w_n| \qquad (3.27)$$

for sufficiently large n, which gives

$$\left| \frac{h_n}{h_n+w_n} \right| < 1 + \frac{|w_n|}{|h_n + w_n|} < 2 \text{ from some n on.} \quad \square \qquad (3.28)$$

From (3.22), (3.26) and (3.28) we obtain the following estimate for the improvement by this method ($f \neq \infty$):

$$\left| \frac{f-S_n(w_n)}{f-S_n(0)} \right| < \frac{2\rho}{|w_n|-\rho} \quad \left(\approx \frac{2\rho}{|\sqrt{a_{n+1}}|-\rho}\right) \quad \text{from some n on.} \qquad (3.29)$$

We also get the following truncation error bounds:

<u>Theorem</u> 3.5. <u>Let</u> $K(a_n/1)$ <u>satisfy the conditions of Theorem</u> 3.1B, <u>and let</u> $a_n \in E_n$ <u>for all</u> $n \geq 1$, <u>where</u> E_n <u>is given by</u> (3.8) <u>with</u> $C_n = w_n$ <u>and</u> $\rho < |1+w_n|$ <u>for all</u> n. <u>Then</u>

$$|f - S_n(w_n)| < 2\rho \prod_{j=1}^{n} \frac{|a_j|}{(|1+w_j|-\rho)^2}, \quad n=1,2,3, \ldots \qquad (3.30)$$

<u>Proof:</u> Since $f = S_n(f^{(n)})$, where $f^{(n)} \in V_n$, we have

$$|f - S_n(w_n)| < \mathrm{diam}\, S_n(V_n), \quad n = 1,2,3, \ldots . \qquad (3.31)$$

Moreover, it is well known that S_n can be written as the composition

$$S_n(z) = s_1 \circ s_2 \circ \ldots \circ s_n(z), \qquad \text{where } s_k(z) = \frac{a_k}{1+z}, \text{ [3]}. \qquad (3.32)$$

Hence we can find $S_n(V_n)$ by successive mappings by the n linear fractional transformations s_k, where $s_k(V_k) \subseteq V_{k-1}$ by (3.10). Clearly, $s_n(V_n)$ is a disk contained in V_{n-1}, $s_{n-1} \circ s_n(V_n)$ is a disk contained in V_{n-2}, etc. In general, mapping the disk

$$W = \{v \in \mathbb{C};\ |v - C| < r\} \subseteq V_k \qquad (3.33)$$

by s_k gives

$$s_k(W) = \{v\in\mathbb{C};\ |v - \frac{(1+\bar{C})a_k}{|1+C|^2-r^2}| < \frac{|a_k|r}{|1+C|^2-r^2}\} \subseteq V_{k-1}. \quad (3.34)$$

That is

$$\frac{\operatorname{diam} s_k(W)}{\operatorname{diam} W} = \frac{|a_k|}{|1+C|^2-r^2} | < \frac{|a_k|}{(|1+w_k|-\rho)^2} . \quad (3.35)$$

Since

$$\operatorname{diam} S_n(V_n) = \operatorname{diam} V_n \cdot \prod_{k=1}^{n} \frac{\operatorname{diam} s_k(s_{k+1}\circ \dots \circ s_n(V_n))}{\operatorname{diam} s_{k+1}\circ \dots \circ s_n(V_n)}, [5] \quad (3.36)$$

the conclusion (3.30) follows. (Convention for k=n: Denom. = diam(V_n).) □

<u>Remarks</u> 3.6.

(i) If $a_n \in E_n$ only from some n on, say $n > N$, we get

$$|f - S_{N+n}(w_{N+n})| = |S_N(f^{(N)}) - S_N(S_n^{(N)}(w_{N+n}))|$$

$$= \frac{|f_N - f_{N-1}||h_N|}{|h_N + f^{(N)}||h_N + S_n^{(N)}(w_{N+n})|} \cdot | f^{(N)} - S_n^{(N)}(w_{N+n})|$$

$$< \frac{|f_N - f_{N-1}|\,|h_N|}{(|h_N+ w_N|-\rho)^2} . 2\rho \prod_{k=N+1}^{N+n} \frac{|a_k|}{(|1+w_k|-\rho)^2}, \quad n = 1, 2, 3, \dots (3.37)$$

if $f \neq \infty$ and $|h_N + w_N|>\rho$.

(ii) The expression (3.30) can be simplified. We have for instance

$$|1+w_k| - \rho = |1/2 + \sqrt{a_{k+1}+ 1/4}| - \rho$$

$$\geq \operatorname{Re}(\tfrac{1}{2} e^{-i\varphi_{k+1}/2} + \sqrt{|a_{k+1}| + \tfrac{1}{4} e^{-i\varphi_{k+1}}}) - \rho , \quad (3.38)$$

where $\varphi_{k+1} = \arg a_{k+1} \to 2\alpha$. Hence, if $\rho < \frac{1}{2} \cos \frac{\varphi_{k+1}}{2}$ for all k,

we get

$$|1+w_k| - \rho > \begin{cases} \sqrt{|a_{k+1}|} & \text{if } |\varphi_{k+1}| < \pi/2, \\ \sqrt{|a_{k+1}|} - 1/4 & \text{if } |\varphi_{k+1}| > \pi/2 ,\ |a_{k+1}| > 1/4 \end{cases} \quad (3.39)$$

which gives

$$|f - S_n(w_n)| \leq \begin{cases} 2\rho|a_1|/|a_{n+1}| \text{ if } |\varphi_{k+1}| \leq \pi/2 \text{ for } 1\leq k\leq n, \\ 2\rho \prod_{k=1}^{n} \frac{|a_k|}{|a_{k+1}| - \frac{1}{4}} \text{ if } |a_{k+1}| > 1/4 \text{ for } 1\leq k\leq n. \end{cases} \tag{3.40}$$

4. A numerical example

Choosing a = 1/2 in the incomplete Γ-function (2.1) gives the complementary error function

$$\begin{aligned} \operatorname{erfc}(z) &= \frac{1}{\sqrt{\pi}} \Gamma(1/2, z^2) \\ &= \frac{1}{\sqrt{\pi}} z e^{-z^2} \{ \frac{z^{-2}}{1} + \frac{\frac{1}{2} z^{-2}}{1} + \frac{\frac{2}{2} z^{-2}}{1} + \frac{\frac{3}{2} z^{-2}}{1} + \ldots \} \end{aligned} \tag{4.1}$$

for $|\arg z| < \pi/2$. Its value at z=1 is given by

$$\operatorname{erfc}(1) = 0.15729920705029$$

where all the 14 digits are significant. Computing the ordinary and the modified approximants of (4.1) gives the following table:

n	$\|f - S_n(0)\|$	$\|f - S_n(w_n)\|$	$\left\|\frac{f - S_n(w_n)}{f - S_n(0)}\right\|$
2	$9 \cdot 10^{-3}$	$4 \cdot 10^{-4}$	0.04
5	$2 \cdot 10^{-3}$	$7 \cdot 10^{-5}$	0.04
10	$2 \cdot 10^{-4}$	$3 \cdot 10^{-6}$	0.01
20	$5 \cdot 10^{-6}$	$4 \cdot 10^{-8}$	0.01
50	$4 \cdot 10^{-9}$	$1 \cdot 10^{-11}$	0.003
90	$4 \cdot 10^{-12}$	$4 \cdot 10^{-15}$	0.001

References

1. Gill, J., Convergence Acceleration for Continued Fractions $K(a_n/1)$ with $\lim a_n = 0$, Lecture Notes in Math., Springer-Verlag No 932 (1982), 67 - 70.
2. Jacobsen, L., Thron, W. J. , Oval Convergence regions and Circular limit regions for continued fractions $K(a_n/1)$. In preparation.
3. Jones, W.B. and Thron, W. J., Continued Fractions. Analytic Theory and Applications, Encyclopedia of Mathematics and Its Applications No 11, Addison-Wesley, Reading, Mass. 1980.
4. Jones, W. B. and Thron, W. J., On the Computation of Incomplete Gamma Functions in the Complex Domain, J. Comp. and Appl. Math. 12 & 13 (1985), 401-407.
5. Rye, Egil and Waadeland, H., Reflections on Value Regions, Limit Regions and Truncation Errors for Continued Fractions, Numer. Math. 1985. In print.
6. Thron, W. J., On Parabolic Convergence Regions for Continued Fractions, Math. Zeitschr. Bd 69 (1958), 173 - 182.
7. Thron, W. J. and Waadeland, H., Accelerating Convergence of Limit Periodic Continued Fractions $K(a_n/1)$, Numer. Math. 34 (1980), 155 - 170.

PERRON-CARATHÉODORY CONTINUED FRACTIONS

William B. Jones*†
University of Colorado
Boulder, Colorado 80309, USA

Olav Njåstad
University of Trondheim-NTH
7034 Trondheim, Norway

W. J. Thron*
University of Colorado
Boulder, Colorado 80309, USA

Abstract. Perron-Carathéodory continued fractions (PC-fractions) have recently been investigated in connection with the trigonometric moment problem and Szegö polynomials (orthogonal on the unit circle) [5] and with Wiener's linear prediction method used in digital signal processing [4]. Further properties of PC-fractions are developed here. These include: fast algorithms for computing PC-fractions, connections with other strong moment problems (Stieltjes and Hamburger) and the relationship to the more general class of Perron continued fractions.

1. Introduction. A continued fraction of the form

$$b_0 + \frac{a_1}{1} + \frac{1}{b_2 z} + \frac{a_3 z}{b_3} + \frac{1}{b_4 z} + \frac{a_5 z}{b_5} + \frac{1}{b_6 z} + \cdots \tag{1.1a}$$

where z is a complex variable and the a_{2n+1} and b_n are complex constants satisfying

$$a_{2n-1} \neq 0, \quad \kappa_n := a_{2n+1} + b_{2n} b_{2n+1} \neq 0, \quad n = 1,2,3,\ldots \tag{1.1b}$$

is called a general Perron-Carathéodory continued fraction (or general PC-fraction). If we define $\{\alpha_{2n+1}\}$ and $\{\beta_n\}$ by $\alpha_1 := a_1$, $\beta_0 = b_0$ and

$$\alpha_{2n+1} := \frac{a_{2n+1}}{\kappa_n}, \quad \beta_{2n} := b_{2n} \prod_{j=1}^{n-1} \kappa_j, \quad \beta_{2n+1} := \frac{b_{2n+1}}{\prod_{j=1}^{n} \kappa_j}, \quad n=1,2,3,\ldots, \tag{1.2}$$

then it is readily seen that the continued fraction

*Research supported in part by the U.S. National Science Foundation under Grant No. DMS-8401717.

†Research supported in part by grants from the United States Educational Foundation in Norway (Fulbright-Hays Grant), The Norwegian Marshall Fund and the University of Colorado Council on Research and Creative Work.

$$\beta_0 + \frac{\alpha_1}{1} + \frac{1}{\beta_2 z} + \frac{\alpha_3 z}{\beta_3} + \frac{1}{\beta_4 z} + \frac{\alpha_5 z}{\beta_5} + \frac{1}{\beta_6 z} + \cdots \tag{1.3a}$$

is equivalent to (1.1a) and that

$$\alpha_{2n-1} \neq 0 \quad \text{and} \quad \alpha_{2n+1} + \beta_{2n}\beta_{2n+1} = 1, \quad n = 1,2,3,\ldots . \tag{1.3b}$$

We call (1.3) a (normalized) PC-fraction and note that every general PC-fraction is equivalent to a uniquely determined PC-fraction.

PC-fractions were introduced in [5] where basic correspondence and convergence properties were given and where connections with the trigonometric moment problem and Szegö polynomials (orthogonal on the unit circle) were described. PC-fractions have been shown to be closely related to Wiener filters and digital signal processing in [4]. The purpose of the present paper is to develop further properties of PC-fractions. Section 2 describes connections with a more general class of continued fractions referred to as Perron-fractions. Relationships between PC-fractions and strong moment problems are given in Section 3. For that purpose we introduce a number of important subclasses of PC-fractions and characterize each subclass (Theorem 3.1) in terms of associated Hankel determinants. Connections with the strong Stieltjes moment problem (SSMP) and the strong Hamburger moment problem (SHMP) are described by Theorem 3.3. Section 4 is used to derive two fast algorithms that can be used to compute a PC-fraction in terms of the coefficients of its corresponding formal Laurent series (fLs). The first of these is a quotient-difference algorithm based on the FG-algorithm of McCabe and Murphy. The second is Levinson's algorithm extended to Hermitian PC-fractions. Before proceeding to Section 2, we summarize here, for completeness and later reference, some basic known results about PC-fractions.

It is known (Theorem 1.1) that every PC-fraction corresponds to a uniquely determined pair (L_0, L_∞) of fLs of the form

$$L_0 = c_0^{(0)} + \sum_{k=1}^{\infty} c_k z^k, \quad L_\infty = -c_0^{(\infty)} - \sum_{k=1}^{\infty} c_{-k} z^{-k}, \quad c_0 := c_0^{(0)} + c_0^{(\infty)}. \tag{1.4}$$

To describe that correspondence we note that the Toeplitz and Hankel determinants associated with the double sequence of complex numbers $\{c_k\}_{-\infty}^{\infty}$ are defined, for $m = 0,\pm1,\pm2,\ldots$ and $k = 1,2,3,\ldots$, by

$$T_k^{(m)} := \det(c_{m-\mu+\nu})_{\mu,\nu=0}^{k-1}, \quad H_k^{(m)} := \det(c_{m+\mu+\nu})_{\mu+\nu=0}^{k-1} \tag{1.5}$$

and $T_0^{(m)} := H_0^{(m)} := 1$, respectively. Of particular interest are the determinants

$$\Delta_n := T_{n+1}^{(0)} = (-1)^{\frac{n(n+1)}{2}} H_{n+1}^{(-n)}, \; n = 0,1,2,\ldots \; (\Delta_{-1} := -\Delta_{-2} := 1), \tag{1.6a}$$

$$\Theta_n := T_n^{(-1)} = (-1)^{\frac{n(n-1)}{2}} H_n^{(-n)}, \quad n = 0,1,2,\ldots, \tag{1.6b}$$

$$\Phi_n := T_n^{(1)} = (-1)^{\frac{n(n-1)}{2}} H_n^{(-n+2)}, \quad n = 0,1,2,\ldots . \tag{1.6c}$$

Jacobi's identities can then be expressed by

$$\Delta_{n-1}^2 = \Delta_n \Delta_{n-2} + \Theta_n \Phi_n, \quad n = 1,2,3,\ldots . \tag{1.7}$$

Theorem 1.1. [5, Theorems 2.1 and 2.2] (A) A given PC-fraction (1.3) with n^{th} numerator $P_n(z)$ and denominator $Q_n(z)$ corresponds to a uniquely determined pair (L_0, L_∞) of fLs (1.4) in the sense that, for $n = 0,1,2,\ldots$,

$$L_0 - \Lambda_0 \left[\frac{P_{2n}(z)}{Q_{2n}(z)}\right] = \beta_{2n+2} \prod_{j=0}^{n} \alpha_{2j+1} z^{n+1} + O(z^{n+2}), \tag{1.8a}$$

$$L_\infty - \Lambda_\infty \left[\frac{P_{2n+1}(z)}{Q_{2n+1}(z)}\right] = \frac{-\beta_{2n+3} \prod_{j=0}^{n} \alpha_{2j+1}}{z^{n+1}} + O\left(\left(\frac{1}{z}\right)^{n+2}\right) \tag{1.8b}$$

and

$$Q_{2n}(z)L_0 - P_{2n}(z) = \beta_{2n+2} \prod_{j=0}^{n} \alpha_{2j+1} z^{n+1} + O(z^{n+2}) \tag{1.9a}$$

$$Q_{2n}(z)L_\infty - P_{2n}(z) = \prod_{j=0}^{n} \alpha_{2j+1} + O\left(\frac{1}{z}\right) \tag{1.9b}$$

$$Q_{2n+1}(z)L_0 - P_{2n+1}(z) = -\prod_{j=0}^{n} \alpha_{2j+1} z^n + O(z^{n+1}) \tag{1.9c}$$

$$Q_{2n+1}(z)L_\infty - P_{2n+1}(z) = \frac{-\beta_{2n+3} \prod_{j=0}^{n} \alpha_{2j+1}}{z} + O\left(\left(\frac{1}{z}\right)^2\right) . \tag{1.9d}$$

Moreover,

$$\Delta_n \neq 0, \quad n = 0,1,2,\ldots, \tag{1.10}$$

and for $n = 1,2,3,\ldots$,

$$\alpha_1 = -\Delta_0, \quad \alpha_{2n+1} = \Delta_n \Delta_{n-2} / \Delta_{n-1}^2, \tag{1.11a}$$

$$\beta_{2n} = (-1)^n \Phi_n / \Delta_{n-1}, \quad \beta_{2n+1} = (-1)^n \Theta_n / \Delta_{n-1}, \tag{1.11b}$$

and

$$Q_{2n}(z) = \frac{1}{\Delta_{n-1}} \begin{vmatrix} c_0 & c_1 & \cdots & c_n \\ c_{-1} & c_0 & \cdots & c_{n-1} \\ \vdots & \vdots & & \vdots \\ c_{-n+1} & c_{-n+2} & \cdots & c_1 \\ z^n & z^{n-1} & \cdots & 1 \end{vmatrix},$$

$$Q_{2n+1}(z) = \frac{1}{\Delta_{n-1}} \begin{vmatrix} c_0 & c_{-1} & \cdots & c_{-n} \\ c_1 & c_0 & \cdots & c_{-n+1} \\ \vdots & \vdots & & \vdots \\ c_{n-1} & c_{n-2} & \cdots & c_{-1} \\ 1 & z & \cdots & z^n \end{vmatrix} \tag{1.12a}$$

$$Q_0(z) = Q_1(z) = 1 . \tag{1.12b}$$

(B) Conversely, let (L_0,L_∞) be a given pair of fLs (1.4) such that (1.10) holds. Let $\{\alpha_{2n+1}\}$ and $\{\beta_n\}$ be defined by (1.11). Then (1.3b) holds and hence (1.3a) is a PC-fraction. Moreover, (1.3a) corresponds to (L_0,L_∞) in the sense that (1.8) and (1.9) hold. The denominators $Q_n(z)$ of the PC-fraction also satisfy (1.12).

We note that the symbol $O(z^r)$ employed in (1.8) and (1.9) denotes a fLs in increasing powers of z, starting with a power not less than r. If R is a rational function of z, then $\Lambda_0(R)$ ($\Lambda_\infty(R)$) denotes the Laurent expansion of R about $z = 0$ ($z = \infty$). It is readily shown that P_{2n}, Q_{2n}, P_{2n+1} and Q_{2n+1} are polynomials in z of degree at most n, $Q_{2n}(0) = 1$ and $Q_{2n+1}(z) = z^n$ + lower powers of z.

Let n, r, t be non-negative integers satisfying

$$r + t > 2n + 1$$

and let P and Q be polynomials in z of degrees at most n. Then P/Q is called the weak (n,n) two-point Padé approximant of order (r,t) for the pair of fLs (L_0,L_∞) of the form (1.4) if

$$QL_0 - P = O(z^r) \quad \text{and} \quad QL_\infty - P = O\left(\left(\tfrac{1}{z}\right)^{t-n}\right).$$

If

$$L_0 - \Lambda_0(P/Q) = O(z^r) \quad \text{and} \quad L_\infty - \Lambda_\infty(P/Q) = O\left(\left(\tfrac{1}{z}\right)^{t}\right)$$

then P/Q is called the (n,n) two-point Padé approximant of order (r,t) for (L_0,L_∞). Weak two-point Padé approximants always exist and are unique. A ("strong") (n,n) two-point Padé approximant may or may not exist but, if it exists, then it equals the corresponding weak approximant [9, Theorems 6 and 8]. It can be seen from Theorem 1.1 (A) that P_{2n}/Q_{2n} and P_{2n+1}/Q_{2n+1} are the weak (n,n) two-point Padé approximants of orders (n+1,n) and (n,n+1), respectively, for (L_0,L_∞). The situation in which P_{2n}/Q_{2n} and/or P_{2n+1}/Q_{2n+1} are (strong) two-point Padé approximants is described by Theorem 1.4.

A PC-fraction is called <u>positive</u> (<u>PPC-fraction</u>) if, in addition to (1.3b), its coefficients satisfy

$$\alpha_1 = -2\beta_0 < 0, \quad \beta_{2n} = \bar{\beta}_{2n+1} \quad \text{and} \quad |\beta_{2n}| < 1, \quad n = 1,2,3,\ldots \; . \tag{1.13}$$

Clearly (1.3b) and (1.13) imply $\alpha_{2n+1} > 0$, $n \geq 1$. If we set $\delta_n :=$ β_{2n+1}, then the PPC-fraction can be expressed in the equivalent form

$$\beta_0 - \frac{2\beta_0}{1} \; \frac{1}{+ \; \bar{\delta}_1 z \; +} \; \frac{(1-|\delta_1|^2)z}{\delta_1} \; \frac{1}{+ \; \bar{\delta}_2 z \; +} \; \frac{(1-|\delta_2|^2)z}{\delta_2} \; \frac{\cdots}{+} \tag{1.14a}$$

where

$$\beta_0 > 0, \quad |\delta_n| < 1, \quad n = 1,2,3,\ldots \; . \tag{1.14b}$$

Positive PC-fractions are characterized by the following:

<u>Theorem 1.2.</u> [5, Theorems 3.1 and 3.2] (A) (<u>Correspondence</u>) <u>A PC-fraction</u> (1.3) <u>is positive if and only if</u>

$$\Delta_n > 0, \quad n = 0,1,2,\ldots \tag{1.15a}$$

<u>and</u>

$$c_0^{(0)} = c_0^{(\infty)} > 0, \quad c_n = \bar{c}_{-n}, \quad n = 0,1,2,\ldots \; . \tag{1.15b}$$

<u>Here the</u> c_k <u>are the coefficients of the corresponding fLs</u> (1.4).

(B) (<u>Convergence</u>) <u>Let</u> (1.14) <u>be a given PPC-fraction with</u> n^{th} <u>numerator</u> P_n <u>and denominator</u> Q_n. <u>Then</u>:

(B1) <u>For</u> $|z| < 1$, $\{P_{2n}(z)/Q_{2n}(z)\}$ <u>converges to a holomorphic function</u> $f(z)$ <u>with Taylor series at</u> $z = 0$ <u>given by</u> L_0, <u>and</u>

$$\text{Re}\, P_{2n}(z)/Q_{2n}(z) > 0, \quad \text{Re}\, f(z) \geq 0 \; . \tag{1.16}$$

<u>The convergence is uniform on compact subsets of</u> $|z| < 1$.

(B2) <u>For</u> $|z| > 1$, $\{P_{2n+1}(z)/Q_{2n+1}(z)\}$ <u>converges to a holomorphic function</u> $g(z)$ <u>whose Laurent series at</u> $z = \infty$ <u>is</u> L_∞, <u>and</u>

$$\text{Re}\, P_{2n+1}(z)/Q_{2n+1}(z) < 0, \quad \text{Re}\, g(z) \leq 0 \; . \tag{1.17}$$

<u>The convergence is uniform on compact subsets of</u> $|z| > 1$.

Let $\Psi_\infty(a,b)$ denote the set of all functions $\psi(t)$ bounded and non-decreasing with infinitely many points of increase on $-\infty \leq a < t < b \leq +\infty$.

Theorem 1.2 can be used to solve the trigonometric moment problem: Let $\{\mu_n\}_{-\infty}^{\infty}$ be a given double sequence of complex numbers; find necessary and sufficient conditions for the existence of a function $\psi \in \Psi_\infty(-\pi,\pi)$ such that

$$\mu_n = \frac{1}{2\pi}\int_{-\pi}^{\pi} e^{-in\theta}\, d\psi(\theta), \quad n = 0,\pm1,\pm2,\ldots . \tag{1.18}$$

Theorem 1.3. [5, Theorem 3.3] Let $\{\mu_n\}_{-\infty}^{\infty}$ be a given double sequence of complex numbers. Let (L_0,L_∞) be the pair of fLs (1.4) where $c_0^{(0)} := c_0^{(\infty)} := \mu_0$, $c_n := 2\mu_n$, $n = 0,\pm1,\pm2,\ldots$. Then the following three statements are equivalent:

(A) There exists $\psi \in \Psi_\infty(-\pi,\pi)$ such that (1.18) holds.

(B)

$$\mu_n = \bar{\mu}_{-n} \quad \text{and} \quad \mathcal{D}_n := \det(\mu_{-i+j})_{i,j=0}^{n} > 0, \quad n = 0,1,2,\ldots . \tag{1.19}$$

(C) There exists a positive PC-fraction (1.14) corresponding to (L_0,L_∞).

Finally, we recall the connection between PC-fractions and general T-fractions given in the following:

Theorem 1. 4. [5, Theorem 4.1] Let (1.3) be a given PC-fraction; let P_n and Q_n denote its nth numerator and denominator, respectively; and let (L_0,L_∞) denote the pair of fLs (1.4) to which (1.3) correponds.

(A) If

$$\beta_{2n} \neq 0, \quad n = 1,2,3,\ldots, \tag{1.20}$$

then

$$L_0 - \Lambda_0\left[\frac{P_{2n}}{Q_{2n}}\right] = \beta_{2n}\prod_{j=0}^{n}\alpha_{2j+1} z^{n+1} + O(z^{n+2}), \quad n = 0,1,2,\ldots \tag{1.21a}$$

and

$$L_0 - \frac{P_0}{Q_0} = \alpha_1 + O(\tfrac{1}{z}), \quad L_\infty - \Lambda_\infty\left[\frac{P_{2n}}{Q_{2n}}\right] = \frac{\prod_{j=0}^{n}\alpha_{2j+1}}{\beta_{2n} z^n} + O\left(\left(\tfrac{1}{z}\right)^{n+1}\right), \tag{1.21b}$$

$$n = 1,2,3,\ldots .$$

Hence, for $n > 0$, P_{2n}/Q_{2n} is the (n,n) two-point Padé approximant of order (n+1,n) for (L_0,L_∞) and is also the nth approximant of the general T-fraction

$$\beta_0 + \frac{F_1 z}{1+G_1 z} + \frac{F_2 z}{1+G_2 z} + \frac{F_3 z}{1+G_3 z} + \cdots \tag{1.22a}$$

where, for n = 2,3,4,...

$$F_1 := \alpha_1\beta_2 = H_1^{(1)}, \quad F_n := -\alpha_{2n-1}\frac{\beta_{2n}}{\beta_{2n-2}} = -\frac{H_{n-2}^{(-n+3)}H_n^{(-n+2)}}{H_{n-1}^{(-n+2)}H_{n-1}^{(-n+3)}}, \tag{1.22b}$$

$$G_1 := \beta_2 = -\frac{H_1^{(1)}}{H_1^{(0)}}, \quad G_n := \frac{\beta_{2n}}{\beta_{2n-2}} = -\frac{H_{n-1}^{(-n+2)}H_n^{(-n+2)}}{H_n^{(-n+1)}H_{n-1}^{(-n+3)}} \tag{1.22c}$$

(B) If

$$\beta_{2n+1} \neq 0, \quad n = 1,2,3,... \tag{1.23}$$

then

$$L_0 - \Lambda_0\left[\frac{P_{2n+1}}{Q_{2n+1}}\right] = -\frac{\prod_{j=0}^{n}\alpha_{2j+1}}{\beta_{2n+1}} z^n + O(z^{n+1}), \ n = 0,1,2,..., \tag{1.24a}$$

$$L_\infty - \Lambda_\infty\left[\frac{P_{2n+1}}{Q_{2n+1}}\right] = -\frac{\beta_{2n+3}\prod_{j=0}^{n}\alpha_{2j+1}}{z^{n+1}} + O\left(\left(\frac{1}{z}\right)^{n+2}\right), \ n = 0,1,2,... . \tag{1.24b}$$

Hence, for $n \geq 0$, P_{2n+1}/Q_{2n+1} is the (n,n) two-point Padé approximant of order (n,n+1) for (L_0,L_∞) and is also the nth approximant of the M-fraction.

$$\beta_0 + \alpha_1 + \frac{U_1}{V_1+z} + \frac{U_2}{V_2+z} + \frac{U_3}{V_2+z} + \cdots \tag{1.25a}$$

where, for n = 2,3,4,...,

$$U_1 := -\alpha_1\beta_3 = -H_1^{(-1)}, \quad U_n := -\alpha_{2n-1}\frac{\beta_{2n+1}}{\beta_{2n-1}} = -\frac{H_{n-2}^{(-n+3)}H_n^{(-n)}}{H_{n-1}^{(-n+2)}H_{n-1}^{(-n+1)}}, \tag{1.25b}$$

$$V_1 := \beta_3 = -\frac{H_1^{(-1)}}{H_1^{(0)}}, \quad V_n := \frac{\beta_{2n+1}}{\beta_{2n-1}} = -\frac{H_{n-1}^{(-n+2)}H_n^{(-n)}}{H_n^{(-n+1)}H_{n-1}^{(-n+1)}} \tag{1.25c}$$

2. Perron-Fractions. A continued fraction of the form

$$\beta_0 + \frac{\alpha_1}{\beta_1} + \frac{1}{\beta_2 z} + \frac{\alpha_3 z}{\beta_3} + \frac{1}{\beta_4 z} + \frac{\alpha_5 z}{\beta_5} + \frac{1}{\beta_6 z} + \cdots, \tag{2.1a}$$

where z is a complex variable and where α_{2n+1} and β_n are complex constants with

$$\alpha_{2n+1} \neq 0, \quad n = 0,1,2,\ldots \tag{2.1b}$$

is called a Perron continued fraction (or Perron-fraction). In this section we explore correspondence properties of Perron-fractions and relations to PC-fractions. The existence of an even and/or an odd part of a Perron-fraction is dealt with in Theorem 2.1. In Theorem 2.2 it is shown that the even approximants always correspond to a fLs L_0 at $z = 0$; the odd approximants correspond weakly to L_0 in the sense of (2.10b) and, under the additional condition (2.11), they correspond (strongly) to L_0. It can be seen from (2.8) that the condition $\beta_1 \neq 0$ is very useful to prove correspondence to a fLs L_∞ at $z = \infty$; if $\beta_1 \neq 0$, then without loss of generality we can set $\beta_1 = 1$ as is done for PC-fractions (1.3). The case in which a Perron-fraction reduces to a PC-fraction is treated in Theorem 2.3.

If P_n and Q_n denote the nth numerator and denominator, respectively, of (2.1) then by the difference equations [6, (2.1.6)]

$$P_0 = \beta_0, \quad P_1 = \alpha_1 + \beta_0\beta_1, \quad Q_0 = 1, \quad Q_1 = \beta_1, \tag{2.2a}$$

and, for $n = 1,2,3,\ldots,$

$$P_{2n} = \beta_{2n}zP_{2n-1} + P_{2n-2}, \quad Q_{2n} = \beta_{2n}zQ_{2n-1} + Q_{2n-2}, \tag{2.2b}$$

$$P_{2n+1} = \beta_{2n+1}P_{2n} + \alpha_{2n+1}zP_{2n-1}, \quad Q_{2n+1} = \beta_{2n+1}Q_{2n} + \alpha_{2n+1}zQ_{2n-1}. \tag{2.2c}$$

We define

$$\kappa_n := \alpha_{2n+1} + \beta_{2n}\beta_{2n+1}, \quad n = 0,1,\ldots\ . \tag{2.3}$$

It is easily verified from (2.2) that P_n and Q_n are polynomials in z of the forms given by (2.2a) and, for $n = 1,2,3,\ldots,$

$$P_{2n}(z) = \sum_{j=0}^{n} p_{2n,j}z^j = \beta_0 + \ldots + \beta_{2n}\prod_{j=0}^{n-1}\kappa_j z^n, \tag{2.4a}$$

$$Q_{2n}(z) = \sum_{j=0}^{n} q_{2n,j}z^j = 1 + \ldots + \beta_1\beta_{2n}\prod_{j=1}^{n-1}\kappa_j z^n, \tag{2.4b}$$

$$P_{2n+1}(z) = \sum_{j=0}^{n} p_{2n+1,j}z^j = \beta_0\beta_{2n+1} + \ldots + \prod_{j=0}^{n}\kappa_j z^n \tag{2.4c}$$

$$Q_{2n+1}(z) = \sum_{j=0}^{n} q_{2n+1,j}z^j = \beta_{2n+1} + \ldots + \beta_1\prod_{j=1}^{n}\kappa_j z^n \tag{2.4d}$$

From this and the determinant formulas [6, (2.1.9)] we obtain

$$P_{2n}Q_{2n-1} - P_{2n-1}Q_{2n} = -\prod_{j=0}^{n-1}\alpha_{2j+1}z^{n-1}, \quad n = 1,2,3,\ldots, \tag{2.5a}$$

$$P_{2n+1}Q_{2n} - P_{2n}Q_{2n+1} = \prod_{j=0}^{n} \alpha_{2j+1} z^n, \quad n = 0,1,2,\ldots, \tag{2.5b}$$

$$P_{2n}Q_{2n-2} - P_{2n-2}Q_{2n} = \beta_{2n} \prod_{j=0}^{n-1} \alpha_{2j+1} z^n, \quad n = 1,2,3,\ldots, \tag{2.5c}$$

$$P_{2n+1}Q_{2n-1} - P_{2n-1}Q_{2n+1} = -\beta_{2n+1} \prod_{j=0}^{n-1} \alpha_{2j+1} z^{n-1}, \quad n = 1,2,3,\ldots . \tag{2.5d}$$

From (2.5c,d) we obtain immediately

Theorem 2.1. (A) *A Perron-fraction* (2.1) *has an even part if and only if*

$$\beta_{2n} \neq 0, \quad n = 1,2,3,\ldots . \tag{2.6}$$

(B) *A Perron-fraction* (2.1) *has an odd part if and only if*

$$\beta_{2n+1} \neq 0, \quad n = 1,2,3,\ldots . \tag{2.7}$$

From (2.2a) and (2.5) we obtain

$$\frac{P_{2n}}{Q_{2n}} - \frac{P_{2n-1}}{Q_{2n-1}} = \frac{-\prod_{j=0}^{n-1} \alpha_{2j+1} z^{n-1}}{\beta_{2n-1} + \ldots + \beta_1^2 \beta_{2n} \prod_{j=1}^{n-1} \kappa_j^2 z^{2n-1}}, \quad n = 1,2,3,\ldots, \tag{2.8a}$$

$$\frac{P_1}{Q_1} - \frac{P_0}{Q_0} = \frac{\alpha_1}{\beta_1} \tag{2.8b}$$

$$\frac{P_{2n+1}}{Q_{2n+1}} - \frac{P_{2n}}{Q_{2n}} = \frac{\prod_{j=0}^{n} \alpha_{2j+1} z^n}{\beta_{2n+1} + \ldots + \beta_1^2 \beta_{2n} \kappa_n \prod_{j=1}^{n-1} \kappa_j^2 z^{2n}}, \quad n = 1,2,3,\ldots, \tag{2.8c}$$

$$\frac{P_2}{Q_2} - \frac{P_0}{Q_0} = \frac{\alpha_1 \beta_2 z}{1 + \beta_1 \beta_2 z} \tag{2.8d}$$

$$\frac{P_{2n}}{Q_{2n}} - \frac{P_{2n-2}}{Q_{2n-2}} = \frac{\beta_{2n} \prod_{j=0}^{n-1} \alpha_{2j+1} z^n}{1 + \ldots + \beta_1^2 \beta_{2n-2} \beta_{2n} \kappa_{n-1} \prod_{j=1}^{n-2} \kappa_j^2 z^{2n-1}}, \quad n = 2,3,4,\ldots, \tag{2.8e}$$

$$\frac{P_3}{Q_3} - \frac{P_1}{Q_1} = \frac{-\beta_3 \alpha_1}{\beta_1 \beta_3 + \beta_1^2 \kappa_1 z} \tag{2.8f}$$

$$\frac{P_{2n+1}}{Q_{2n+1}} - \frac{P_{2n-1}}{Q_{2n-1}} = \frac{-\beta_{2n+1} \prod_{j=0}^{n-1} \alpha_{2j+1} z^{n-1}}{\beta_{2n-1} \beta_{2n+1} + \ldots + \beta_1^2 \kappa_n \prod_{j=1}^{n-1} \kappa_j^2 z^{2n-1}}, \quad n = 2,3,4,\ldots . \tag{2.8g}$$

By using (2.8d,e,f,g) we can readily prove the following:

Theorem 2.2. Let (2.1) be a given Perron-fraction. (A) Then there exists a unique fLs $L_0 = \beta_0 + \sum_1^\infty c_k z^k$ such that, for $n = 0,1,2,\ldots$,

$$L_0 - \Lambda_0\left[\frac{P_{2n}}{Q_{2n}}\right] = \beta_{2n+2} \prod_{j=0}^{n} \alpha_{2j+1} z^{n+1} + O(z^{n+2}), \tag{2.9}$$

and

$$Q_{2n}L_0 - P_{2n} = \beta_{2n+2} \prod_{j=0}^{n} \alpha_{2j+1} z^{n+1} + O(z^{n+2}), \tag{2.10a}$$

$$Q_{2n+1}L_0 - P_{2n+1} = -\prod_{j=0}^{n} \alpha_{2j+1} z^{n} + O(z^{n+1}). \tag{2.10b}$$

(B) If, in addition,

$$\beta_{2n+1} \neq 0, \quad n = 0,1,2,\ldots \tag{2.11}$$

then

$$L_0 - \frac{P_{2n+1}}{Q_{2n+1}} = \frac{-\prod_{j=0}^{n} \alpha_{2j+1} z^n}{\beta_{2n+1}} + O(z^{n+1}) \tag{2.12}$$

Under the conditions of Theorem 2.2(B) we see that the Perron-fraction corresponds to a fLs L_0 at $z = 0$. This is the case considered by Perron [10, p. 176-178] and it motivates the nomenclature, "Perron-fraction", for the more general continued fractions (2.1). Our next theorem gives sufficient conditions for the existence of a fLs L_∞ to which the odd order approximants correspond at $z = \infty$. It follows directly from (2.8d,e,f,g).

Theorem 2.3. Let (2.1) be a Perron-fraction such that

$$\beta_1 \neq 0 \quad \text{and} \quad \kappa_j \neq 0, \quad j = 1,2,3,\ldots . \tag{2.13}$$

Then: (A) There exists a fLs $L_\infty = -c_0^{(\infty)} - \sum_1^\infty c_{-k} z^{-k}$ such that, for $n = 0,1,2,\ldots$,

$$L_\infty - \Lambda_\infty\left[\frac{P_{2n+1}}{Q_{2n+1}}\right] = \frac{-\beta_{2n+3} \prod_{j=0}^{n} \alpha_{2j+1}}{\beta_1^2 \kappa_{n+1} \prod_{j=1}^{n} \kappa_j^2 z^{n+1}} + O\left(\left(\frac{1}{z}\right)^{n+2}\right) \tag{2.14}$$

and

$$Q_{2n+1}L_\infty - P_{2n+1} = \frac{-\beta_{2n+3} \prod_{j=0}^{n} \alpha_{2j+1}}{\beta_1 \prod_{j=1}^{n+1} \kappa_j z} + O\left(\left(\frac{1}{z}\right)^2\right), \tag{2.15a}$$

$$Q_{2n}L_\infty - P_{2n} = \frac{\prod_{j=0}^{n} \alpha_{2j+1}}{\beta_1 \prod_{j=1}^{n} \kappa_j} + O(\tfrac{1}{z}). \tag{2.15b}$$

Thus a Perron-fraction satisfying (2.13) is a PC-fraction, since it is equivalent to a continued fraction of the form (1.3).

3. Connections with Strong Moment Problems. The trigonometric moment problem (an example of a strong moment problem) was seen (Theorem 1.3) to have a solution $\psi \in \Psi_\infty(-\pi,\pi)$ if and only if there exists a positive PC-fraction corresponding to the pair (L_0, L_∞) of fLs (1.4). In this section we show (Theorem 3.3) that two other strong moment problems can be dealt with in a similar manner by means of other types of PC-fractions (SSPC-fractons and APTPC-fractions). Theorem 3.2 relates these PC-fractions to well-known classes of general T-fractions. We also consider in this section several other subclasses of PC-fractions associated in a natural way with general T-fractions and M-fractions. A characterization of the correspondence of each of these special PC-fractions is described by Theorem 3.1 in terms of Hankel determinants. We begin with some definitions.

A PC-fraction (1.3) is called a TPC-fraction (Thron PC-fraction) if

$$\beta_{2n} \neq 0, \quad n = 1,2,3,\ldots; \tag{3.1}$$

(1.3) is called an MPC-fraction (Murphy PC-fraction) if

$$\beta_{2n+1} \neq 0, \quad n = 1,2,3,\ldots; \tag{3.2}$$

(1.3) is called an HPC-fraction (Hermitian PC-fraction) if

$$\beta_{2n} = \overline{\beta_{2n+1}}, \quad n = 1,2,3,\ldots; \tag{3.3}$$

(1.3) is called an HMTPC-fraction (Hermitian, Murphy, Thron PC-fraction) if

$$\beta_{2n} = \overline{\beta_{2n+1}} \neq 0, \quad n = 1,2,3,\ldots . \tag{3.4}$$

(1.3) is called an SSPC-fraction (strong Stieltjes PC-fraction if

$$\alpha_1 > 0, \quad \alpha_{2n+1} < 0, \quad \beta_{2n} > 0, \quad n = 1,2,3,\ldots . \tag{3.5}$$

(1.3) is called an APTPC-fraction (alternating positive term PC-fraction) if

$$\begin{gathered} \alpha_1 > 0, \quad \alpha_3\beta_4 < 0, \quad \alpha_{4n+1} < 0, \quad \alpha_{4n+3}\frac{\beta_{4n+4}}{\beta_{4n}} < 0, \\ \alpha_{2n+1} \in \mathbb{R}, \quad \beta_n \in \mathbb{R}, \quad \beta_{2n} \neq 0, \quad n = 1,2,3,\ldots; \end{gathered} \tag{3.6}$$

in all of the above cases it is tacitly assumed that (1.3b) holds. Our first result here characterizes each of the above types of PC-fractions in terms of the associated Hankel determinants.

Theorem 3.1. For each pair (X,Y) given below, there exists an X-fraction corresponding to a pair (L_0,L_∞) of fLs (1.4) in the sense of (1.8) and (1.9) if and only if condition Y holds. In all cases $H_k^{(m)}$ denotes the Hankel determinant (1.5) associated with (L_0,L_∞).

X	Y
PC	$H_n^{(-n+1)} \neq 0, \quad n \geq 1$
HPC	$H_n^{(-n+1)} \neq 0, \quad H_n^{(-n)} = \overline{H_n^{(-n+2)}}, \quad n \geq 1$
PPC	$(-1)^{\frac{n(n+1)}{2}} H_n^{(-n+1)} > 0, \quad H_n^{(-n)} = \overline{H_n^{(-n+2)}}, \quad n \geq 1$
TPC	$\underbrace{H_{2n}^{(-2n+1)} \neq 0, \; H_{2n+1}^{(-2n)} \neq 0}_{(H_n^{(-n+1)} \neq 0)}, \quad \underbrace{H_{2n+1}^{(-2n+1)} \neq 0, \; H_{2n}^{(-2n+2)} \neq 0}_{(H_n^{(-n+2)} \neq 0)}, \quad n \geq 0$ (equivalent to above)
MPC	$H_n^{(-n+1)} \neq 0, \quad H^{(-n)} \neq 0, \quad n \geq 0$
HMTPC	$H_n^{(-n+1)} \neq 0, \quad H_n^{(-n)} = \overline{H_n^{(-n+2)}} \neq 0, \quad n \geq 1$
SSPC	$H_{2n}^{(-2n+1)} > 0, \quad H_{2n+1}^{(-2n)} < 0, \quad H_{2n+1}^{(-2n+1)} > 0, \quad H_{2n}^{(-2n+2)} > 0, \quad n \geq 0$ $c_m \in \mathbb{R}$ for all m
APTPC	$H_{2n}^{(-2n+1)} \neq 0, \quad H_{2n+1}^{(-2n)} < 0, \quad H_{2n+1}^{(-2n+1)} \neq 0, \quad H_{2n}^{(-2n+2)} > 0, \quad n \geq 0$ $c_m \in \mathbb{R}$ for all m

Proof. The assertions for PC and PPC are restatements of results given by Theorems 1.1 and 1.2(A), making use of (1.11b) and (1.5). The assertion for HPC follows from (3.3), (1.1b) and (1.5). Assertions for TPC, MPC and HMPC follow from (1.5) and (1.11b) applied to (3.10, (3.2) and (3.3), respectively.

APTPC: Suppose that the Hankel determinant condition holds. Then by PC there exists a PC-fraction corresponding to (L_0,L_∞) and $\alpha_{2n+1} \in \mathbb{R}$, $\beta_n \in \mathbb{R}$, $n > 1$. $\beta_{2n} \neq 0$ for $n \geq 1$ follows from (1.5), (1.11b) and $H_n^{(-n+2)} \neq 0$, $n \geq 0$. From (1.11a) we obtain

$$\alpha_1 = - H_1^{(0)} > 0$$

and

$$\alpha_{4n+1} = \frac{(-1)^{\frac{2n(2n+1)}{2}} H_{2n+1}^{(-2n)} (-1)^{\frac{(2n-2)(2n-1)}{2}} H_{2n-1}^{(-2n+2)}}{(\Delta_{2n-1})^2} < 0, \quad n \geq 1$$

since $H_{2n+1}^{(-2n)} < 0$, $n \geq 0$. Also by (1.11) and (1.5)

$$\alpha_{4n+3} \frac{\beta_{4n+4}}{\beta_{4n}} = \left(\frac{\Delta_{2n-1}}{\Delta_{2n}}\right)^2 \frac{(-1)^{\frac{(2n+2)(2n+1)}{2}} H_{2n+2}^{(-2n)}}{(-1)^{\frac{2n(2n-1)}{2}} H_{2n}^{(-2n+2)}} < 0, \quad n \geq 1$$

since $H_{2n}^{(-2n+2)} > 0$, $n \geq 0$. Finally

$$\alpha_3 \beta_4 = - \frac{H_2^{(0)}}{(\Delta_0)^2} < 0 \quad \text{since} \quad H_2^{(0)} > 0 \ .$$

Conversely, by similar methods we can prove that the Hankel determinant conditions are satisfied, if the corresponding PC-fraction is an APTPC-fraction.

SSPC: Suppose the Hankel determinant conditions hold. Then the corresponding PC-fraction is at least an APTPC-fraction. Thus it suffices to show that

$$\alpha_{4n+3} < 0 \quad \text{and} \quad \beta_{2n} > 0, \quad n = 1,2,3,\dots \ .$$

From (1.11a) and (1.5) we obtain

$$\alpha_{4n+3} = \frac{(-1)^{\frac{(2n+1)(2n+2)}{2}} H_{2n+2}^{(-2n-1)} (-1)^{\frac{(2n-1)(2n)}{2}} H_{2n}^{(-2n+1)}}{(\Delta_{2n})^2} < 0, \quad n \geq 1$$

since $H_{2n}^{(-2n+1)} > 0$, $n \geq 0$. Also

$$\beta_{2n} = (-1)^n \frac{H_n^{(-n+2)}}{H_n^{(-n+1)}} > 0, \quad n \geq 1,$$

since $H_{2m}^{(-2m+2)} > 0$, $H_{2m}^{(-2m+1)} > 0$, $H_{2m+1}^{(-2m+1)} > 0$ and $H_{2m+1}^{(-2m)} < 0$ for $m \geq 0$. Conversely, by similar methods one can easily show that the Hankel determinant conditions are satisfied if the corresponding PC-fraction is an SSPC-fraction. □

It is readily seen from Theorems 3.1 and 1.4 that the even part of a TPC-fraction is a general T-fraction (1.22a) and the odd part of an MPC-fraction is an M-fraction (1.25). The following theorem deals with the even parts of SSPC-fractions and APTPC-fractions. We recall the following definitions. A general T-fraction

$$\beta_0 + \frac{F_1 z}{1+G_1 z} + \frac{F_2 z}{1+G_2 z} + \frac{F_3 z}{1+G_3 z} + \cdots, \quad F_n \neq 0, \quad n \geq 1 \tag{3.7a}$$

is called a <u>positive T-fraction</u> if

$$F_n > 0, \quad G_n > 0, \quad n = 1,2,3,\ldots . \tag{3.7b}$$

A general T-fraction (3.7a) is called an <u>APT-fraction</u> if

$$F_n \in \mathbb{R}, \ 0 \neq G_n \in \mathbb{R}, \quad F_{2n-1}F_{2n} > 0, \quad F_{2n-1}/G_{2n-1} > 0, \ n = 1,2,3,\ldots . \tag{3.7c}$$

<u>Theorem 3.2</u>. (A) <u>The even part of an APTPC-fraction is an APT-fraction</u>.

(B) <u>The even part of an SSPC-fraction is a positive T-fraction</u>.

<u>Proof</u>. An immediate consequence of Theorems 3.1 and 1.4(A). □

The <u>strong Stieltjes moment problem</u> (SSMP) for a double sequence of real numbers $\{c_n\}_{-\infty}^{\infty}$ is to determine whether or not there exists a function $\psi \in \Psi_\infty(0,\infty)$ such that

$$c_n = -\int_0^\infty (-t)^{-n} d\psi(t), \quad n = 0,\pm1,\pm2,\ldots . \tag{3.8}$$

The <u>strong Hamburger moment problem</u> (SHMP) for a given double sequence of real numbers $\{c_n\}_{-\infty}^{\infty}$ is to determine whether or not there exists a function $\psi \in \Psi_\infty(-\infty,\infty)$ such that

$$c_n = -\int_{-\infty}^\infty (-t)^{-n} d\psi(t), \quad n = 0,\pm1,\pm2,\ldots . \tag{3.9}$$

<u>Theorem 3.3</u>. <u>Let</u> $\{c_n\}_{-\infty}^{\infty}$ <u>be a given double sequence of real numbers and let</u> (L_0,L_∞) <u>be the pair of fLs</u> (1.4). <u>Then</u> (A): <u>There exists a solution</u> $\psi \in \Psi_\infty(0,\infty)$ <u>to the</u> SSMP <u>for</u> $\{c_n\}$ <u>if and only if there exists a SSPC-fraction corresponding to</u> (L_0,L_∞).

(B) <u>If there exists an APTPC-fraction corresponding to</u> (L_0,L_∞), <u>then there exists a solution</u> $\psi \in \Psi_\infty(-\infty,\infty)$ <u>to the SHMP for</u> $\{c_n\}$.

Proof. We make use of Theorem 3.1 and well-known results on strong moment problems.

(A): It was shown in [6, Theorem 6.3] that for a given double sequence $\{c_n^*\}_{-\infty}^{\infty}$ of real numbers, there exists a function $\psi \in \Psi_\infty(0,\infty)$ such that

$$c_n^* = \int_0^\infty (-t)^n d\psi(t), \quad n = 0,\pm1,\pm2,\ldots \tag{3.10}$$

if and only if

$$H_{2n}^{(-2n+1)*} > 0, \quad H_{2n+1}^{(-2n)*} > 0, \quad H_{2n+1}^{(-2n+1)*} < 0, \quad H_{2n}^{(-2n)*} > 0, \quad n \geq 0. \tag{3.11}$$

Here $H_k^{(m)*}$ denotes the Hankel determinants associated with $\{c_n^*\}$. If we set

$$c_n^* = -c_{-n}, \quad n = 0,\pm 1,\pm 2,\ldots \tag{3.12}$$

then (3.10) is equivalent to (3.8) and

$$H_k^{(m)*} = (-1)^k H_k^{(-m-2k+2)}, \quad H_k^{(m)} = (-1)^k H_k^{(-m-2k+2)*}. \tag{3.13}$$

It follows that

$$H_{2n}^{(-2n+1)*} = H_{2n}^{(-2n+1)}, \quad H_{2n+1}^{(-2n)*} = -H_{2n+1}^{(-2n)}, \quad H_{2n+1}^{(-2n+1)*} = -H_{2n+1}^{(-2n+1)},$$
$$\text{and} \quad H_{2n}^{(-2n)*} = H_{2n}^{(-2n+2)}, \quad n = 0,1,2,\ldots . \tag{3.14}$$

Thus (3.11) is equivalent to the determinant condition for SSPC-fractions in Theorem 3.1. This proves (A).

(B): In [3, Theorem 7.2] it is shown that for a given double sequence of real numbers $\{c_n\}_{-\infty}^{\infty}$ there exists a $\psi \in \Psi_\infty(-\pi,\pi)$ such that

$$c_n^* = \int_{-\infty}^{\infty} (-t)^n d\psi(t), \quad n = 0,\pm 1,\pm 2,\ldots \tag{3.15}$$

if and only if

$$H_{2n+1}^{(-2n)*} > 0, \quad H_{2n}^{(-2n)*} > 0, \quad n = 0,1,2,\ldots . \tag{3.16}$$

It follows from (3.14) that (3.16) is equivalent to

$$H_{2n+1}^{(-2n)} < 0, \quad H_{2n}^{(-2n+2)} > 0, \quad n = 0,1,2,\ldots . \tag{3.17}$$

The assertion follows since (3.15) is equivalent to (3.9) and (3.17) is implied by the determinant condition in Theorem 3.1 for APTPC-fractions. □

4. Algorithms for PC-fractions. We begin by considering a given pair (L_0,L_∞) of fLs (1.4) and suppose that each coefficient c_r has a decomposition

$$c_r = c_r^{(0)} + c_r^{(\infty)}, \quad r = 0,\pm 1,\pm 2,\ldots . \tag{4.1}$$

Then for each $r = 0, \pm 1, \pm 2, \pm 3,\ldots,$ we define a pair $(L_0^{(r)},L_\infty^{(r)})$ of fLs

$$L_0^{(r)} := c_r^{(0)} + \sum_{k=1}^{\infty} c_{k+r}z^k, \quad L_\infty^{(r)} := -c_r^{(\infty)} - \sum_{k=1}^{\infty} c_{-k+r}z^{-k} . \tag{4.2}$$

It follows that: for $r = 0$,

$$(L_0^{(0)},L_\infty^{(0)}) = (L_0,L_\infty); \tag{4.3a}$$

for $r \geq 1$,

$$L_0^{(r)} = \frac{L_0-(c_0^{(0)}+\sum_{k=1}^{r-1} c_k z^k + c_r^{(\infty)} z^r)}{z^r}, \tag{4.3b}$$

$$L_\infty^{(r)} = \frac{L_\infty-(c_0^{(0)}+\sum_{k=1}^{r-1} c_k z^k + c_r^{(\infty)} z^r)}{z^r}; \tag{4.3c}$$

and for $r \leq -1$,

$$L_0^{(r)} = \frac{L_0+(c_0^{(\infty)}+\sum_{k=1}^{-r-1} c_{-k} z^{-k} + c_r^{(0)} z^r)}{z^r}, \tag{4.3d}$$

$$L_\infty^{(r)} = \frac{L_\infty+(c_0^{(\infty)}+\sum_{k=1}^{-r-1} c_{-k} z^{-k} + c_r^{(0)} z^r)}{z^r}. \tag{4.3e}$$

Associated with each pair $(L_0^{(r)},L_\infty^{(r)})$ of fLs (4.2) are the determinants

$$\Delta_n^{(r)} := T_{n+1}^{(r)}, \quad \Theta_n^{(r)} := T_n^{(r-1)}, \quad \Phi_n^{(r)} := T_n^{(r+1)} \tag{4.4}$$

analogous to (1.5). If and only if

$$\Delta_n^{(r)} \neq 0, \quad n = 0,1,2,\ldots \tag{4.5}$$

then there exists a PC-fraction

$$\beta_0^{(r)} + \frac{\alpha_1^{(r)}}{1} + \frac{1}{\beta_2^{(r)} z} + \frac{\alpha_3^{(r)} z}{\beta_3^{(r)}} + \frac{1}{\beta_r^{(r)} z} + \frac{\alpha_5^{(r)} z}{\beta_5^{(r)}} + \cdots \tag{4.6}$$

corresponding to $(L_0^{(r)},L_\infty^{(r)})$ in the sense analogous to that in Theorem 1.1; the coefficients $\alpha_{2n+1}^{(r)}$, $\beta_n^{(r)}$ are expressed in terms of the determinants (4.4) in a manner analogous to formulas (1.11). If

$$\beta_{2n}^{(r)} \neq 0, \quad n = 1,2,3,\ldots \tag{4.7}$$

then the even part of (4.6) is the general T-fraction

$$c_r^{(0)} + \frac{c_{r+1} z}{1+G_1^{(r)} z} + \frac{F_2^{(r)} z}{1+G_2^{(r)} z} + \frac{F_3^{(r)} z}{1+G_3^{(r)} z} + \cdots, \tag{4.8a}$$

where, for $n = 2,3,4,\ldots$,

$$F_n^{(r)} = \alpha_{2n-1}^{(r)} \frac{\beta_{2n}^{(r)}}{\beta_{2n-2}^{(r)}} = -\frac{H_{n-2}^{(r-n+3)} H_n^{(r-n+2)}}{H_{n-1}^{(r-n+2)} H_{n-1}^{(r-n+3)}}, \tag{4.8b}$$

$$G_1^{(r)} = \beta_2^{(r)} = -\frac{H_1^{(r+1)}}{H_1^{(r)}}, \quad G_n^{(r)} = \frac{\beta_{2n}^{(r)}}{\beta_{2n-2}^{(r)}} = -\frac{H_{n-1}^{(r-n+2)}H_n^{(r-n+2)}}{H_n^{(r-n+1)}H_{n-1}^{(r-n+3)}} \tag{4.8c}$$

It follows from [6, (7.3.48)] that, for $r = 0,\pm 1,\pm 2,\ldots$, if we set

$$F_1^{(r)} := 0, \tag{4.9a}$$

then

$$G_1^{(r)} = -\frac{c_{r+1}}{c_r} \tag{4.9b}$$

and, for $n = 1,2,3,\ldots$

$$F_{n+1}^{(r)} + G_n^{(r)} = F_n^{(r+1)} + G_n^{(r+1)} \tag{4.9c}$$

$$F_{n+1}^{(r)}G_{n+1}^{(r+1)} = F_{n+1}^{(r+1)}G_n^{(r)} \; . \tag{4.9d}$$

Equations (4.9) are called the F,G-relations and can be used to form an algorithm for computing the $F_n^{(r)}$ and $G_n^{(r)}$ [6, Algorithm 7.3.1]. Once these are computed one can easily compute the $\alpha_{2n+1}^{(r)}$ and $\beta_n^{(r)}$ by

$$\beta_0^{(r)} = c_r^{(0)}, \quad \beta_2^{(r)} = G_1^{(r)}, \quad \beta_{2n}^{(r)} = G_n^{(r)}\beta_{2n-2}^{(r)}, \quad n = 2,3,4,\ldots, \tag{4.10a}$$

$$\beta_{2n+1}^{(r)} = 1 + F_{n+1}^{(r)}/G_{n+1}^{(r)}, \quad n = 1,2,3,\ldots, \tag{4.10b}$$

$$\alpha_1^{(r)} = -c_r, \quad \alpha_{2n+1}^{(r)} = 1 - \beta_{2n}^{(r)}\beta_{2n+1}^{(r)}, \quad n = 1,2,3,\ldots \; . \tag{4.10c}$$

Finally, to obtain the PC-fraction (1.3) corresponding to (L_0,L_∞) we set

$$\alpha_{2n+1} = \alpha_{2n+1}^{(0)} \quad \text{and} \quad \beta_n = \beta_n^{(0)}, \quad n = 0,1,2,\ldots \; . \tag{4.11}$$

The algorithm based on (4.9), (4.10) and (4.11) can be used provided conditions (4.5) and (4.7) hold. Another algorithm similar to this has been given by McCabe [2] (see also [1]).

Let (L_0,L_∞) be a given pair of fLs (1.4) satisfying

$$\Delta_n \neq 0 \quad \text{and} \quad c_n = \bar{c}_{-n}, \quad n = 0,1,2,\ldots \; .$$

Associated with the double sequence $\{c_n\}_{-\infty}^{\infty}$ we define, on the space Λ of Laurent polynomials

$$\sum_{k=m}^{n} a_k z^k, \quad a_k \in \mathbb{C}, \quad -\infty < m \leq n < \infty ,$$

the linear functional

$$\mathcal{M}\Big(\sum_{k=m}^{n} a_k z^k\Big) := \sum_{k=m}^{n} a_k c_{-k} \tag{4.12}$$

and the pseudo-inner product

$$\langle P,Q\rangle := \mathcal{M}\left(P(z)\overline{Q}(1/z)\right), \quad P,Q \in \Lambda . \tag{4.13}$$

Here $\overline{Q}(z) := \sum_{k=m}^{n} \bar{a}_k z^k$ if $Q(z) = \sum_{k=m}^{n} a_k z^k$. By Theorem 3.1 there exists an HPC-fraction (1.14a) corresponding to (L_0,L_∞). We shall derive an algorithm to compute the δ_n of (1.14a). Let $Q_n(z)$ denote the nth denominator of (1.14a) and let

$$\rho_n(z) := Q_{2n+1}(z), \quad \rho_n^*(z) := Q_{2n}(z), \quad n = 0,1,2,\dots . \tag{4.14}$$

By (1.12) it follows that, for $n \geq 1$, $\rho_n(z)$ is a monic polynomial in z,

$$\rho_n^*(z) = z^n \bar{\rho}_n(1/z) , \tag{4.15}$$

$$\langle \rho_n, z^m\rangle = \begin{cases} 0, & 0 \leq m \leq n-1 \\ \Delta_n/\Delta_{n-1}, & m = n, \end{cases} \tag{4.16a}$$

$$\langle \rho_n^*, z^m\rangle = \begin{cases} \Delta_n/\Delta_{n-1}, & m = 0 \\ 0, & 1 \leq m \leq n . \end{cases} \tag{4.16b}$$

From the difference equations (2.2) (also satisfied by the Q_n) one can easily show that

$$\rho_0(z) = \rho_0^*(z) = 1 , \tag{4.17a}$$

$$\rho_n(z) = z\rho_{n-1}(z) + \delta_n \rho_{n-1}^*(z), \quad n = 1,2,3,\dots, \tag{4.17b}$$

$$\rho_n^*(z) = \bar{\delta}_n z\rho_{n-1}(z) + \rho_{n-1}^*(z), \quad n = 1,2,3,\dots . \tag{4.17c}$$

From (4.16) and (4.17b) we obtain

$$\delta_n = -\frac{\langle z\rho_{n-1},1\rangle}{\langle \rho_{n-1}^*,1\rangle} = -\frac{\sum_{j=0}^{n-1} r_{n-1,j}c_{-j-1}}{\sum_{j=0}^{n-1} \bar{r}_{n-1,j}c_{j+1-n}} \tag{4.18}$$

where for $m = 1,2,3,\dots$

$$\rho_m(z) =: \sum_{j=0}^{m} r_{m,j}z^j, \quad \rho_m^*(z) = \sum_{j=0}^{m} \bar{r}_{m,j}z^{m-j}, \quad r_{m,m} = 1 . \tag{4.19}$$

Levinson's algorithm for computing the δ_n consists of alternately applying (4.17) and (4.18). The algorithm was originally derived for Wiener filters [8].

References

1. Bultheel, Adhemar, Algorithms to compute the reflection coefficients of digital filters, Numerical Methods of Approximation Theory, Vol. 7, (eds. L. Collatz, G. Meinardus, H. Werner) Birkhäuser Verlag, Basel (1984), 33-50.

2. McCabe, J.H., On the even extension of an M-fraction, in: Padé Approximation and Its Applications (eds. M.G. deBruin and H. van Rossum), Lecture Notes in Mathematics 888, Springer-Verlag, New York (1980), 290-299.

3. Jones, William B., Olav Njåstad and W.J. Thron, Orthogonal Laurent polynomials and the strong Hamburger moment problem, J. Math. Anal. and Applic., 98, No. 2 (February 1984), 528-554.

4. Jones, William B., Olav Njåstad and W.J. Thron, Continued fractions associated with Wiener's linear prediction method, Proceedings of the International Symposium on the Mathematical Theory of Networks and Systems, Stockholm, Sweden in June 1985, North Holland, to appear.

5. Jones, William B., Olav Njåstad and W.J. Thron, Continued fractions associated with the trigonometric and other strong moment problems, submitted.

6. Jones, William B. and Thron, W.J., Continued fractions: Analytic theory and applications, Encyclopedia of Mathematics and Its Applications, No. 11, Addison-Wesley Publishing Company, Reading, Mass. (1980).

7. Jones, William B., W.J. Thron, and H. Waadeland, A strong Stieltjes moment problem, Trans. Amer. Math. Soc., V. 261, No. 2 (October, 1980), 503-528.

8. Levinson, Norman, The Wiener RMS (root mean square) error criterion in filter design and prediction, J. Math. and Physics 25 (1947), 261-278.

9. Magnus, Arne, On the structure of the two-point Padé table, Analytic theory of continued fractions, (eds. W.B. Jones, W.J. Thron, H. Waadeland), Lecture Notes in Mathematics 932, Springer-Verlag, New York (1982), 176-193.

10. Perron, O., Die Lehre von den Kettenbrüchen, Band II, B.G. Teubner, Stuttgart (1957).

ON APPROXIMATION OF FUNCTIONS BY TWO-DIMENSIONAL CONTINUED FRACTIONS

Kh. I. Kuchminskaya
Institute of Applied Problems
of Mechanics and Mathematics
Ukrainian Academy of Sciences
Lvov, 290047, U.S.S.R.

1. Introduction. The first approximation and interpolation formulas for multivariable functions in the branched continued fraction form are given in [1]. These formulas are not very convenient for the practical use. At first, there is no error formula for approximation of functions by such fractions and moreover we have some troubles with the location of interpolation points. For the second, it is difficult to find an easy connection between approximants of branched continued fraction and corresponding multiple power series.

Therefore, we were looking for a particular form of branched continued fractions given in [1] which will be useful in practice. Such forms are given in [4], [8] and [9]. In this paper we deal with branched continued fractions defined in [4] and [8]. We restrict our considerations to the two-dimensional case.

By a two-dimensional continued fraction /shortly TDCF/ we mean an expression of the form

$$K_0 + \sum_{i=1}^{\infty} \frac{a_i\, xy|}{|\, K_i} \qquad /1.1/$$

or

$$\frac{a_0}{K_0} + \sum_{i=1}^{\infty} \frac{a_i\, xy|}{|\, K_i} \qquad /1.2/$$

where

$$K_i = b_i + \sum_{p=i}^{\infty} \frac{a_{p+1,i}\, x|}{|b_{p+1,i}} + \sum_{p=i}^{\infty} \frac{a_{i,p+1}\, y|}{|b_{i,p+1}} .$$

The n-th approximant of TDCF /1.1/ is defined as

$$f_n = \frac{A_n}{B_n} = K_0^{(n)} + \sum_{i=1}^{[n/2]} \frac{a_i\, xy|}{|K_i^{n-2i}} \qquad /1.3/$$

while for TDCF /1.2/ as

$$f_n = \frac{A_n}{B_n} = \frac{a_0|}{|K_0^{(n-1)}} + \sum_{i=1}^{\left[\frac{n-1}{2}\right]} \frac{a_i\, xy\ |}{|K_i^{(n-2i-1)}} \quad , \quad \frac{A_1}{B_1} = \frac{a_0}{b_0} \quad , \qquad /1.4/$$

where

$$K_i^{(m)} = b_i + \sum_{p=i}^{m} \frac{a_{p+1,i}\ x|}{|b_{p+1,i}} + \sum_{p=i}^{m} \frac{a_{i,p+1}\ y|}{|b_{i,p+1}} .$$

Here $[x]$ denotes the integral part of x, $K_i^{(0)} = b_i$.

TDCF-s introduced in [4] and [8] generalize in the natural way constructions based on ordinary continued fractions and can be used for solving two-variable interpolation and approximation problems. The following properties of TDCF-s can be easily checked:

/i/ Defining Property: Coefficients of the TDCF corresponding to the given two-variable power series are uniquely defined. The n-th approximant of the given TDCF is a rational two-variable function and its power series agrees with the given power series up to terms of the total power n.

/ii/ Symmetry Property: The construction of TDCF corresponding to the given power series do not depend on the ordering of variables /examples of unsymmetrical branched continued fractions are given in [3] and [10] /.

/iii/ Projection Property: If we put x=0 or y=0 in TDCF /1.1/ /respectively in TDCF /1.2/ / then it reduces to the ordinary continued fraction corresponding to the reduced one variable series.

/iv/ Reciprocal Property: TDCF corresponding to the reciprocal of the given power series is the reciprocal of a TDCF corresponding to that series.

2. Convergence of TDCF.

Problem of a pointwise convergence of TDCF-s leads to the investigation of convergence of TDCF-s with constant coefficients

$$\frac{a_0|}{|K_0} + \sum_{i=1}^{\infty} \frac{a_i|}{|K_i} \quad , \quad K_i = b_i + \sum_{m=i}^{\infty} \frac{a_{m+1,i}|}{|b_{m+1,i}} + \sum_{m=i}^{\infty} \frac{a_{i,m+1}|}{|b_{i,m+1}} \qquad /2.1/$$

TDCF /2.1/ is said to converge if the limit of its sequence of approximants $\lim_{n\to\infty} f_n$ exists and is finite. The value of TDCF is defined to be the limit of its sequence of approximants.

TDCF /2.1/ is said to converge absolutely if

$$\sum_{n=0}^{\infty} \left| f_{n+1} - f_n \right| < \infty .$$

Defining

$$Q_i^{(s-1-2i)} = K_i^{(s-1-2i)} + \sum_{j=i+1}^{\left[\frac{s-1}{2}\right]} \frac{a_j |}{|K_j^{s-1-2j}} , \quad Q_i^{(0)} = K_i^{(0)} ,$$

$$Q_i^{(-1)} = \infty , \quad i=0,1,\ldots,\left[\frac{s-1}{2}\right] , \quad s=1,2,\ldots$$

and using mathematical induction, it is not diffucult to obtain a formula for the difference between r-th and m-th approximants. We have

$$\frac{A_r}{B_r} - \frac{A_m}{B_m} = \sum_{i=0}^{\left[\frac{r-1}{2}\right]} \frac{(-1)^{i+1} \left[K_i^{(r-1-2i)} - K_i^{(m-1-2i)}\right] \prod_{j=0}^{i} a_i}{\prod_{j=0}^{i} Q_j^{(r-1-2j)} \; Q_j^{(m-1-2j)}} +$$

$$+ \frac{(-1)^{\left[\frac{r-1}{2}\right]} \prod_{i=0}^{\left[\frac{r-1}{2}\right]+1} a_i}{\prod_{j=0}^{\left[\frac{r-1}{2}\right]} Q_j^{(r-1-2j)} \prod_{j=0}^{\left[\frac{r-1}{2}\right]+1} Q_j^{(m-1-2j)}} . \qquad /2.2/$$

TDCF /2.1/ with complex coefficients is called the majorant for TDCF

$$\sum_{i=0}^{\infty} \frac{a_i(z) |}{|K_i(z)} \qquad /2.3/$$

where $a_i(z)$, $b_i(z)$, $a_{ij}(z)$, $b_{ij}(z)$ are arbitrary complex valued functions defined in the domain $D \subset \mathbb{C}^2$, $z=(z_1,z_2)$ if there exist an integer s and a positive constant M , such that for arbitrary integers $m,n \geq s$

$$| \tilde{f}_n - \tilde{f}_m | \leq M | f_n - f_m | ,$$

where $\tilde{f}_p$ is the p-th approximant of TDCF /2.3/.

TDCF /2.3/ converges uniformly if there exists the majorant fraction and it is convergent.

If all elements of TDCF /2.1/ are functions defined in the common domain D then we say that this TDCF converges uniformly in D if its sequence of approximants converges uniformly in D .

<u>Theorem 2.1.</u> If for TDCF /2.1/

$$| b_i | \geq | a_i | + 3$$
$$| b_{ij} | \geq | a_{ij} | + 1 , \quad i \neq j, \; i,j=0,1,\ldots \qquad /2.4/$$

then this TDCF converges absolutely and its values belongs to the disc

$$| z | \leq 1 . \qquad /2.5/$$

Theorem 2.1 is an analogue of Pringsheim's theorem for ordinary continued fractions [2] . Its proof follows easily from the difference formula /2.2/ [6] .

It should be noted that it is difficult to find recurence relationsfor obtaining the n-th numerator and denominator of TDCF. Therefore the well known methods of investigation of ordinary continued fractionscan not be applied in the case of TDCF-s.

Other convergence properties can be found in [6] and [7].

Theorem 2.2. Let the function $f(x,y)$, defined in the rectangle $D\subset\mathbb{R}^2$ be analytic in the neighbourhood of the point $(x_0,y_0)\in D$ and let there exist TDCF of the form /1.1/ corresponding to this function. The following error formula is fulfilled

$$f(x,y)-\frac{A_m}{B_m}=R_m(f,f_m)=\sum_{k=0}^{m+1}\left.\frac{\partial^{m+1}F(x,y)}{\partial x^k\,\partial y^{m+1-k}}\right|_{\substack{x=\xi\\ y=\eta}}\frac{(x-x_0)^k(y-y_0)^{m+1-k}}{k!\,(m+1-k)!\,B_m(x,y)} \qquad /2.6/$$

where $F(x,y)=f(x,y)B_m(x,y)-A_m(x,y)$ and $(\xi,\eta)\in U(x_0,y_0)$.

It is not difficult to prove this theorem using the auxiliary function

$$\varphi(t)=F(x,y)-\sum_{k+p\geq 0}^{m}\frac{\partial^{k+p}}{\partial x^k\partial y^p}F(xt,yt)\frac{x^ky^p}{k!\;p!}(1-t)^{k+p}-$$
$$-\;R_m(1-t)^m$$

and Lagrange theorem [3], [5] where $(x_0,y_0)=(0,0)$.

Now let us consider branched continued fraction of the form

$$f(x_0,y_0)+\sum_{p=1}^{\infty}\frac{y-y_0|}{|\,a_{0p}}+\cfrac{x-x_0}{a_{10}+\sum_{p=1}^{\infty}\frac{y-y_0|}{|\,a_{1p}}+\cfrac{x-x_0}{a_{20}+\sum_{p=1}^{\infty}\frac{y-y_0|}{|\,a_{2p}}+}} \qquad /2.7/$$

where

$$a_{kp}=p_y'\left[{}^{(p-1)}_{\;y}\left[k_x'\left[{}^{(k-1)}_{\;x}\;f(x,y)\right]\right]\right]_{(x,y)=(x_0,y_0)}$$

$$a_{0p}=p_y'\left[{}^{(p-1)}_{\;y}\;f(x,y)\right]_{(x,y)=(x_0,y_0)},$$

$$a_{p0}=p_x'\left[{}^{(p-1)}_{\;x}\;f(x,y)\right]_{(x,y)=(x_0,y_0)}\qquad,\;k,p=1,2,\dots$$

are partial reciprocal derivatives of $f(x,y)$ in the point (x_0,y_0) [3] . The n-th approximant of /2.7/ is defined as

$$\frac{P_n(x,y)}{Q_n(x,y)} = f(x_0,y_0) + \sum_{p=1}^{n-1} \frac{y-y_0|}{|a_{0p}} + \sum_{k=1}^{n-1} \frac{x - x_0 \quad |}{|\, a_{k0} + \sum_{p=1}^{n-k-1} \frac{y-y_0|}{|a_{kp}}}.$$

The above form of TDCF was introduced independently by W. Siemaszko [10].

Theorem 2.3. Let the function $f(x,y)$ be defined in the rectangle $D=\{x\in[0,1], y\in[0,1]\}$ and moreover let

$$\max_{0\leq x\leq 1}\left|\frac{\partial^i}{\partial x^i} f(x,y)\right| \leq M < \infty \qquad , i=1,2,\ldots$$

$$K \geq \left| p_y'\left[\,_y^{(p-1)}\left[k_x'\left[\,_x^{(k-1)} f(x,y)\right]\right]\right]_{(x,y)=(x_0,y_0)}\right| \geq 4 ,$$

$$K \geq \left| p_{x_i}'\left[\,_{x_i}^{(p-1)} f(x,y)\right]_{(x,y)=(x_0,y_0)}\right| \geq 4 ,$$

$p,k=1,2,\ldots$, $i=1,2$, $x_1=x$, $x_2=y$, M,K are positive constants. Then for any integer $n \geq 3$ we have the following inequality

$$\left| f(x,y) - \frac{P_n(x,y)}{Q_n(x,y)} \right| < C \frac{L^{n-1}}{(n+2)^{n/2}}$$

where L and C are positive constants.

Proof. Let all partial denominators in /2.7/ are positive. Otherwise, one can always do it by multiplying corresponding numerators and denominators by (-1).

If we denote as

$$B_{k,n-1} = a_{k0} + \sum_{p=1}^{n-k-1} \frac{y-y_0|}{|a_{kp}} \qquad , k=1,2,\ldots, n-1$$

then

$$f(x,y) - \frac{P_n(x,y)}{Q_n(x,y)} = f(x,y) - B_{0,n-1}(y) - \cfrac{x-x_0}{B_{1,n-1}(y) + \cfrac{x-x_0}{B_{2,n-1}(y) + \cdots + \cfrac{x-x_0}{B_{n-1,n-1}}}}$$

$$B_{n-1,n-1} = a_{n-1,0} .$$

If we take y arbitrary constant, $y\in(0,1)$ then there exists $\xi\in(0,1)$ such that

$$f(x,y) - \frac{P_n(x,y)}{Q_n(x,y)} = \frac{(x-x_0)^n}{n!\,\tilde{Q}_n(x,y)} \frac{\partial^n}{\partial x^n} f(\xi,y)\tilde{Q}_n(\xi,y) ,$$

where $\tilde{Q}_n(x,y)$ is the denominator of the n-th approximant of the ordinary continued fraction

$$B_{0,n-1}(y) + \sum_{p=1}^{n-1} \frac{x-x_0 \,|}{|\, B_{p,n-1}(y)} \ .$$

Since continued fractions $B_{k,n-1}(y)$, $k=1,\dots,n-2$ satisfy the Worpitsky condition, we have

$$a_{k0}\left[1 - \cfrac{1/16}{1 - \cfrac{1/16}{1 - \cfrac{1/16}{1 - \ddots}}}\right] \leqslant B_{k,n-1} \leqslant a_{k0}\left[1 + \cfrac{1/16}{1 - \cfrac{1/16}{1 - \ddots}}\right]$$

Therefore

$$|(y-y_0)/\, a_{kp}a_{kp-1}| \leqslant 1/16 \ ,$$

$$1 + \frac{1/16|}{|\,1} - \frac{1/16|}{|\,1} - \frac{1/16|}{|\,1} - \dots = (1/2)(3-\sqrt{3}/2) ,$$

$$1 - \frac{1/16|}{|\,1} - \frac{1/16|}{|\,1} - \frac{1/16|}{|\,1} - \dots = (1/2)(1+\sqrt{3}/2) ,$$

and

$$2+\sqrt{3} \leqslant B_{k,n-1} \leqslant K\,\tfrac{1}{2}\,(3-\sqrt{3}/2) \ .$$

Moreover, the following inequalities are fulfilled

$$K \geqslant B_{n-1,n-1} = a_{n-10} \geqslant 4 \ ,$$

$$|B_{0,n-1}| \leqslant |f(x_0,y_0)| + 1/(2+\sqrt{3}) \ .$$

Now we will find estimations for $\tilde{Q}_n(x,y)$. If $(x-x_0) \geqslant 0$, then

$$\tilde{Q}_1(x,y) = 1,$$

$$\tilde{Q}_2(x,y) = B_{1,n-1} \geqslant 2+\sqrt{3} \ ,$$

$$\dots$$

$$\tilde{Q}_{n-1}(x,y) \geqslant (2+\sqrt{3})^{n-2} \ ,$$

$$\tilde{Q}_n(x,y) = B_{n-1,n-1}\,\tilde{Q}_{n-1}(x,y) + (x-x_0)\,\tilde{Q}_{n-2}(x,y) \geqslant (2+\sqrt{3})^{n-1} \ .$$

If $(x-x_0) < 0$, then we willtransform the n-th approximant of /2.7/ to the form

$$B_{0,n-1}(y) + \cfrac{x - x_0}{B_{1,n-1}(y) - 1 + \cfrac{1}{1 + \cfrac{|x - x_0|}{B_{2,n-1}(y) - |x-x_0| - 1 + \cfrac{1}{1 + \ddots}}}}$$

$$\cdot \cdot \cdot + \cfrac{x - x_0}{B_{n-1,n-1} - |x-x_0| - 1 + \frac{1}{1}} \; .$$

Let A_{2n-1} be the denominator of the $(2n-1)$-th approximant of this fraction. Obviously it coincides with $\tilde{Q}_n(x,y)$. Using mathematical induction method we obtain

$$A_{2n-1} > (3+\sqrt{3})(1+\sqrt{3})^{n-2} \quad , \; n=4,5,\dots \; .$$

Therefore, $\tilde{Q}_n(x,y) \geqslant (1+\sqrt{3})^{n-2}(3+\sqrt{3})$, $n \geqslant 3$. Now the inequality /2.8/ follows from the result contained in [5].

3. Interpolational TDCF. Let $f(x,y)$ be a function defined in the rectangle $D \subset \mathbb{R}^2$ and such that $f(x_s,y_p) = f_{sp}$, $s,p=1,\dots,m$.

We define the following partial inverted divided differences [3], [5] :

$$\rho_{00}(x,y) = f(x,y) \; ,$$

$$\rho_{10}(x_1,x,y_1) = \frac{x - x_1}{f(x,y_1) - f(x_1,y_1)} \; ,$$

$$\rho_{01}(x_1,y_1,y) = \frac{y - y_1}{f(x_1,y) - f(x_1,y_1)} \; ,$$

$$\rho_{11}(x_1,x,y_1,y) = (x-x_1)(y-y_1)\{f(x,y) - f(x_1,y) + f(x_1,y_1) - f(x,y_1)\}^{-1}$$

$$\rho_{mk}(x_1,\dots,x_m,x,y_1,\dots,y_k,y) =$$

$$\begin{cases} (x-x_m)\{\rho_{m-1,k}(x_1,\dots,x_{m-1},x,y_1,\dots,y_k,y) - \\ \quad - \rho_{m-1,k}(x_1,\dots,x_m,y_1,\dots,y_k,y)\}^{-1} & ; \; m > k \\ (y-y_k)\{\rho_{m,k-1}(x_1,\dots,x_m,x,y_1,\dots,y_{k-1},y) - \\ \quad - \rho_{m;k-1}(x_1,\dots,x_m,x,y_1,\dots,y_k)\}^{-1} & ; \; k > m \\ (x-x_m)(y-y_m)\{\rho_{m-1,m-1}(x_1,\dots,x_{m-1},x,y_1,\dots,y_{m-1},y) - \\ - \rho_{m-1,m-1}(x_1,\dots,x_m,y_1,\dots,y_{m-1},y) + \rho_{m-1,m-1}(x_1,\dots,x_m,y_1,\dots,y_m) - \\ \quad - \rho_{m-1,m-1}(x_1,\dots,x_{m-1},x,y_1,\dots,y_m)\}^{-1} \end{cases} \qquad /3.1/$$

$$m,k=0,1,\dots \; , \; \rho_{-1,k} = \rho_{m,-1} = \rho_{-1,-1} = 0 \; .$$

Denoting as

$$K_k(x,y) = \rho_{kk}(x_1,\dots,x_{k+1},y_1,\dots,y_{k+1}) \quad +$$

$$+ \sum_{t=k+1}^{m-1} \frac{x - x_t \quad |}{|\rho_{tk}(x_1,\dots,x_{t+1},y_1,\dots,y_{k+1})} +$$

$$+ \sum_{t=k+1}^{m-1} \frac{y - y_t \quad |}{|\rho_{kt}(x_1,\dots,x_{k+1},y_1,\dots,y_{t+1})} ,$$

$k=0,\dots,m-1$, we d efine the interpolational TDCF for m^2 interpolation nodes (x_s, y_p), $s,p=1,\dots,m$ as a branched continued fraction of the form

$$K_0(x,y) + \cfrac{(x-x_1)(y-y_1)}{K_1(x,y) + \cfrac{(x-x_2)(y-y_2)}{K_2(x,y) + \cfrac{}{\ddots + \cfrac{(x-x_{m-1})(y-y_{m-1})}{K_{m-1}(x,y)}}}} \qquad /3.2/$$

It is not diffucult to show that the continued fraction /3.2/ interpolates $f(x,y)$ in corresponding nodes (x_s, y_p), $s,p=1,\dots,m$.

<u>Theorem 3.1</u>. /[3],[5]/. Let the function $f(x,y)$ possess continuous partial derivatives $\frac{\partial^{k+s} f\ x,y}{\partial x^k\ \partial y^s}$, $k,s=0,\dots,m$ in the domain D = $\{ a_1 \leqslant x \leqslant b_1$, $a_2 \leqslant y \leqslant b_2$, containing the interpolation nodes. Then we have the following formula for the error of interpolation with TDCF of the form /3.2/

$$R_m(x,y) = \frac{1}{m!\ Q_m(x,y)} \left\{ \prod_{i=1}^{m} (x-x_i) \frac{\partial^m F(\xi,y)}{\partial x^m} + \right. \qquad /3.3/$$

$$+ \prod_{i=1}^{m} (y-y_i) \frac{\partial^m F(x,\eta)}{\partial y^m} - \frac{1}{m!} \prod_{i=1}^{m} (x-x_i)(y-y_i) \frac{\partial^{2m} F(\bar{\xi},\bar{\eta})}{\partial x^m \partial y^m},$$

where $F(x,y) = f(x,y)Q_m(x,y) - P_m(x,y)$, $a_1 < \xi, \bar{\xi} < b_1, a_2 < \eta, \bar{\eta} < b_2$, and $P_m(x,y)/Q_m(x,y)$ is the m-th approximant of /3.2/.

<u>Proof.</u> Let us consider the function

$$F(x,y) = f(x,y)Q_m(x,y) - P_m(x,y) . \qquad /3.4/$$

In all points (x_i, y_j), $i,j=1,\dots,m$ we have $F(x_i, y_j) = 0$. From the Newton interpolation formula for function $F(x,y)$ we obtain

$$F(x,y) = \sum_{\nu=1}^{m} \sum_{\mu=1}^{m} F(x_1,\dots,x_\nu,y_1,\dots,y_\mu) \prod_{k=1}^{\nu-1}(x-x_k)\prod_{k=1}^{\mu-1}(y-y_k) +$$

$$+ \frac{\partial^m F(\xi,y)}{\partial x^m \cdot m!} \prod_{k=1}^{m}(x-x_k) + \frac{\partial^m F(x,\eta)}{\partial y^m\, m!} \prod_{k=1}^{m}(y-y_k) -$$

$$- \frac{\partial^{2m} F(\bar{\xi},\bar{\eta})}{(m!)^2 \partial x^m \partial y^m} \prod_{k=1}^{m} (x-x_k)(y-y_k) \quad ,$$

$a_1 < \xi,\ \bar{\xi} < b_1$; $a_2 < \eta,\ \bar{\eta} < b_2$, and $F(x_1,\dots,x_\nu,y_1,\dots,y_\mu)$ are double divided differences for function $F(x,y)$ and we put

$$\prod_{k=1}^{0}(x-x_k) = \prod_{k=1}^{0}(y-y_k) = 1 \quad .$$

Calculating divided differences for $F(x,y)$ and then returning to the formula /3.4/ we have

$$f(x,y) - \frac{P_m(x,y)}{Q_m(x,y)} = R_m(x,y) ,$$

where $R_m(x,y)$ is defined in /3.3/.

References

[1] P. I. Bodnarchuk, Ch. J. Kuchminskaja, Interpolation and approximationformulas for multivariate functions in branched continued fraction forms, Mat.Metody i Fiz.-Mech.Polya,2 /1975/,/in Russian/

[2] W. B. Jones, W. J. Thron,Continued Fractions :Analytic Theory and Applications,Addison Wesley, Reading, London /1980/

[3] Kh. I. Kuchminskaya , Approximation and interpolation of Functions by continued and branched continued fractions, Thesis,L'vov, /1976/, /in Russian/

[4] Kh. I. Kuchminskaya , Corresponding and associated branched continued fractions for double power series, Doklady Akad.Nauk Ukr.SSR, Ser.A,7 /1978/,614-617, /in Russian/

[5] Kh. I. Kuchminskaya On approximation of functions by continued and branched continued fractions, Mat.Metody i Fiz.-Mech. Polya, 12/1980/, 3-10, /in Russian/

[6] Kh. I. Kuchminskaya, On sufficient conditions of the absolute convergence of two dimensional continued fractions, Mat.Metody i Fiz.-Mech. Polya, 20 /1984/, 19-23, /in Russian/

[7] Kh. I. Kuchminskaya , On the convergence of two dimensional continued fractions, Constructive Theory of Functions '84, Publi-

shing House of the Bulgarian Academy of Sciences, Sofia /1984/, 501-506

[8] J.A. Murphy, M. R. O'Donohoe, A two variable generalization of the Stieltjes-type continued fraction, J.Comp.Appl.Math.,4 /3/ /1978/, 181-190

[9] W. Siemaszko, Branched continued fractions for double power series, J. Comp.Appl.Math.,6 /2/, /1980/, 121-125

[10] W. Siemaszko , Thiele-type branched continued fractions for two-variable functions, J. Comp.Appl.Math., 9 /1983/, 137-153

ON THE CONVERGENCE OF THE MULTIDIMENSIONAL LIMIT-PERIODIC CONTINUED FRACTIONS

V.I. PARUSNIKOV

Keldysh Institute of Applied Mathematics

Miusskaya Sq. 4, 125047 Moscow A-47, USSR

1. INTRODUCTION

Many usual functions admit expansions in the so called limit-periodic continued fractions, i.e. continued fractions

$$p_0 + \frac{q_1|}{|p_1} + \cdots + \frac{q_j|}{|p_j} + \cdots ,$$

for which there exists a positive integer T (period), such that

$t=1, \dots ,T$: $\lim_{k\to\infty} q_{t+kT} = \mathcal{q}_t$, $\qquad \lim_{k\to\infty} p_{t+kT} = \mathcal{p}_t$.

Some interesting results in the expansion of Markov functions into limit-periodic continued P-fractions were obtained by Widom [10]. The convergence theorems for the above fractions were first obtained by Van Vleck [9] and Pringsheim [6]; the generalization of those results to arbitrary period was performed by Szász [8]. One can mention also the Perron theorem (see [2]) about the asymptotic behaviour of the solutions of the finite- difference equation with limit-periodic coefficients. The criterion of convergence of some limit-periodic multidimensional continued fractions with period T=1 may be deduced from the latter theorem. However, Perron's result does not include the case when T >1, or the case when, among the roots of the characteristic equation, some of them have the same moduli. In this paper we will try to make up for this deficiency.

2. MAIN DEFINITIONS

Let $\vec{a} = (a_1, \dots ,a_n) \in \mathbb{C}^n$, $n \geqslant 1$ and $a_n \neq 0$. Define the inverse of $\vec{a}$ by

$$\frac{1}{\vec{a}} := \left(\frac{1}{a_n}, \frac{a_1}{a_n}, \frac{a_2}{a_n}, \dots , \frac{a_{n-1}}{a_n}\right) \tag{1}$$

which allows the introduction of the following (infinite) multidimensional continued fraction

$$\vec{p}_0 + \frac{q_1|}{|\vec{p}_1} + \cdots + \frac{q_j|}{|\vec{p}_j} + \cdots, \tag{2}$$

where $\vec{p}_j=(p_{j,1}, \ldots, p_{j,n})\in\mathbb{C}^n$, $n \geqslant 1$, and $q_j\in\mathbb{C}^*$. Then the j-th approximant to the continued fraction (2) is

$$j\in\mathbb{N}: \quad \vec{\alpha}_j := \vec{p}_0 + \frac{q_1|}{|\vec{p}_1} + \cdots + \frac{q_j|}{|\vec{p}_j} = (\alpha_{j,1}, \ldots, \alpha_{j,n}). \tag{3}$$

The components of the vectors $\vec{\alpha}_j$ are represented by the quotient of numerators $A_{j,i}$ $(i=1,2, \ldots ,n)$ and denominators $A_{j,0}$ of the approximants

$$\alpha_{j,i} = A_{j,i}/A_{j,0} \tag{4}$$

and are linear with respect to the components of the vectors $\vec{p}_l$ and to the numbers $1/q_l$ $(l=0,1, \ldots ,j)$. The values $A_{j,i}$ are determined by the recurrence relations

$$j\in\mathbb{N}: \qquad A_{i,j} = \sum_{l=0}^{n} p_{j,l}A_{j+l-n-1}, \qquad p_{j,0}:= 1/q_{j+1} \tag{5}$$

with the following initial conditions

$$A_{j,j+n+1} = 1, \qquad A_{j,i} = 0 \quad \text{if} \quad i \neq j+n+1, \qquad j=-1,-2, \ldots ,-n-1.$$

Even in the unidimensional case the j-th approximants of the continued fraction are not always defined (as a element of $\mathbb{C}$), but only where $A_{j,0}\neq 0$. To get rid of this restriction one can enlarge the value domain to the projective space $\overline{\mathbb{C}} = \mathbb{C}\mathbb{P}^1$. Analogously, in the case n<1 we can consider the <u>n-dimensional projective space $\mathbb{C}\mathbb{P}^n$</u> instead of $\mathbb{C}^n$. Remark that the projective vectors, elements of $\mathbb{C}\mathbb{P}^n$, are the equivalence classes in $\mathbb{C}^{n+1}\backslash\{\vec{0}\}$ modulo the proportionality relation "$\sim$": $\vec{f}\sim\vec{g}$ if $\vec{f}=k\vec{g}$ where $k\in\mathbb{C}^*$. Let us denote by $\bar{d}=(d_0: d_1: \ldots :d_n)$ the projective vector (a bar above denotes the elements of $\mathbb{C}\mathbb{P}^n$) defined by the class containing $(d_0, d_1, \ldots ,d_n) \in \mathbb{C}^{n+1}\backslash\{\vec{0}\}$.

The space $\mathbb{C}\mathbb{P}^n$ is a compact n-dimensional manifold, whose atlas consists of the (n+1) maps $\mathcal{H}_i= \{(d_0:d_1: \ldots :d_n) \in \mathbb{C}\mathbb{P}^n \mid d_i\neq 0 \}$; the ratios d_k/d_i $(k=0,\ldots,i-1,i+1,\ldots,n)$ are the local coordinates of the map $\mathcal{H}_i$. Identify $\mathbb{C}^n$ and $\mathcal{H}_0$ by the means of the submergence $\varphi : \mathbb{C}^n\to\mathbb{C}\mathbb{P}^n$; $\varphi((a_1, \ldots ,a_n))=(1:a_1: \ldots :a_n)$. Finally, let

$$M\bar{d} = \left(\sum_{i=0}^{n} (M)_{0,i}d_i : \ldots : \sum_{i=0}^{n} (M)_{n,i}d_i \right)$$

be a result of application of the fractional-linear transformation M with $\det(M)\neq 0$ to the projective vector $\bar{d} = (d_0:d_1: \ldots :d_n)$. The elements of the matrix M are designated by $(M)_{i,j}$; those of the vector $\vec{f}$: by $(f)_i$ or f_i.

Three successive operations are necessary to obtain an approximant of the continued fraction: inverse a vector $\vec{a}$ in the mean of (1), multiply the vector $\vec{a}$ by the scalar $q_j \neq 0$ and add a vector $\vec{p} \in \mathbb{C}^n$ to $\vec{a} \in \mathbb{C}^n$. These operations are represented in $\mathbb{C}\mathbb{P}^n$ by the following matrices (in $\mathcal{H}_0 \cap \mathcal{H}_n$ in the first case and in $\mathcal{H}_0$ in the two others)

$$\begin{bmatrix} 0 & 0 & . & 0 & 1 \\ 1 & 0 & . & 0 & 0 \\ 0 & 1 & . & 0 & 0 \\ . & . & . & . & . \\ 0 & 0 & . & 1 & 0 \end{bmatrix}, \begin{bmatrix} 1/q & 0 & 0 & . & 0 \\ 0 & 1 & 0 & . & 0 \\ 0 & 0 & 1 & . & 0 \\ . & . & . & . & . \\ 0 & 0 & 0 & . & 1 \end{bmatrix}, \begin{bmatrix} 1 & 0 & 0 & . & 0 \\ (\vec{p})_1 & 1 & 0 & . & 0 \\ (\vec{p})_2 & 0 & 1 & . & 0 \\ . & . & . & . & . \\ (\vec{p})_n & 0 & 0 & . & 1 \end{bmatrix}$$

All these fractional-linear transformations, which are characterised by the same matrices, can be extended to the whole $\mathbb{C}\mathbb{P}^n$ space.

Denote by P_j (j=0,1, ...) the following product related to the n-fraction (2):

$$P_j = \begin{pmatrix} 1 & 0 & . & 0 \\ p_{j,1} & 1 & . & 0 \\ . & . & . & . \\ p_{j,n} & 0 & . & 1 \end{pmatrix} \times \begin{pmatrix} p_{j,0} & 0 & . & 0 \\ 0 & 1 & . & 0 \\ . & . & . & . \\ 0 & 0 & . & 1 \end{pmatrix} \times \begin{pmatrix} 0 & . & 0 & 1 \\ 1 & . & 0 & 0 \\ . & . & . & . \\ 0 & . & 1 & 0 \end{pmatrix} =$$

$$= \begin{pmatrix} 0 & . & 0 & 0 & p_{j,0} \\ 1 & . & 0 & 0 & p_{j,1} \\ . & . & . & . & . \\ 0 & . & 1 & 0 & p_{j,n-1} \\ 0 & . & 0 & 1 & p_{j,n} \end{pmatrix}. \qquad (6)$$

and define the j-th projective approximant to (2) by

$$\overline{\alpha}_j = (A_{j,0}:A_{j,1}: \dots :A_{j,n}) = P_0 P_1 \dots P_j \overline{e}_n \qquad (7)$$

where $\overline{e}_n=(0: \dots :0:1) \in \mathbb{C}\mathbb{P}^n$. This definition is compatible with the definitions (3) and (4) of the approximant $\vec{\alpha}_j$: if $\overline{\alpha}_j \in \mathcal{H}_0$ then $\vec{\alpha}_j = \varphi^{-1}(\overline{\alpha}_j)$. By convergence of the continued fraction (2) in $\mathbb{C}^n$ we mean the convergence of the sequence $\{\vec{\alpha}_j\}_{j=0}^{\infty}$ to some vector $\vec{a} \in \mathbb{C}^n$, and by convergence in $\mathbb{C}\mathbb{P}^n$ we mean the convergence of the sequence $\{\overline{\alpha}_j\}_{j=0}^{\infty}$ to some projective vector $\overline{a} \in \mathbb{C}\mathbb{P}^n$. If, in addition, the components of the vectors $\vec{\alpha}_j$ or the local coordinates of projective vectors in the map $\mathcal{H}_i$ converge geometrically to some limit with the rate θ $(0 \leq \theta < 1)$, i.e. if

$i=1, \dots ,n$: $\quad \overline{\lim}_{j\to\infty} |(\vec{\alpha}_j-\vec{a})_i|^{1/j} \leq \theta$

we say that the <u>rate of convergence</u> of the continued fraction (2) in $\mathbb{C}^n$ or $\mathbb{C}\mathbb{P}^n$ is equal to θ.

The n-fraction (2) is called limit-periodic with period T (T > 0) if the following limits exist

$$i=0, \dots, n: \qquad \lim_{k\to\infty} p_{t+kT,i} = \mathcal{P}_{t,i}. \tag{8}$$

and if

$$\mathcal{P}_{t,0} \neq 0.$$

The second condition assumes that the limits $\lim_{k\to\infty} q_{t+kT}$ are finite and non zero, because $p_{j+1,0}=1/q_j$. For our purposes it is convenient to define $\mathcal{P}_{t,i}$ for all integer t. We also introduce the $(n+1)\times(n+1)$ matrices (cf.(6))

$$\mathcal{P}_t = \begin{bmatrix} 0 & 0 & \dots & 0 & \mathcal{P}_{t,0} \\ 1 & 0 & \dots & 0 & \mathcal{P}_{t,1} \\ \dots & \dots & \dots & \dots & \dots \\ 0 & 0 & \dots & 1 & \mathcal{P}_{t,n} \end{bmatrix} \tag{9}$$

$$\mathcal{M}_t = \mathcal{P}_t \, \mathcal{P}_{t+1} \cdots \mathcal{P}_{t+T-1} \tag{10}$$

and the characteristic polynomial of the continued fraction

$$Q(\lambda) = \det(\mathcal{M}_t - \lambda E) = \sum_{i=0}^{n} q_i \lambda^i \tag{11}$$

where E is the unit matrix. The expression (11) does not depend on t, because all the matrices $\mathcal{M}_t$ are similar. Let the zeros of the polynomial Q be arranged as follows $|\lambda_0| \geq |\lambda_1| \geq \dots \geq |\lambda_n|$.

3. CONVERGENCE RESULTS

Our main purpose is to prove the following proposition.

Theorem 1.

Let the limit-periodic continued fraction (2) *satisfy the following conditions:*

i) $|\lambda_1| = \theta_0 |\lambda_0|, \quad \theta_0 < 1,$

ii) *for* $t=1,2,\dots,T$ *the vector* $(0,0,\dots,1) \in \mathbb{C}^{n+1}$ *is not contained in the maximal subspace of* $\mathbb{C}^{n+1}$ *invariant by* $\mathcal{M}_t$ *and not containing the eigenvector of* $\mathcal{M}_t$ *corresponding to* λ_0.

Then:

i) *this n-fraction converges in* $\mathbb{CP}^n$ *to some projective vector* $\bar{a}$ *with the rate of convergence* $\theta_0^{1/T}$;

ii) *if for some* $i \in \{0,1,\dots,n\}$ $\bar{a} \in \mathcal{R}_i$, *then for a large enough* j *one has*

$$A_{j,i} \neq 0$$

and $$\lim_{j\to\infty} (A_{j+T,i} / A_{j,i}) = \lambda_0.$$

Remark 1 The conditions of the theorem imply that any given matrix S_t, which reduces $\mathcal{M}_t$ to the form $\mathcal{N}_t=S_t^{-1}\mathcal{M}_t S_t$, where $(\mathcal{N}_t)_{0,0}=\lambda_0$, $(\mathcal{N}_t)_{0,i}=(\mathcal{N}_t)_{i,0}=0$, $i\neq 0$ has the property $(S_t^{-1})_{0,n}\neq 0$.

Remark 2 These conditions are close to those necessary for the convergence of the limit-periodic continued fractions. It may be proved that if for a periodic continued fraction (2) either only the first condition holds, or if instead of the first condition the following relations

$$|\lambda_0|=|\lambda_1|=\ldots=|\lambda_k|>|\lambda_{k+1}|\geqslant\ldots\geqslant|\lambda_n|,\qquad \lambda_i\neq\lambda_j \text{ if } i\neq j,\ i,j=0,\ldots,k$$

are valid, then the n-fraction (2) diverges in $\mathbb{C}\mathbb{P}^n$ and hence in $\mathbb{C}^n$.

Proof of the Theorem 1. Let t $(t\in\{1,\ldots,T\})$ and θ $(\theta_0<\theta<1)$ be fixed. Consider a matrix C which reduces the matrix M t to the special Jordan form

$$J=C^{-1}\mathcal{M}_t C. \tag{12}$$

All elements of the upper row and of the left column of J are zero except $(J)_{0,0}=\lambda_0$. The moduli of the extradiagonal elements of J are not greater than $(\theta-\theta_0)|\lambda_0|$. The block structure of J implies the following inequalities for the k-th degrees of J

$$(J^k)_{0,0}=\lambda_0^{\,k},\qquad |(J^k)_{i,l}|\leqslant\theta^k|\lambda_0|^k,\qquad i\neq 0 \text{ or } l\neq 0. \tag{13}$$

Consider the matrices

$$s,k\in\mathbb{N}:\qquad R_{s,k}:=C^{-1}\,(P_{sT+t}P_{sT+t+1}\cdots P_{(s+k)T+t-1})\ C \tag{14}$$

From (6),(8) and (9) it follows that

$$\lim_{s\to\infty}R_{s,k}=J^k.$$

Denote by σ_k the number, starting with which $(s\geqslant\sigma_k)$ the inequality $|(R_{s,k}-J^k)_{i,l}|\leqslant\theta^k|\lambda_0|^k$ $(i,l=0,\ldots,n)$ holds. Then for $s\geqslant\sigma_k$ condition (13) implies

$$(1-2\theta^k)\,|\lambda_0|^{\,k}\leqslant|(R_{s,k})_{0,0}|\leqslant(1+2\theta^k)\,|\lambda_0|^{\,k},$$
$$|(R_{s,k})_{\,i,l}|\leqslant 2\theta^k)\,|\lambda_0|^{\,k},\qquad i\neq 0 \text{ or } l\neq 0 \tag{15}$$

Now, because the matrix $C_k=(\mathcal{P}_t\mathcal{P}_{t+1}\cdots\mathcal{P}_{t+k})^{-1}C$, $k\in\{0,\ldots,T-1\}$ reduces the matrix $\mathcal{M}_{t+k+1}$ to the form $J=C_k^{-1}\mathcal{M}_{t+k+1}C_k$, then according to the remark 1 the value $(C^{-1}\mathcal{P}_t\cdots\mathcal{P}_{t+k+1})_{0,n}$ does not vanish. Consequently the value

$$D=\max\left(\frac{5}{2},\ 2\max_{\substack{l=1,\ldots,n\\ k=0,\ldots,T-1}}\left|\frac{(C^{-1}\mathcal{P}_t\cdots\mathcal{P}_{t+k+1})_{j,n}}{(C^{-1}\mathcal{P}_t\cdots\mathcal{P}_{t+k+1})_{0,n}}\right|\right)$$

is finite. In $\mathbb{C}\mathbb{P}^n$ (in $\mathcal{R}_0$) let $\mathcal{D}$ be the image by φ of the ball of radius D:

$$\{(x_1, \dots ,x_n) \in \mathbb{C}^n \Big| \sum_{i=1}^n x_i^2 \leqslant D^2\}$$

Now fix the natural number r_0 to be large enough that

$$2\theta^{r_0} < 1/4nD \leqslant 1/10n. \tag{16}$$

We say that the sequence $\{h_\nu\}_{\nu=0}^\infty$ of integers is <u>admissible</u> if

$$h_0 = \max(\sigma_{r_0}, \sigma_{r_0+1}, \dots ,\sigma_{2r_0-1})$$

and if the first differences r_ν satisfy the following conditions

$$r_\nu := h_\nu - h_{\nu-1} \geqslant r_0, \quad \sigma_{r_\nu} \leqslant h_{\nu-1}.$$

Note that any number $u \geqslant h_0 + r_0$ is a term of some admissible sequence, for which the numbers r_ν can be chosen from the set $\{r_0, r_0+1, \dots , 2r_0-1\}$. Let the integer ϱ_u be the smallest of the numbers r such that for $u \geqslant h_0+r_0$ both u and u-r are the neighbour terms in one admissible sequence, i.e. $u=h_\nu$ and $u-r=h_{\nu-1}$. Because $u \geqslant \max(h_0,\sigma_r)+r_0+r$ then the integer ϱ_u exceeds r and we have

$$\lim_{u\to\infty} \theta^{\varrho_u} = 0.$$

To any fixed admissible sequence $\{h_\nu\}_{\nu=0}^\infty$ we associate the matrices

$$\nu\in\mathbb{N}\setminus\{0\}:\ H^{(0)} = \begin{vmatrix} 1 & 0 & \dots & 0 \\ 1 & 2\theta^{r_0} & \dots & 0 \\ \dots & \dots & \dots & \dots \\ 1 & 0 & \dots & 2\theta^{r_0} \end{vmatrix}, \qquad H^{(\nu)} = H^{(\nu-1)} R_{h_{\nu-1},r_\nu} = H^{(\nu-1)} R_{h_0,h_\nu-h_0} \tag{17}$$

The following properties of these matrices can be easily established by induction using (15)-(17) for i=0, ... ,n

$$|(H^{(\nu)})_{i,0}| \geqslant |\lambda_0|^{h_\nu-h_0} (3/5)^{\nu/2}, \tag{18}$$

$$|(H^{(\nu)})_{i,l}/(H^{(\nu)})_{i,0}| \leqslant 4\theta^{r_\nu}, \qquad l=1, \dots ,n. \tag{19}$$

Taking ϱ_u we can rewrite the inequality (19) as follows ($u \geqslant h_0 + r_0$):

$$|(H^{(0)}R_{h_0,u})_{i,l} / (H^{(0)}R_{h_0,u})_{i,0} \leqslant 4\,\theta^{\varrho_u}.$$

Introduce the following short notation:

$$i,j,k,l=0, \dots ,n: \qquad H^{(\nu)}_{i,j,k,l} = (H^{(\nu)})_{i,j} (H^{(\nu)})_{k,l} - (H^{(\nu)})_{i,l} (H^{(\nu)})_{k,j}.$$

From the obvious property

$$\max \left| H^{(0)}_{i,j,k,l} \right| = 2\,\theta^{r_0}$$

and the recurrent relations

$$H^{(\nu)}_{i,j,k,l} = \sum_{s,u=0}^{n} (R_{h_{\nu-1},r_\nu})_{s,j} (R_{h_{\nu-1},r_\nu})_{u,l} H^{(\nu-1)}_{i,s,k,u}$$

(see(17)) one may get, using the inequalities (15) and (16), the following estimation

$$\max_{i,j,k,l}|H^{(\nu)}{}_{i,j,k,l}| \leqslant |\lambda_0|^{2r_\nu}(4n\theta^{r_\nu}+4n(n-1)\theta^{2r_\nu}) \max_{i,j,k,l}|H^{(\nu-1)}{}_{i,j,k,l}| \leqslant$$
$$\leqslant 2\,\theta^{r_0}|\lambda_0|^{2(h_\nu-h_0)}\prod_{\mu=1}{}^{\nu}(6n\theta^{r_\mu}) \qquad (20)$$

Consider now an arbitrary sequence $\bar{d}_\nu=(d_{\nu,0}:\dots:d_{\nu,n})$, $\nu\in\mathbb{N}$, of projetive vectors of D. We prove that the projective vectors $H^{(\nu)}\bar{d}_\nu$ belong to $\mathcal{R}_0$ and that the above sequence converges to some projective vector belonging also to $\mathcal{R}_0$. The first assertion is evident because the "0,i"-th uniform coordinates $\sum^n_{i=0}(H^{(\nu)})_{0,i}d_{\nu,i}$ of vectors $H^{(\nu)}\bar{d}_\nu$ are not equal to 0:

$$\left|\sum_{i=1}{}^n(H^{(\nu)})_{0,i}d_{\nu,i}\right| \leqslant 4n\theta^{r_\nu}D\left|(H^{(\nu)})_{0,0}\right|\left|d_{\nu,0}\right| < \left|(H^{(\nu)})_{0,0}\right|\left|d_{\nu,0}\right|/2.$$

In order to establish the convergence of the local coordinates of $H^{(\nu)}\bar{d}_\nu$ in the map $\mathcal{R}_0$ it will be necessary to prove the convergence of the following series (i=1, ... ,n):

$$\frac{(H^{(0)}\bar{d}_0)_i}{(H^{(0)}\bar{d}_0)_0}+\sum_{\vartheta=0}^{\infty}\left(\frac{(H^{(\vartheta+1)}\bar{d}_{\vartheta+1})_i}{(H^{(\vartheta+1)}\bar{d}_{\vartheta+1})_0}-\frac{(H^{(\vartheta)}\bar{d}_\vartheta)_i}{(H^{(\vartheta)}\bar{d}_\vartheta)_0}\right)=\frac{(H^{(0)}\bar{d}_0)_i}{(H^{(0)}\bar{d}_0)_0}+$$
$$+\sum_{\vartheta=0}^{\infty}\frac{\sum_{s,k,l=0}^{n}(H^{(\vartheta)}_{i,s,0,l}(R_{h_{\vartheta-1},r_\vartheta})_{s,k}\,d_{\vartheta+1,k}\,d_{\vartheta,l}}{\sum_{k=0}^{n}(H^{(\vartheta+1)})_{0,k}d_{\vartheta+1,k}\sum_{l=0}^{n}(H^{(\vartheta)})_{0,l}d_{\vartheta,l}} \qquad (21)$$

Estimating the numerators of the above series terms with the help of (20) and (15), and the corresponding denominators with the help of (18), one can deduce that the module of the ν-th term does not exceed the following value

$$c_0\prod_{\mu=1}{}^{\nu}(10n\theta^{r_\mu}) \leqslant c_0(1/2)^\nu \qquad (22)$$

where the constant $c_0=80(n+1)^3D^2\theta^{r_0}/\sqrt{15}$ does not depend either on the admissible sequence $\{h_\nu\}_{\nu=0}{}^\infty$, or on projective vectors $\bar{d}_\nu\in D$. Then the series (21) converges to the limit b_i independent on the sequence $\{\bar{d}_\nu\}_{\nu=0}{}^\infty$:

$$\lim_{\nu\to\infty} H^{(\nu)}\bar{d}_\nu=\bar{b}=(1:b_1:\dots:b_n).$$

For every two admissible sequences there exists a third one containing an infinite number of terms of both previous sequences. This implies that the limits b_i do not depend on the admissible sequence $\{h_\nu\}_{\nu=0}{}^\infty$, which was used to built the matrices $H^{(\nu)}$.

Remember that an arbitrary integer $u \geq h_0+r_0$ may be chosen as a term of some admissible sequence. Then we find that the whole sequence $\{H^{(0)}R_{h_0,u}\bar{d}_u\}_{u=0}^{\infty}$, $\bar{d}_u \in D$, converges to $\bar{b}$, and not only its subsequences $\{H^{(\nu)}\bar{d}_{h_\nu-h_0}\}_{\nu=0}^{\infty} = \{H^{(0)}R_{h_0,h_\nu-h_0}\}_{\nu=0}^{\infty}$, i.e. not only when u tends to infinity on sequences $\{h_\nu-h_0\}_{u=0}^{\infty}$, where $\{h_\nu\}_{\nu=0}^{\infty}$ is some admissible sequence. Now consider as $\{\bar{d}_u\}_{u=0}^{\infty}$ the following sequences

$$\{C^{-1}P_{(h_0+u)T+t}P_{(h_0+u)T+t+1}\cdots P_{(h_0+u)T+t+k}\bar{e}_n\}_{u=0}^{\infty}$$

($\bar{e}_n=(0:\ldots:0:1)$; $k=0,\ldots,T-1$). From (8) and the choice of D it follows that for a large enough u the terms of these sequences belong to D. Therefore all T subsequences of the sequence

$$\{H^{(0)}C^{-1}P_{h_0T+t}P_{h_0T+t+1}\cdots P_{h_0T+t+u}\bar{e}_n\}_{u=0}^{\infty} \tag{23}$$

composed by terms numbered by the arithmetic progression with step T, converge to $\bar{b}$. Then, the sequence (23) converges to $\bar{b}$.

The sequence of projective approximants $\{\bar{\alpha}_{h_0+t-1+u}\}_{u=0}^{\infty}$ having been obtained by the application of a non degenerated (det (F) $\neq$ 0) fractional-linear transformation

$$F=(P_0P_1\cdots P_{h_0T+t-1})^{-1}C(H^{(0)})^{-1},$$

to the sequence (23), we can conclude that the considered continued fraction converges in $\mathbb{CP}^n$ to $\bar{a} = F\bar{b}$.

To estimate the rate of convergence, we recall the estimation (22) of the term of the series (21). For each integer $u \geq h_0+r_0$ there exists an admissible sequence $\{h_\nu\}_{\nu=0}^{\infty}$ containing u, say $h_k=u$. Let $k=k(u)$ be the smallest index k in the sequences in question. Prove that

$$\lim_{u\to\infty} k(u)/u = 0. \tag{24}$$

Let $r \geq r_0$, $r\in\mathbb{Z}$, $u_0 = \max(\sigma_r,\sigma_{r+1},h_0+r_0)$. Then every integer $u \geq u_0+r^2-r$ is contained in the admissible sequence such that $u_0=h_{k(u_0)}$, and the differences $h_{\nu+1}-h_\nu$ for $\nu \geq k(u_0)$ are equal to r or to r+1. Then $k(u) \leq k(u_0)+(u-u_0)/r$ and $\overline{\lim}_{u\to\infty} k(u)/u \leq 1/r$. Because r is arbitrarily large, the property (24) is proved.

Now, the existence of the limit of the sequence (23) and the estimation (22) of its rate of convergence allow the following inequality to be written

$$\left|b_i - \frac{(H^{(0)}C^{-1}P_{h_0T+t}P_{h_0T+t+1}\cdots P_{h_0T+t+v}\bar{e}_n)_i}{(H^{(0)}C^{-1}P_{h_0T+t}P_{h_0T+t+1}\cdots P_{h_0T+t+v}\bar{e}_n)_0}\right| \leq 2c_0(10n)^{k(u)}\,\Theta^{u-h_0},$$

where u is the integral part of the number v/T. Bearing in mind (24), we conclude that

$$\overline{\lim_{v\to\infty}} \left| b_i - \frac{(H^{(0)}C^{-1}P_{h_0T+t} \cdots P_{h_0T+t+v}\bar{e}_n)_i}{(H^{(0)}C^{-1}P_{h_0T+t} \cdots P_{h_0T+t+v}\bar{e}_n)_0} \right|^{1/v} \leqslant \theta^{1/T},$$

i.e. the rate of convergence of (23) is $\theta^{1/T}$. The rate of convergence of $\{\bar{\alpha}_j\}_{j=h_0T+t}^{\infty} = \{F\,H^{(0)}C^{-1}\,P_{h_0T+t}\,P_{h_0T+t+1}\ldots P_j\bar{e}_n\}_{j=h_0T+t}^{\infty}$ is the same. Because the number $\theta > \theta_0$ is arbitrary, the rate of convergence of the considered limit-periodic continued fraction is $\theta_0^{1/T} = |\lambda_1/\lambda_0|^{1/T}$.

To prove the second statement of Theorem 1 we fix again $t\in\{1, \ldots ,T\}$ and some θ, $\theta_0 < \theta < 1$. Consider again the matrices $H^{(0)}$ and $R_{h_0,u}$, $u\in\mathbb{N}$ (see (17),(14)), and denote by

$$S^{(u)}=H^{(0)}R_{h_0,u}, \qquad R^{(u)}=R_{h_0+1,u}= C^{-1}P_{(h_0+u)T+t}\ldots P_{(h_0+u)T+t+T-1}C, \quad u\in\mathbb{N}.$$

Then for $j=(h_0+u)T+t$, $u\in\mathbb{N}$ we find from (7) that

$$A_{j,k}= (F\,H^{(0)}R_{h_0,u}C^{-1})_{k,n}= (F\,S^{(u)}C^{-1})_{k,n} \text{ and } A_{j+T,k}= (F\,S^{(u)}R^{(u)}C^{-1})_{k,n}\ (k=0, \ldots ,n).$$

The convergence of the continued fraction in $\mathbb{C}\mathbb{P}^n$ to $\bar{a}\in\mathcal{R}_i$ means that beginning from some index $j=j_0$ the homogenous coordinates $A_{j,i}$ of $\bar{\alpha}_j$ are not zeros. Then for $j \geqslant \max(j_0,h_0T+t)$ the equality

$$\frac{A_{j+T,i}}{A_{j,i}} = \frac{\sum_{k,l,r=0}^{n} (F)_{i,k}(S^{(u)})_{k,l}(\mathcal{R}^{(u)})_{l,r}(C^{-1})_{r,n}}{\sum_{k,l=0}^{n} (F)_{i,k}(S^{(u)})_{k,l}(C^{-1})_{l,n}} \tag{25}$$

holds. Observe that $(C^{-1})_{0,n}\neq 0,\ a_i = \sum_{k=0}^{n} (F)_{i,k}b_k\neq 0$

and that for $k, r=0, \ldots ,n;\ l=1,\ldots ,n$ we have

$$\lim_{u\to\infty} (S^{(u)})_{k,0}/(S^{(u)})_{0,0}=b_k, \quad \lim_{u\to\infty}(S^{(u)})_{k,l}/(S^{(u)})_{k,0}=0, \quad \lim_{u\to\infty}(R^{(u)})_{l,r}=(J)_{l,r};$$

then we can turn to the limits in the items of the sums in (25):

$$\lim_{j\to\infty} \frac{A_{j+T,i}}{A_{j,i}} = \frac{\sum_{k=0}^{n} (F)_{i,k}b_k(J)_{0,0}(C^{-1})_{0,n}}{\sum_{k=0}^{n} (F)_{i,k}b_k(C^{-1})_{0,n}} = \lambda_0. \tag{26}$$

The equality (26) holds for arbitrary $t\in\{1, \ldots ,T\}$, thus Theorem 1 is proved ∎

Note that neither Theorem 1, nor any theorem of convergence of the n-fractions in $\mathbb{C}\mathbb{P}^n$

can answer in general questions about the convergence of the n-fraction in $\mathbb{C}^n$. One can build up the continued fraction with new "storeys" in such a way that a convergent in $\mathbb{C}^n$ n-fraction becomes divergent, and vice versa.

To draw a conclusion about the convergence of the n-fractions in $\mathbb{C}^n$ it is necessary to have additional information about the structure of their elements. The so called n-dimensional P-fractions (n-P-fractions) i.e. the n-fractions (2), whose elements $p_{j,i}$ are polynomials in $z \in \mathbb{C}$, $q_j = 1$ and

$$\deg(p_{j,n}) \geqslant 1 + \max_{i=1,\dots,n-1} (0, \deg(p_{i,j})), \tag{27}$$

can be used as an example illustrating this possibility.

The n-dimensional vectors, whose components are formal power series $\sum_{j=r}^{\infty} c_j z^{-j}$, $r \in \mathbb{Z}$, $c_j \in \mathbb{C}$,(see [3]) can be expanded to the n-P-fractions by means of the Jacobi-Perron algorithm ([5], [1], [7]).

We say that the n-P-fraction is limit-priodic with period T, if for each $z \in \mathbb{C}$ there exist limits

$$i=1, \dots, n;\ t \in \mathbb{Z}: \qquad \lim_{k\to\infty} p_{t+kT,i} = \rho_{t,i}$$

which are also the polynomials in z and the degrees of which satisfy the inequalities of type (27):

$$\deg(\rho_{t,n}) \geqslant 1 + \max_{i=1,\dots,n-1}(0, \deg(\rho_{t,i}))$$

In other words the n-fraction with the elements $\vec{\rho}_t = (\rho_{t,1}, \dots, \rho_{t,n})$ should be n-P-fraction.

Let, for the limit-periodic n-P-fraction, I_0 be a set, where the first condition of the Theorem 1 is false, and Δ be a set , where the first condition is true, but the second one does not hold. The set Δ is a finite point set, and I_0 consists of a finite number of the compact pieces of algebraic curves and divides the whole complex plane $\mathbb{C}$ by a finite number of open connected components.

With the help of the same methods as those used for the proof of the Theorem 1 and considering the analyticity of the elements of the n-P-fraction and of other functions of z which are related to them, we have improved Theorem 1 in the following way [4]:

Theorem 2.

The limit-periodic n-P-fraction converges in $\mathbb{C}^n$ *(except, at most the countable set of points) at all points of a unique unbounded connected component of the set* $\mathbb{C}\setminus(I_0 \cup \Delta)$ *and also at all points of those components where this continued fraction converges in* $\mathbb{C}^n$*, even at only one point.*

■ ■ ■ ■ ■

REFERENCES

[1] BERNSTEIN L., The Jacobi-Perron algorithm. Its theory and application., Lecture Notes in Mathematics, **207**, Springer-Verlag, (1971).

[2] GELFOND A.O., KUBENSKAYA I.M., On Perron theorem in the theory of finite -difference equations (in russian), Izv.Akad.Nauk SSSR, Ser. Mat., **17**,83-86,(1953).

[3] PARUSNIKOV V.I., The Jacobi-Perron algorithm and simultaneous approximation of functions (in russian), Mat. Sbornik, **114** (156), 322-333, (1981).

[4] PARUSNIKOV V.I., Limit-periodic multidimensional continued fractions (in russian), preprint, M.V. Keldysh Inst. of Appl. Math., Akad. Nauk SSSR, Moscow, N°**62**, (1983).

[5] PERRON O., Gründlagen für eine Theorie des Jacobischen Kettenbruch Algorithmus., Math. Ann., **64**,1-76, (1907).

[6] PRINGSHEIM A., Über Konvergenz und functionen-theoretischen Character gewisser Limitar-periodische Kettenbrüche, Sitzungsber., Bayer. Akad. Wiss., München, Math.-Phys. **6**, 1-52, (1910).

[7] SCHWEIGER F., The Metrical Theory of Jacobi-Perron Algorithm., Lecture Notes in Mathematics, **334**, Springer-Verlag, (1973).

[8] **SZÁSZ** O., Über die Erhaltung der Konvergenz unendlicher Kettenbrüche bei independenter Veränderlichkeit aller ihrer Elemente, J. Reine Angew. Math., **147**,(1917).

[9] VAN VLECK E.B., On the convergence of algebraic continued fractions whose coefficients have limiting values, Trans. Am. Math. Soc., **5**, 253-262, (1904).

[10] WIDOM H., Extremal polynomials assiociated with a system of curves in the complex plane, Adv. Math., **3**,127-232, (1969).

QUELQUES GENERALISATIONS DE LA REPRESENTATION DE REELS PAR DES FRACTIONS CONTINUES

Stefan Paszkowski
Institute of Low Temperature and Structure Research
PO Box 937, 50-950 Wrocław, Poland

0. Introduction. Chaque réel s'exprime sous forme d'une fraction continue arithmétique (ayant les numérateurs égaux à 1 et les dénominateurs partiels naturels). Cette représentation a des avantages bien connus (rappelons-nous la propriété de la meilleure approximation), mais aussi certains défauts: des algorithmes des opérations arithmétiques sur fractions continues sont compliqués et les dénominateurs partiels peuvent être arbitrairement grands. Cet article démontre qu'en modifiant la notion de fraction continue arithmétique on peut éviter ses certains défauts sans perdre ses principaux avantages. On utilise à cet effet des transformations homographiques convenablement choisies.

1. Définitions, exemples et propriétés. Chaque réel $x \in (0,1)$ se développe en fraction continue

(1.1) $$x = \frac{1|}{|d_1} + \frac{1|}{|d_2} + \dots$$

dont les dénominateurs partiels d_n sont naturels. Si x est un rationnel, alors (pour assurer l'unicité du développement) on admet que le dernier dénominateur soit supérieur à 1. Dans [1], [2], [5] on proposait des généralisations du développement (1.1). Une classe des cas utiles dans le calcul peut être définie comme suit:

On appelle système canonique $\mathbb{K}$ une suite contenant au moins deux couples (I_j, φ_j) où $I_1, I_2, \dots$ sont des intervalles non vides, deux à deux disjoints et tels que

$$\bigcup_j I_j = (0,1)$$

et où φ_j est une transformation homographique, à coefficients entiers, d'intervalle I_j dans $(0,1]$ ou $[0,1)$. Si $\varphi_j(x) = (ex-f)/(g-hx)$ où les entiers e, f, g, h n'ont aucun diviseur commun supérieur à 1, alors le nombre $\det(\varphi_j) := eg-fh$ est dit déterminant de l'homographie φ_j. Par définition, $\det(\varphi_j) \neq 0$. Le déterminant de la composition de deux homographies φ, ψ est égal à $\det(\varphi)\det(\psi)$ ou à un diviseur de ce produit.

A chaque réel $x \in (0,1)$ correspond une suite des homographies τ_1, τ_2, ... et une suite des réels x_1, $x_2, \ldots$, définies pour $x_0 := x$ de la façon suivante. Pour $k = 0, 1, \ldots$, s'il existe un j tel que $x_k \in I_j$ (ce qui est nécessairement vrai pour $k = 0$), alors on pose

$$\tau_{k+1} := \varphi_j, \quad x_{k+1} := \tau_{k+1}(x_k).$$

Lorsque $x_k = 0$ ou 1, les suites $\{\tau_n\}$, $\{x_n\}$ se terminent par τ_k et x_k, respectivement. La suite $\{\tau_n\}$ s'appelle <u>représentation de x en système canonique</u> $\mathbb{K}$:

(1.2) $$x = (\tau_1, \tau_2, \ldots)_{\mathbb{K}}.$$

Une représentation finie (1.2) permet de redécouvrir exactement, au moyen d'un nombre fini d'opérations arithmétiques, le réel x. En effet, soit $\tau_k = \varphi_j$ la dernière transformation dans (1.2). La valeur $x_k := \varphi_j(x_{k-1})$ est connue; elle est égale ou à 0 ou bien à 1. Il suffit donc de calculer x de la formule

$$x = (\tau_k^{-1} \circ \tau_{k-1}^{-1} \circ \ldots \circ \tau_1^{-1})(x_k).$$

Notons que les transformations inverses τ_n^{-1} sont des homographies à coefficients entiers. Le symbole $\circ$ désigne une composition des transformations effectuées à partir de celle de gauche:

$$x = \tau_1^{-1}(\ldots(\tau_{k-1}^{-1}(\tau_k^{-1}(x_k)))\ldots).$$

Pour une représentation infinie (1.2) on peut en déduire une suite descendante des intervalles contenant x:

$$(\tau_{n-1}^{-1} \circ \ldots \circ \tau_1^{-1})(I_{j_n}) \quad (n = 2, 3, \ldots)$$

où j_n est tel que $\tau_n = \varphi_{j_n}$; elle converge vers x.

Les deux systèmes canoniques sont bien connus:

<u>Exemple 1.1</u>. Pour un entier $b > 1$ soit

$$\delta_{b,j}(x) := bx - j \quad (j = 0, 1, \ldots, b-1).$$

La représentation (1.2) de x en système canonique fini

$$\mathbb{K}_0(b) := \left\{\left(0, \tfrac{1}{b}\right), \delta_{b,0}; \left[\tfrac{1}{b}, \tfrac{2}{b}\right), \delta_{b,1}; \ldots; \left[\tfrac{b-1}{b}, 1\right), \delta_{b,b-1}\right\}$$

équivaut au développement de x suivant la base b:

$$x = c_1 b^{-1} + c_2 b^{-2} + \ldots \Leftrightarrow x = (\delta_{b,c_1}, \delta_{b,c_2}, \ldots)_{\mathbb{K}_0(b)}. \quad \square$$

<u>Exemple 1.2</u>. Soit

$$\rho_j(x) := \frac{1}{x} - j \quad (j = 1, 2, \ldots).$$

La représentation (1.2) de x en système canonique infini

$$\mathbb{K}_1 := \left\{\left(\tfrac{1}{2}, 1\right), \rho_1; \left(\tfrac{1}{3}, \tfrac{1}{2}\right], \rho_2; \left(\tfrac{1}{4}, \tfrac{1}{3}\right], \rho_3; \ldots\right\}$$

équivaut au développement (1.1):

$$x = \frac{1|}{|d_1} + \frac{1|}{|d_2} + \ldots \Leftrightarrow x = (\rho_{d_1}, \rho_{d_2}, \ldots)_{\mathbb{K}_1}. \quad \square$$

Un défaut essentiel des fractions continues (1.1), c'est-à-dire celui du système $\mathbb{K}_1$, est que les dénominateurs partiels d_n peuvent être arbitrairement grands ce qui rend difficile leur codage. Il est cependant facile de remplacer $\mathbb{K}_1$ par un système équivalent (dans un certain sens) à $\mathbb{K}_1$ et fini, donc tel que dans (1.2) chaque position ne peut contenir qu'un nombre fini de transformations différentes.

Exemple 1.3. Pour un entier $b \geqslant 1$ soit

$$\gamma_{1,b}(x) := \frac{x}{1 - bx},$$

$$\mathbb{K}_2(b) := \left\{\left(0, \frac{1}{b+1}\right], \gamma_{1,b}; \left(\frac{1}{2}, 1\right), \rho_1; \left(\frac{1}{3}, \frac{1}{2}\right], \rho_2; \ldots; \left(\frac{1}{b+1}, \frac{1}{b}\right], \rho_b\right\}$$

(cf. [4], § 1, où $b = 1$). On vérifie facilement que pour $l \geqslant 0$, $1 \leqslant m \leqslant b$

$$\rho_{lb+m}(x) = \begin{cases} (\underbrace{\gamma_{1,b} \circ \cdots \circ \gamma_{1,b}}_{l \text{ fois}} \circ \rho_m)(x) & \text{si } x \in \left(\frac{1}{lb+m+1}, \frac{1}{lb+m}\right) \\ & \text{ou } x = \frac{1}{lb+m} \text{ et } m > 1, \\ (\underbrace{\gamma_{1,b} \circ \cdots \circ \gamma_{1,b}}_{l \text{ fois}})(x) & \text{si } x = \frac{1}{lb+1}. \end{cases}$$

Il en résulte que si

$$x = (\rho_{d_1}, \rho_{d_2}, \ldots)_{\mathbb{K}_1} \tag{1.3}$$

alors

$$x = (\underbrace{\gamma_{1,b}, \ldots, \gamma_{1,b}}_{l_1 \text{ fois}}, \rho_{m_1}, \underbrace{\gamma_{1,b}, \ldots, \gamma_{1,b}}_{l_2 \text{ fois}}, \rho_{m_2}, \ldots)_{\mathbb{K}_2(b)} \tag{1.4}$$

où $d_k = l_k b + m_k$, $l_k \geqslant 0$, $1 \leqslant m_k \leqslant b$. Si une représentation (1.3) est finie et son dernier élément est égal à d_k (où $d_k > 1$) alors la représentation (1.4) se termine par les l_k éléments $\gamma_{1,b}, \ldots, \gamma_{1,b}$ (pour $m_k = 1$) ou par les l_k+1 éléments $\gamma_{1,b}, \ldots, \gamma_{1,b}, \rho_{m_k}$ (pour $m_k > 1$). $\square$

Même pour des rationnels simples la suite (1.4) peut être très longue. Dans un cas extrême, pour $x = 1/p$ (p naturel), la représentation (1.4) se compose de $[(p+1)/b]$ éléments. Il existe des systèmes canoniques finis permettant de coder les rationnels d'une façon plus économique.

Exemple 1.4. Pour des entiers b, c tels que

$$b > 1, \quad 0 \leqslant c < b-1 \tag{1.5}$$

soit

$$\gamma_{a,c}(x) := \frac{ax}{1-cx},$$

$$\mathbb{K}_3(b,c) := \left\{\left(0,\tfrac{1}{b}\right],\gamma_{b-c,c};\left(\tfrac{1}{2},1\right),\rho_1;\left(\tfrac{1}{3},\tfrac{1}{2}\right],\rho_2;\ldots;\left(\tfrac{1}{b},\tfrac{1}{b-1}\right],\rho_{b-1}\right\}$$

(cf. [4], § 2, où $b = 2$, $c = 0$). En permettant la valeur $c = b-1$ on obtiendrait $\mathbb{K}_3(b,b-1) = \mathbb{K}_2(b-1)$.

Dans les cas $(b,c) = (2,0)$, $(3,0)$, $(3,1)$, $(4,0)$, $(4,1)$, $(4,2)$ on a trouvé les représentations en $\mathbb{K}_3(b,c)$ de tous les rationnels les plus simples. Soit, pour des b, c fixés, $\lambda(n)$ (resp., $\Lambda(n)$) la longueur moyenne (resp., maximale) des représentations calculée pour l'ensemble de tous les rationnels l/m tels que l, m sont relativement premiers et que $0 < l < m \leq n$. On a obtenu, entre autres, les résultats suivants:

b	c	$\lambda(50)$	$\Lambda(50)$	$\lambda(100)$	$\Lambda(100)$
2	0	10.49	21	12.95	28
3	0	7.03	14	8.58	17
3	1	7.08	17	8.69	23
4	0	5.71	12	6.98	14
4	1	6.10	14	7.41	18
4	2	5.34	10	6.53	12

On en déduit que pour $b = 3$ (resp., $b = 4$) les longueurs moyennes et maximales des suites d'homographies sont les plus petites si $c = 0$ (resp., $c = 2$). Remarquons aussi que le nombre des bits nécessaires pour coder les suites d'homographies pour tous les rationnels l/m $(0 < l < m \leq 100)$ est égal pour $b = 2, 3, 4$ à 28, 27 ($17 \operatorname{Log}_2 3 \approx 26.9$) et 24, respectivement, et qu'il est minimal pour $b = 4$ (ce qui n'est point evident). Il y a les 3043 rationnels de la forme considérée et pour les coder il faut au moins 12 bits. □

On sait que la représentation en $\mathbb{K}_1$ d'un nombre $x \in (0,1)$ est périodique si et seulement si ce nombre est un irrationnel quadratique

$$\frac{i\sqrt{j}+l}{m} \tag{1.6}$$

(j - un naturel non carré, i, l, m - des entiers, $i \neq 0$, $m \neq 0$). En vertu de (1.3), (1.4) le même est vrai pour $\mathbb{K}_2(b)$. Pour d'autres systèmes canoniques il n'est évident que si la représentation de x est périodique alors x est un irrationnel quadratique. Il semble que la représentation en $\mathbb{K}_3(b,c)$ (sous l'hypothèse (1.5)) et, plus généralement, en chaque système canonique $\mathbb{K}$ dans lequel au moins une homographie φ_j a le déterminant différent de ± 1, peut être apériodique. Cette conjecture cependant sera sans doute difficile à démontrer (sauf le cas très particulier de $\mathbb{K}_2(b)$).

Exemple 1.5. Pour chaque irrationnel (1.6) tous les x_n sont, eux aussi, des irrationnels quadratiques et on peut les calculer exactement. On a vérifié qu'en $\mathbb{K}_3(4,2)$, en particulier,

$$\sqrt{5}-2 = (\overline{0,1,0,3}),$$
$$\sqrt{6}-2 = (2,\overline{0,1}),$$
$$\sqrt{8}-2 = (1,0,1,\overline{2}),$$
$$\sqrt{22}-4 = (1,2,0,1,0,1,\overline{1,0,3,2})$$

où les chiffres 0, 1, 2, 3 remplacent les symboles $\gamma_{2,2}$, ρ_1, ρ_2, ρ_3 et la barre au dessus des chiffres désigne la période de la représentation. Dans tous les cas les nombres x_n sont de la forme (1.6) avec des entiers i, l, m relativement petits. Par contre, pour tous les irrationnels $\sqrt{26}-5$, $\sqrt{32}-5$, $\sqrt{37}-6$, $\sqrt{41}-6$, $\sqrt{50}-7$ la suite $\{x_n\}$ se complique très vite (p. ex., si $x = \sqrt{26}-5$ alors $x_{47} = (4096\sqrt{26}-17652)/64367$) et il semble peu probable qu'elle soit périodique. □

On va démontrer qu'il est possible de généraliser la notion de système canonique (même en comparaison de [1], [2], [5]) de façon à garantir la périodicité de la représentation de tous les irrationnels quadratiques et à assurer d'autres avantages. Notamment,

1° on supprime la condition que les I_j soient deux à deux disjoints,

2° on admet que 0 ou 1 appartiennent à $\bigcup I_j$,

3° on suppose que s'il existe au moins deux intervalles de la famille (I_j) contenant un certain x_k, alors le choix d'un d'eux et de l'homographie correspondante dépend de $\tau_1, \tau_2, \ldots, \tau_k$.

On peut donc opposer la construction <u>non stationnaire</u> des suites $\{\tau_n\}$, $\{x_n\}$ pour un système canonique généralisé à leur construction <u>stationnaire</u> pour un système défini plus tôt. Dans chaque cas particulier il faut évidemment préciser comment on choisit des intervalles.

<u>Exemple 1.6</u>. Pour un naturel $b > 1$ soit

$$\mathbb{K}_4(b) := \Big\{\big(\tfrac{1}{2},1\big],\rho_1;\big(\tfrac{1}{3},\tfrac{1}{2}\big],\rho_2;\ldots;\big(\tfrac{1}{b},\tfrac{1}{b-1}\big],\rho_{b-1};$$
$$\big[0,\tfrac{1}{b}\big),\delta_{b,0};\big[\tfrac{1}{b},\tfrac{2}{b}\big),\delta_{b,1};\ldots;\big[\tfrac{b-1}{b},1\big),\delta_{b,b-1}\Big\}.$$

Si $x_k < 1/b$ alors $\tau_{k+1} = \delta_{b,0}$ et si $x_k > 1/b$ alors τ_{k+1} peut être a priori une homographie ρ_j ou une homographie $\delta_{b,j}$. Il existe maintes critères de leur choix. Un critère qui semble être le plus raisonnable résulte d'une identité tout à fait élémentaire: quel que soit naturel

$$d = (c_l c_{l-1} \ldots c_0)_b$$

(c_i - chiffres du développement de d suivant la base b, $c_l > 0$),

$$(1.7) \qquad \rho_d = \underbrace{\delta_0\circ\ldots\circ\delta_0}_{l \text{ fois}}\circ\rho_{c_l}\circ\delta_{c_{l-1}}\circ\ldots\circ\delta_{c_0}$$

où $\delta_j := \delta_{b,j}$. La suite $\{\tau_n\}$ donc, comme pour $\mathbb{K}_2(b)$, résulte directement de la fraction continue (1.1). Si, p.ex., b = 3 et

$$x = \frac{1|}{|7} + \frac{1|}{|1} + \frac{1|}{|15}$$

alors la suite $\{\tau_n\}$ se compose des éléments

$$\underbrace{\delta_0,\ \rho_2,\ \delta_1}_{7\,=\,(21)_3},\ \underbrace{\rho_1}_{1=(1)_3},\ \underbrace{\delta_0,\ \delta_0,\ \rho_1,\ \delta_2,\ \delta_0}_{15\,=\,(120)_3}.$$

On peut aussi construire $\{\tau_n\}$ sans connaître le développement (1.1). Soit, p.ex., $x = 15/241$ et $b = 3$. Puisque $x_0 := x < 1/b$, il résulte de la définition de $\mathbb{K}_4(b)$ que $\tau_1 = \delta_0$, $x_1 = 45/241$. Pour la même raison $\tau_2 = \delta_0$, $x_2 = 135/241$. On constate que $x_2 > 1/3$. Par conséquent, il est certain que τ_3 est une homographie ρ_j et que τ_4, τ_5 sont des homographies δ_j. Plus précisément, $\tau_3 = \rho_1$, $x_3 = 106/135$, $\tau_4 = \delta_2$, $x_4 = 16/45$, $\tau_5 = \delta_1$, $x_5 = 1/15$. L'identité (1.7) détermine aussi les éléments restants des suites $\{\tau_n\}$ et $\{x_n\}$: $\tau_6 = \delta_0$, $x_6 = 1/5 < 1/3$, $\tau_7 = \delta_0$, $x_7 = 3/5 > 1/3$, donc $\tau_8 = \rho_1$, $x_8 = 2/3$, $\tau_9 = \delta_2$, $x_9 = 0$. Malgré la valeur nulle de x_9 il faut effectuer une transformation de plus, δ_0 bien entendu: $\tau_{10} = \delta_0$, $x_{10} = 0$. Remarquons que pour le nombre $x^* := x_3 = 106/135$ on a $x_1^* = \rho_1(x^*) = 29/106 \neq x_4$. Ici se manifeste le fait que la construction de $\{\tau_n\}$ n'est pas stationnaire.

Il vaut noter en plus que pour tous les irrationnels quadratiques un autre critère du choix des homographies ρ_j et $\delta_{b,j}$ est possible. En effet, pour x de la forme (1.6) chaque x_k s'exprime d'une façon analogue:

$$x_k = \frac{i_k\sqrt{j} + l_k}{m_k}$$

où i_k, l_k, m_k sont des entiers et n'ont aucun diviseur commun supérieur à 1. Considérons, p. ex., le système

$$\mathbb{K}_4(2) = \left\{\left(\tfrac{1}{2},1\right],\rho_1;\left[0,\tfrac{1}{2}\right),\delta_{2,0};\left[\tfrac{1}{2},1\right),\delta_{2,1}\right\}.$$

Il faut donc choisir, si $x_k > \frac{1}{2}$, parmi les homographies ρ_1 et $\delta_{2,1}$. Soit

$$\tau_{k+1} = \begin{cases} \rho_1 & (m_k \text{ impair}), \\ \delta_{2,1} & (m_k \text{ pair}). \end{cases}$$

Une définition analogue est valable pour les rationnels. Notons $\mathbb{K}_4^*(2)$ le système canonique engendré par cette convention et restreint aux rationnels et aux irrationnels quadratiques de l'intervalle (0,1). L'identité (1.7) permet de transformer la fraction continue pour $\sqrt{21}-4$ et d'obtenir la représentation

$$\sqrt{21}-4 = (\rho_1,\overline{\rho_1,\delta_0,\rho_1,\delta_0,\rho_1,\rho_1,\delta_0,\delta_0,\delta_0,\rho_1,\delta_0,\delta_0,\delta_0})_{\mathbb{K}_4(2)}.$$

La période en $\mathbb{K}_4^*(2)$ est plus courte:

$$\sqrt{21}-4 = (\overline{\rho_1,\rho_1,\delta_0,\delta_1})_{\mathbb{K}_4^*(2)}.$$

Pour tous les naturels non carrés $n \leq 1003$ on a vérifié que la représentation de $\sqrt{n}-[\sqrt{n}]$ en $\mathbb{K}_4^*(2)$ est périodique. Dans les deux cas seulement la période est mixte:

$$\sqrt{769}-27 = (\rho_1,\delta_0,\delta_1,\delta_0,\rho_1,\delta_0,\delta_0,\delta_0,\overline{\delta_0,\rho_1,\delta_0,\delta_0,\delta_0,\delta_1,\rho_1,\rho_1,\delta_0,\delta_1,\delta_0,\rho_1})_{\mathbb{K}_4^*(2)},$$

$$\sqrt{865}-29 = (\delta_0,\rho_1,\delta_0,\delta_0,\delta_1,\overline{\delta_1,\delta_0,\rho_1,\delta_0,\delta_0,\delta_0,\rho_1,\delta_1,\delta_1,\rho_1,\delta_0,\delta_0,\delta_0})_{\mathbb{K}_4^*(2)}.$$

On n'a trouvé aucun irrationnel quadratique dont la représentation en $\mathbb{K}_4(2)$ aurait la période plus courte que celle en $\mathbb{K}_4^*(2)$. Il reste cependant à trouver une façon d'étendre les régles définissant $\mathbb{K}_4^*(2)$ sur tous les réels de $(0,1)$. On ne sait non plus comment on pourrait utiliser ces régles en effectuant les opérations sur les représentations (v. § 2). □

2. <u>Changement de système canonique. Opérations sur représentations.</u> Etant donnés deux systèmes canoniques (définis au début du § 1)

$$\mathbb{K} := \{I_1,\varphi_1;I_2,\varphi_2;\dots\}, \quad \bar{\mathbb{K}} := \{\bar{I}_1,\bar{\varphi}_1;\bar{I}_2,\bar{\varphi}_2;\dots\}$$

on pose le problème suivant: en ne sachant d'un réel $x\in(0,1)$ que

$$x = (\tau_1,\tau_2,\dots)_{\mathbb{K}} \tag{2.1}$$

trouver sa représentation en $\bar{\mathbb{K}}$:

$$x = (\bar{\tau}_1,\bar{\tau}_2,\dots)_{\bar{\mathbb{K}}}. \tag{2.2}$$

Un cas particulier du problème consiste à déduire la représentation d'un réel x en système $\mathbb{K}_3(b,c)$ du développement de x en fraction continue (1.1).

Le problème posé admet une solution élémentaire. Soit

$$x_0 := x, \quad x_k := \tau_k(x_{k-1}) = (\tau_1\circ\dots\circ\tau_k)(x).$$

La composition $\tau_1\circ\dots\circ\tau_k$ étant une homographie,

$$x_k = \frac{px-q}{r-sx} \tag{2.3}$$

où p, q, r, s sont des entiers tels que $pr-qs \neq 0$. On calcule les coefficients p, q, r, s (dépendant de k) par récurrence, en utilisant les coefficients des homographies τ_n. Puisque $x_k\in(0,1)$ (si x_k n'est pas le dernier élément de $\{x_n\}$), il résulte de (2.3) que

$$x\in\left(\frac{q}{p},\frac{q+r}{p+s}\right); \tag{2.4}$$

dans telles relations les extremités d'intervalle peuvent être transposés. Chaque homographie τ_n introduit une information supplémentaire sur x. En conséquence, pour un irrationnel x les intervalles de (2.4) (k = 1, 2,...) forment une suite descendante qui converge vers x et il existe des entiers k, j_1 tels que

$$\left(\frac{q}{p},\frac{q+r}{p+s}\right) \subset \bar{I}_{j_1}.$$

Il en résulte que $\bar{\tau}_1 = \bar{\varphi}_{j_1}$. Soit $\bar{x}_1 := \bar{\tau}_1 x$. Si

(2.5) $$x = \bar{\tau}_1^{-1}(\bar{x}_1) = \frac{a\bar{x}_1-b}{c-d\bar{x}_1} \quad \text{(a, b, c, d entiers),}$$

alors

(2.6) $$x_k = \frac{p(a\bar{x}_1-b)/(c-d\bar{x}_1)-q}{r-s(a\bar{x}_1-b)/(c-d\bar{x}_1)} = \frac{p_1\bar{x}_1-q_1}{r_1-s_1\bar{x}_1}$$

où

(2.7) $$\begin{bmatrix} p_1 & q_1 \\ s_1 & r_1 \end{bmatrix} := \begin{bmatrix} p & q \\ s & r \end{bmatrix}\begin{bmatrix} a & b \\ d & c \end{bmatrix}.$$

Le déterminant de l'homographie (2.6) est le produit de ceux de (2.3), (2.5). Plus précisément, s'il existe un facteur commun $v > 1$ des entiers p_1, q_1, r_1, s_1, alors on peut les diviser par v et diminuer le déterminant v^2 fois.

S'il existe un j_2 tel que

$$\left(\frac{q_1}{p_1},\frac{q_1+r_1}{p_1+s_1}\right) \subset \bar{I}_{j_2}$$

alors on sait que $\bar{\tau}_2 = \bar{\varphi}_{j_2}$ et on exprime x_k par $\bar{x}_2$. Si, par contre, l'intervalle a une partie commune avec au moins deux intervalles de la famille $(\bar{I}_j)$, alors pour trouver $\bar{\tau}_2$ il faut prendre en considération les homographies τ_{k+1}, τ_{k+2},... On calcule donc d'abord $x_{k+1} = \tau_{k+1}x_k$. Soit

(2.8) $$\tau_{k+1}y = \frac{ey-f}{g-hy} \quad \text{(e, f, g, h entiers).}$$

Alors

(2.9) $$x_{k+1} = \frac{p_2\bar{x}_1-q_2}{r_2-s_2\bar{x}_1}$$

où

(2.10) $$\begin{bmatrix} p_2 & q_2 \\ s_2 & r_2 \end{bmatrix} := \begin{bmatrix} e & f \\ h & g \end{bmatrix}\begin{bmatrix} p_1 & q_1 \\ s_1 & r_1 \end{bmatrix}.$$

Le déterminant de l'homographie (2.9) est le produit de ceux de (2.6), (2.8) ou son diviseur.

Pour un irrationnel x les représentations (2.1), (2.2) sont infinies et pour passer de (2.1) à (2.2) il suffit d'examiner les intervalles (q/p,(q+r)/(p+s)) et d'appliquer les formules (2.7), (2.10). Si, par contre, x est un rationnel, alors

$$x = (\tau_1, \tau_2, \dots, \tau_k)_{\mathbb{K}}$$

et l'algorithme présenté plus haut permet d'exprimer x_k par $\bar{x}_i$ pour un certain $i \geqslant 0$. Conformément à la définition de représentation, x_k est égal ou à 0 ou bien à 1. En connaissant τ_k on peut déterminer x_k, trouver $\bar{x}_i$ (un rationnel, bien entendu) et compléter la suite (2.2).

Exemple 2.1. Soit $\mathbb{K} = \mathbb{K}_1$, $\bar{\mathbb{K}} = \mathbb{K}_3(4,2)$,

$$x = \sqrt{38}-6 = (\overline{\rho_6, \rho_{12}})_{\mathbb{K}}.$$

On calcule successivement

$$x_1 = \rho_6(x) = \frac{1-6x}{x}$$

d'où $x \in \left(\frac{1}{7}, \frac{1}{6}\right)$, $\bar{\tau}_1 = \gamma_{2,2}$, $x = \gamma_{2,2}^{-1}(\bar{x}_1) = \bar{x}_1/(2+2\bar{x}_1)$,

$$x_1 = \frac{2-4\bar{x}_1}{\bar{x}_1}$$

d'où $\bar{x}_1 \in \left(\frac{2}{5}, \frac{1}{2}\right)$, $\bar{\tau}_2 = \rho_2$, $\bar{x}_1 = \rho_2^{-1}(\bar{x}_2) = 1/(2+\bar{x}_2)$,

$$x_1 = 2\bar{x}_2.$$

L'information que $\bar{x}_2 \in \left(0, \frac{1}{2}\right)$ ne permet pas de trouver $\bar{\tau}_3$ et on tient compte de τ_2:

$$x_2 = \rho_{12}(x_1) = \frac{1-24\bar{x}_2}{2\bar{x}_2}$$

d'où $\bar{x}_2 \in \left(\frac{1}{26}, \frac{1}{24}\right)$, $\bar{\tau}_3 = \gamma_{2,2}$, $\bar{x}_2 = \gamma_{2,2}^{-1}(\bar{x}_3)$,

$$x_2 = \frac{1-11\bar{x}_3}{\bar{x}_3}.$$

En continuant les calculs de la même façon on trouve que

(2.11) $$\sqrt{38}-6 = (0,2,0,0,0,1,\overline{3,2,3,2,1,3,1,3,1,2})$$

(les notations comme dans l'exemple 1.5). ▯

Les opérations arithmétiques sur les représentations de réels en système canonique choisi $\mathbb{K}$ sont, en général, difficiles, même dans le cas $\mathbb{K} = \mathbb{K}_1$. Il est cependant relativement facile de déduire de la représentation $x = (\tau_1, \tau_2, \dots)_{\mathbb{K}}$ la partie entière i et la représentation $(\tilde{\tau}_1, \tilde{\tau}_2, \dots)_{\mathbb{K}}$ de la partie fractionnaire d'une expression

(2.12) $$\frac{jx+l}{mx+n} \quad (j,\ l,\ m,\ n \text{ entiers}).$$

On utilise à cet effet un procédé analogue à celui du changement de système canonique (tous les deux généralisent un algorithme de [3]). On trouve d'abord un k tel que les extrémités de l'intervalle

$$\left(\frac{jq/p+l}{mq/p+n}, \frac{j(q+r)/(p+s)+l}{m(q+r)/(p+s)+n}\right)$$

(contenant, en vertu de (2.4), le nombre (2.12)) aient la même partie entière i. La partie fractionnaire de (2.12) est égale à

$$\tilde{x} := \frac{jx+l}{mx+n} - i.$$

On substitue donc dans (2.3)

$$x = \frac{n\tilde{x}-(l-ni)}{(j-mi)-m\tilde{x}}$$

et on obtient

$$x_k = \frac{\tilde{p}\tilde{x}-\tilde{q}}{\tilde{r}-\tilde{s}\tilde{x}} \quad \text{où} \quad \begin{bmatrix}\tilde{p} & \tilde{q}\\ \tilde{s} & \tilde{r}\end{bmatrix} := \begin{bmatrix}p & q\\ s & r\end{bmatrix}\begin{bmatrix}n & l-ni\\ m & j-mi\end{bmatrix}.$$

Il en résulte, comme précédemment, un intervalle contenant $\tilde{x}$. S'il est suffisamment étroit, alors il désigne une homographie $\tilde{\tau}_1 \in \{\varphi_1, \varphi_2, \ldots\}$. Dans ce cas on pose $\tilde{x}_1 = \tilde{\tau}_1(\tilde{x})$, c'est-à-dire $\tilde{x} = \tilde{\tau}_1^{-1}(\tilde{x}_1)$. Sinon, au moyen de la formule $x_{k+1} = \tau_{k+1}(x_k)$ on exprime x_{k+1} par $\tilde{x}$ et on trouve de nouveau un intervalle contenant $\tilde{x}$.

Exemple 2.2. En utilisant (2.11) nous trouverons la partie entière et la représentation en $\mathbb{K}_3(4,2)$ de la partie fractionnaire du réel $10(\sqrt{38}-6)$. Soit $x := \sqrt{38}-6$. Puisque

$$x_1 = \gamma_{2,2}(x) = \frac{2x}{1-2x} \quad \left(\text{d'où } 10x \in \left(0, \tfrac{5}{2}\right)\right),$$

$$x_2 = \rho_2(x_1) = \frac{1-6x}{2x} \quad \left(\text{d'où } 10x \in \left(\tfrac{5}{4}, \tfrac{5}{3}\right)\right),$$

on trouve que $[10x] = 1$. Posons $\tilde{x} := 10x-1$ d'où

$$x_2 = \frac{2-3\tilde{x}}{1+\tilde{x}} \quad \left(\tilde{x} \in \left(\tfrac{1}{4}, \tfrac{2}{3}\right)\right),$$

$$x_3 = \gamma_{2,2}(x_2) = \frac{4-6\tilde{x}}{7\tilde{x}-3} \quad \left(\tilde{x} \in \left(\tfrac{7}{13}, \tfrac{2}{3}\right) \text{ d'où } \tilde{\tau}_1 = \rho_1,\ \tilde{x} = \tilde{\tau}_1^{-1}(\tilde{x}_1)\right),$$

$$x_3 = \frac{4\tilde{x}_1-2}{4-3\tilde{x}_1} \quad \left(\tilde{x}_1 \in \left(\tfrac{1}{2}, \tfrac{6}{7}\right) \text{ d'où } \tilde{\tau}_2 = \rho_1,\ \tilde{x}_1 = \tilde{\tau}_2^{-1}(\tilde{x}_2)\right), \ldots,$$

$$10(\sqrt{38}-6) = 1+(1,1,1,0,1,0,2,1,3,0,1,3,0,0,0,0,0,0,1,1,0,\ldots)$$

(les notations comme dans l'exemple 1.5). Pour trouver ces 21 premières homographies $\tilde{\tau}_n$ il fallait connaître les 17 premières homographies τ_n. Pour déterminer $\tilde{\tau}_{22}$ on utilise la relation

$$x_{17} = \frac{153\tilde{x}_{21}-26}{62+29\tilde{x}_{21}}$$

qui résulte des calculs précédents. Le déterminant de cette homographie est égal à 10240 et, par conséquent, l'intervalle (26/153,22/31) contenant $\tilde{x}_{21}$ est assez large. C'est n'est pas aléatoire. En effet, en général, chaque homographie <u>donnée</u> τ_m égale à $\gamma_{2,2}$ et chaque homographie <u>calculée</u> $\tilde{\tau}_n$ égale à $\gamma_{2,2}$ provoque le doublement du déterminant d'homographie. Dans deux cas seulement le déterminant diminue deux fois par rapport à celui d'une homographie

$$x_m = \frac{p\tilde{x}_n - q}{r - s\tilde{x}_n},$$

notamment si 1° r, s sont pairs et on effectue une homographie $x_{m+1} = \gamma_{2,2}(x_m)$, 2° p, s sont pairs et on substitue $\tilde{x}_n = \gamma_{2,2}^{-1}(\tilde{x}_{n+1}) = \tilde{x}_{n+1}/(2+2\tilde{x}_{n+1})$. □

L'exemple 2.2 permet de supposer que même pour un système canonique fini (p. ex., le système $\mathbb{K}_4(b)$ qui semble être d'ailleurs le plus prometttant) les opérations décrites plus haut provoquent inévitablement une croissance illimitée des coefficients d'homographies. Ce phénomène complique sans aucun doute les calculs. La seule exception font, peut-être, certains systèmes canoniques tels que, quel que soit j, $\det(\varphi_j) = \pm 1$. Tel est le cas du système $\mathbb{K}_2(b)$ dans lequel cependant les représentations sont peu économiques. Le système $\mathbb{K}_1$ a un autre défaut: les dénominateurs partiels de fractions continues (1.1) et, par conséquent, les coefficients d'homographies auxiliaires peuvent être arbitrairement grands. Il faut donc toujours chercher des systèmes canoniques plus avantageux, même au-dehors de la classe définie au début du § 1.

<u>Références</u>

[1] B.H. Bissinger, A generalization of continued fractions, Bull. Amer. Math. Soc. 50(1944), 868-876.

[2] C.J. Everett, Representations for real numbers, Bull. Amer. Math. Soc. 52(1946), 861-869.

[3] S. Paszkowski, Transformations de fractions continues et calcul de leurs valeurs (en polonais), Université de Wrocław, Institut d'Informatique, 1979.

[4] S. Paszkowski, Sur des fractions continues binaires, Université des Sciences et Techniques de Lille, U.E.R. d'I.E.E.A., Publ. ANO-142, Septembre 1984.

[5] B.K. Swartz, B. Wendroff, Continued function expansions of real numbers, Proc. Amer. Math. Soc. 11(1960), 634-639.

LOCAL PROPERTIES OF CONTINUED FRACTIONS

Haakon Waadeland
Department of Mathematics
and Statistics
University of Trondheim
N-7055 Dragvoll
Norway

1. Introduction

In the present paper we shall discuss continued fractions of the form

$$\underset{n=1}{\overset{\infty}{K}} \frac{z_n}{1} = \frac{z_1}{1} + \frac{z_2}{1} + \ldots + \frac{z_n}{1} + \ldots \quad . \tag{1.1}$$

For a given sequence $\{E_n\}$ of convergence regions the c. f. type (1.1) defines a function F from a subset of $\mathbb{C}^{\infty}$ to $\hat{\mathbb{C}}$ by

$$F(z_1,z_2,\ldots,z_n\ldots) := \underset{n=1}{\overset{\infty}{K}} \frac{z_n}{1} \tag{1.2}$$

$$F: \underset{n=1}{\overset{\infty}{X}} E_n \to \hat{\mathbb{C}}$$

We shall illustrate, on some simple examples, how a study of such a function can lead to results in the analytic theory of continued fractions. First some words on notation:

For a fixed $(a_1,a_2,\ldots,a_n,\ldots) \in \underset{n=1}{\overset{\infty}{X}} E_n$ we define for $N \geq 0$:

$$f^{(N)} := F(a_{N+1},a_{N+2},\ldots), \text{ \underline{value of Nth tail} } \underset{n=N+1}{\overset{\infty}{K}} \frac{a_n}{1} \;, \tag{1.3}$$

$f := f^{(0)}$ being the value of the continued fraction $\underset{n=1}{\overset{\infty}{K}}(a_n/1)$ itself.

We shall assume in the following that all continued fraction values in question, as well as all tail values, are <u>finite</u>.

The set

$$L := \{F(z_1, z_2, \ldots) \mid z_n \in E_n\}$$

is the range of the function (1.2). More generally we define for all integers $N \geq 0$:

$$L^{(N)} := \{F(z_{N+1}, z_{N+2}, \ldots) \mid z_n \in E_n\} \quad , \tag{1.4}$$

$L = L^{(0)}$. $\{L^{(N)}\}$ is called <u>a sequence of limit regions</u> for (1.1) corresponding to the sequence $\{E_n\}$ of element regions, actually <u>the</u> sequence of <u>best</u> limit regions corresponding to $\{E_n\}$. [3], [7], [9, Sec. 2]. (The term <u>region</u> is here used loosely to mean any subset of $\hat{\mathbb{C}}$, as in [8, Sec. 4.2].)

Knowledge of the sequence $\{L^{(N)}\}$ or a dominant sequence $\{V^{(N)}\}$, $V^{(N)} \supseteq L^{(N)}$, is of great importance in connection with computation of values of continued fractions $K(a_n'/1)$ with $a_n' \in E_n$ for all n, since it can be used to obtain <u>a priori</u> truncation error estimates, see e.g. [9, Sec. 1 and 2] and [8, Ch. 8].

In the particular case when all E_n are equal, $E_1 = E_2 = \ldots = E_n = E$, E is called a <u>simple convergence region</u>. In this case all $L^{(N)}$ are equal to an L, a <u>simple limit region</u> (<u>the best</u> limit region). As <u>one</u> of the illustrations of results in the analytic theory of continued fractions, obtained by studying (1.2), we shall describe a procedure to determine a good approximation to L under certain conditions.

For "small" convergence regions E_n the number f can, under certain conditions, be regarded as an approximation to the value of $\overset{\infty}{\underset{n=1}{K}}(z_n/1)$, where $z_n \in E_n$ for all n, and $f^{(N)}$ as an approximation to the tail value $\overset{\infty}{\underset{n=N+1}{K}}(z_n/1)$. Results of this type are proved in e.g. [3], [6], and may be regarded as being theorems on the continuity of (1.2) $(d(\bar{z},\bar{a}) = \sup |z_i - a_i|)$. This has then been used for convergence acceleration, see e.g. [10], [4], [5]. In the present paper we shall give a better approximation, based upon differentiability. This will in turn provide a tool for improved convergence acceleration. This is <u>the other</u> illustration of results in the analytic theory of continued fractions, obtained by studying the function (1.2), to be included in the paper.

2. An improved approximation to the value of a continued fraction

For any fixed n the function (1.2) is a rational function of z_n. Let $(\quad)_o$ indicate evaluation at $(z_1,z_2,\ldots) = (a_1,a_2\ldots)$. Then the following formula holds:

Lemma 1

$$\left(\frac{\partial F}{\partial z_{n+1}}\right)_o = \frac{f}{a_{n+1}} \prod_{k=1}^{n} \left(\frac{-f^{(k)}}{1+f^{(k)}}\right) \tag{2.1}$$

The proof is given in [12] in two versions. The most direct one uses merely the chain rule and the recurrence relation between the tails. The formula (4.20) in the unpublished paper [1] contains the expression (2.1) in a different version, without reference to partial derivatives.

The simplest case is when all a_n are equal to $a \notin (-\infty, -\frac{1}{4})$. (Diagonal case in C^∞, periodic case in continued fraction theory.) In this case all tails $f^{(k)}$ are equal to

$$\Gamma = K(a/1) = \tfrac{1}{2}[\sqrt{1+4a} - 1].\ \mathrm{Re}\sqrt{} > 0. \tag{2.2}$$

In this case the formula (2.1) takes the form

$$\left(\frac{\partial F}{\partial z_{n+1}}\right)_o = \frac{1}{1+\Gamma} \cdot \left(\frac{-\Gamma}{1+\Gamma}\right)^n . \tag{2.3}$$

(Keep in mind that $\left|\frac{\Gamma}{1+\Gamma}\right| < 1$.) On this background it is tempting to guess that for a continued fraction

$$\tilde{f} = \mathop{K}_{n=1}^{\infty} \frac{a+\delta_n}{1}, \quad \delta_n \text{ "small"},$$

we must have

$$\tilde{f} \approx \Gamma + \frac{1}{1+\Gamma} \sum_{n=0}^{\infty} \left(\frac{-\Gamma}{1+\Gamma}\right)^n \cdot \delta_{n+1} .$$

We shall justify this, and later illustrate the two applications mentioned in the previous section.

We shall need the following trivial observation: Let $(a_1,a_2,\ldots, a_n,\ldots)$ be such that $F(a_1,a_2,\ldots,a_n,\ldots)$ exists and is finite. Then, to any n there is a δ, such that

$$F(z_1,z_2,\dots,z_n,a_{n+1},a_{n+2},\dots) = \frac{z_1}{1} \; \frac{z_2}{+\,1} \; + \dots + \frac{z_n}{1+f^{(n)}} \tag{2.4}$$

has a finite value for any $(z_1,z_2,\dots,z_n)$, such that $\max_{1\leq i\leq n} |z_i-a_i| \leq \delta$. (2.4) is a rational function of $z_1,z_2,\dots z_n$. Let in particular $z_1,z_2,\dots,z_n$ be holomorphic functions of one complex variable z in a neighborhood of z = 0, and let $z_i(0) = a_i$ for $i = 1,2,\dots,n$. Then with

$$g_n(z) = F(z_1(z),z_2(z),\dots,z_n(z),a_{n+1},a_{n+2},\dots) \quad , \tag{2.5}$$

we have, from (2.1)

$$g_n'(0) = f \,.\, \sum_{\nu=0}^{n-1} \Big(\prod_{k=1}^{\nu} \Big(\frac{-f^{(k)}}{1+f^{(k)}}\Big)\Big) \frac{z_{\nu+1}'(0)}{a_{\nu+1}} \quad . \tag{2.6}$$

Defining for $i > n$ $z_i(z) := a_i$ we may even replace the upper summation limit $n - 1$ by ∞, since $z_i'(0) = 0$ for all $i > n$.

From now on we shall proceed under far more special conditions than needed, in order to present the idea undisturbed by technical details.

Definition For $a \in \mathbb{C}$, $\varepsilon > 0$, $R > 0$ the triple (a,ε,R) satisfies condition C iff

(α) The disk D, given by

$$|w-a| \leq \varepsilon$$

is a convergence region for $K(z_n/1)$.

(β) The disk

$$|w-\Gamma| \leq R$$

contains all possible values of $K(z_n/1)$, $z_n \in D$.

(γ) For any continued fraction $K(z_n/1)$, $z_n \in D$ we have

$$\lim_{N\to\infty} \Big(\frac{z_1}{1} \; \frac{z_2}{+\,1} \; + \dots \frac{z_N}{1+\Gamma}\Big) = \mathop{K}_{n=1}^{\infty} \frac{z_n}{1} \quad .$$

Here $\Gamma = K(a/1) = \frac{1}{2}[\sqrt{1+4a}-1]$, $\mathrm{Re}\sqrt{} > 0$ (see (2.2)).

Remark Under very mild conditions (γ) follows from (α). From formula (1.2) in [10] and the argument used in the proof of Lemma 2.3 in the same paper it follows for instance that for all sufficiently small ε, depending upon a, (γ) follows from (α). For simplicity we keep (γ) as part of the definition of condition C.

Theorem 2 Let (a,ε,R) satisfy condition C, let $0 < \varrho < \varepsilon$ and $|\delta_n| \leq \varrho$ for all n. Then

$$\left| \mathop{K}_{n=1}^{\infty} \frac{a+\delta_n}{1} - \Gamma - \frac{1}{1+\Gamma} \cdot \sum_{n=0}^{\infty} \left(\frac{-\Gamma}{1+\Gamma}\right)^n \cdot \delta_{n+1} \right| \leq \frac{R\left(\frac{\varrho}{\varepsilon}\right)^2}{1-\frac{\varrho}{\varepsilon}} . \tag{2.7}$$

Proof Let G be holomorphic in the unit disk $|z| < 1$, and let $|G(z) - G(0)| \leq R$ there. Let $0 < r < 1$. Then

$$|G(z)-G(0)-G'(0).z| \leq \frac{Rr^2}{1-r} \quad \text{in } |z| \leq r. \tag{2.8}$$

This follows immediately from Cauchy's integral formula.

Let furthermore for any z in the closed unit disk and any $n \geq 1$

$$G_n(z) := \frac{a+\delta_1(\frac{\varepsilon}{\varrho}z)}{1} + \frac{a+\delta_2(\frac{\varepsilon}{\varrho}z)}{1} + \dots + \frac{a+\delta_n(\frac{\varepsilon}{\varrho}z)}{1+\Gamma} . \tag{2.9}$$

The conditions imply pointwise convergence of $\{G_n\}$ to $K((a+\delta_n(\frac{\varepsilon}{\varrho})z/1))$ for all z in $|z| \leq 1$. Since all G_n are rational functions with values in $|w-\Gamma| \leq R$ they are all holomorphic and uniformly bounded. Thus $\{G_n\}$ is normal. By Stieltjes-Vitali's theorem $\{G_n\}$ converges uniformly on compact subsets of $|z| < 1$ to the holomorphic function G,

$$G(z) = \mathop{K}_{n=1}^{\infty} \frac{a+\delta_n(\frac{\varepsilon}{\varrho}z)}{1} , \tag{2.10}$$

in particular $G(\frac{\varrho}{\varepsilon}) = K((a+\delta_n)/1)$. From (2.9) and (2.6) follows

$$G_n'(0) = \frac{1}{1+\Gamma} \sum_{\nu=0}^{n-1} \left(\frac{-\Gamma}{1+\Gamma}\right)^\nu \delta_{\nu+1} \cdot \frac{\varepsilon}{\varrho} ,$$

and hence

$$G'(0) = \frac{1}{1+\Gamma} \sum_{\nu=0}^{\infty} \left(\frac{-\Gamma}{1+\Gamma}\right)^\nu \delta_{\nu+1} \frac{\varepsilon}{\varrho} ,$$

With $z = r = \frac{\varrho}{\varepsilon}$ in (2.8) we get (2.7), and Theorem 2 is thus proved.

Remarks: For fixed a, ε, R the bound in (2.7) depends only upon ϱ, not the individual continued fraction. The bound may be replaced by $K\varrho^2$. In the Worpitzky case, i.e. $a = \Gamma = 0$, $\varepsilon = \frac{1}{4}$, $R = \frac{1}{2}$, any value > 8 will do as K, for all sufficiently small ϱ. It is well known, that for any $a \notin (-\infty, -\frac{1}{4}]$ there is an $\varepsilon > 0$ and an R, such that the condition C is satisfied [3]. From [3] follows also rather simply an extension of Theorem 2 to the case when the elements are periodically located in certain disks.

3. Approximation of limit regions

Let E be a convergence region for $K(z_n/1)$. Assume that for some $a \in E$ the set E is contained in a closed ϱ-disk centered at a, and that (a, ε, R) satisfies the C-condition for some $\varepsilon > \varrho$. The following procedure, based upon Theorem 2, can be used to determine a set L^*, such that

$$L^* \approx L \text{ (such that error estimate is known)},$$

and

$$L^* \supseteq L .$$

We recall that L is the set of values of $K(z_n/1)$, $z_n \in E$.

Step 1: Compute $\Gamma = \frac{1}{2}[\sqrt{1+4a}-1]$ and $\frac{-\Gamma}{1+\Gamma}$.
Determine the set

$$H := \{ \sum_{n=0}^{\infty} \left(\frac{-\Gamma}{1+\Gamma}\right)^n \delta_{n+1} \mid a+\delta_{n+1} \in E\} . \tag{3.1}$$

Step 2: Determine the set

$$L_* = \Gamma + \frac{1}{1+\Gamma} H \tag{3.2}$$

Step 3: Cover L_* by all $K\varrho^2$-disks, centered at points in L_*, $K\varrho^2$ being a fixed upper bound for the error term in Theorem 2. Then L^* = the union of all the disks.

To determine H is in most cases non-trivial, but is computationally much simpler than direct continued fraction approaches (substantially fewer operations).

We shall illustrate step 1 and 2, i.e. the determination of L_*, on two examples.

<u>Example 1</u> E = closed ϱ-disk, centered at a, completely contained in the complement of $(-\infty, -\frac{1}{4}]$.

A simple argument on (3.1) shows that H is the disk

$$|w| \leq \frac{\varrho}{1-|\frac{\Gamma}{1+\Gamma}|},$$

and L_* is the disk

$$|w-\Gamma| \leq \frac{\varrho}{|1+\Gamma| - |\Gamma|}.$$

<u>Example 2</u> E is the line segment from a to $a + \varrho e^{i\alpha}$, completely in the complement of $(-\infty, -\frac{1}{4}]$. In this case H is the set of values $\varrho e^{i\alpha} \cdot \sum_{n=0}^{\infty} (\frac{-\Gamma}{1+\Gamma})^n t_{n+1}$, where all t_k vary independently in the interval [0,1]. We shall restrict ourselves to three very simple cases, to indicate the flavor of the problem.

<u>Case 1</u> $0 < \frac{-\Gamma}{1+\Gamma} < 1$

Here $-\frac{1}{2} < \Gamma < 0$, and $-\frac{1}{4} < a < 0$. A straightforward argument shows that L_* is the line segment from

$$\Gamma \text{ to } \Gamma + \frac{\varrho e^{i\alpha}}{1+2\Gamma}.$$

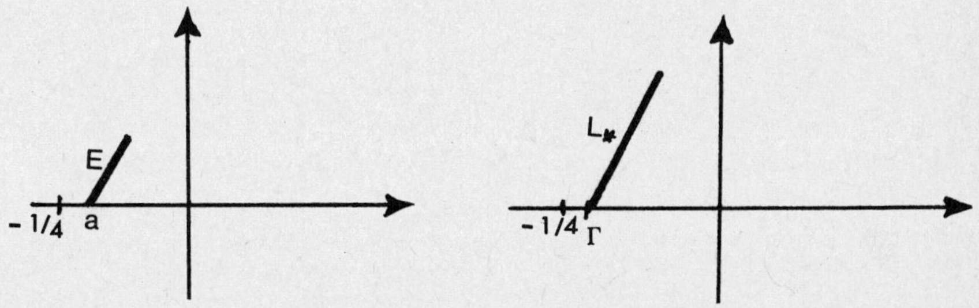

<u>Case 2</u> $\quad -1 < \frac{-\Gamma}{1+\Gamma} < 0$

Here $\Gamma > 0$ and $a > 0$. In this case it is of advantage to split the H-sum into two parts

$$\sum_{m=o}^{\infty} \left(\frac{-\Gamma}{1+\Gamma}\right)^{2m} . t_{2m+1} + \sum_{m=o}^{\infty} \left(\frac{-\Gamma}{1+\Gamma}\right)^{2m+1} . t_{2m+2} \quad ,$$

with ranges $[0, \frac{1}{1-\left(\frac{-\Gamma}{1+\Gamma}\right)^2}]$ and $[\frac{\frac{-\Gamma}{1+\Gamma}}{1-\left(\frac{-\Gamma}{1+\Gamma}\right)^2}, 0]$.

L_* is the line segment from $\Gamma - \frac{\Gamma}{1+2\Gamma}\varrho e^{i\alpha}$ to $\Gamma + \frac{1+\Gamma}{1+2\Gamma}\varrho e^{i\alpha}$.

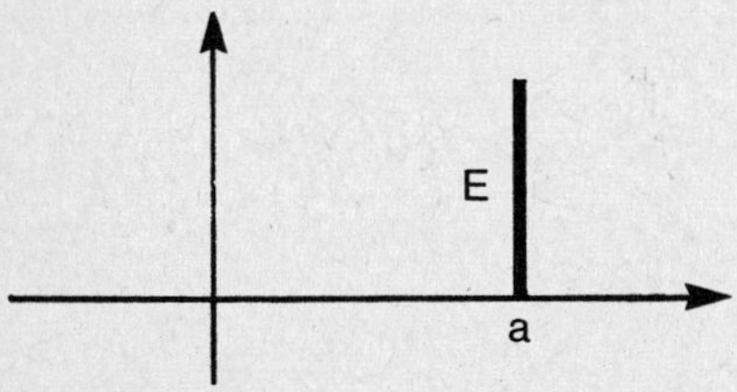

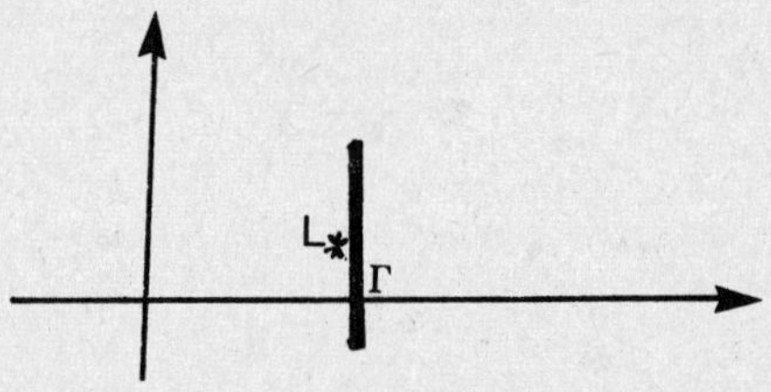

Observe that L_* here has length ϱ, the same as E.

<u>Case 3</u> $\quad \frac{-\Gamma}{1+\Gamma} = re^{\frac{2\pi i}{3}} \quad , \quad 0 < r < 1$

In this case it is of advantage to split the H-sum into three parts, and a simple argument shows that with $H = H'\rho e^{i\alpha}$ we have

$$H' = \{\frac{\tau_o + \tau_1 re^{\frac{2\pi i}{3}} + \tau_2 r^2 e^{\frac{4\pi i}{3}}}{1-r^3} \mid 0 \leqslant \tau_k \leqslant 1\} \quad .$$

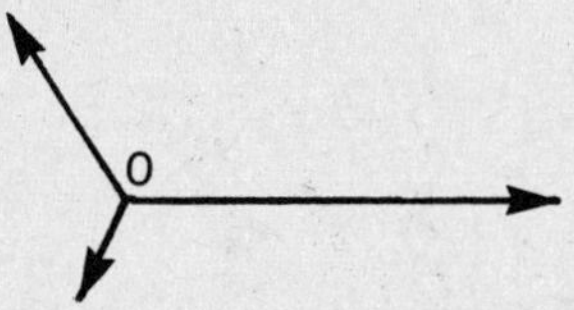

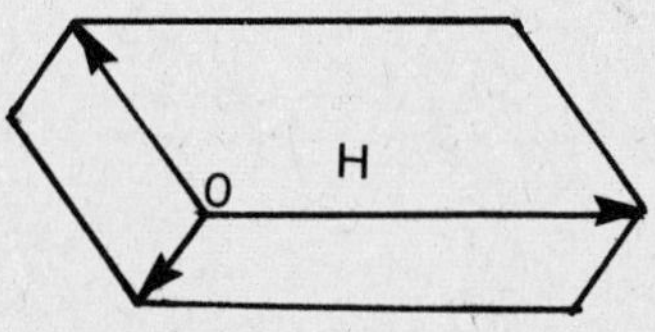

$$L_* = \Gamma + \frac{\varrho e^{i\alpha}}{1+\Gamma} H'$$

In the illustration below the computer drawing of L is <u>not</u> based upon Theorem 2, but on a certain strategy for computing limit regions [2].

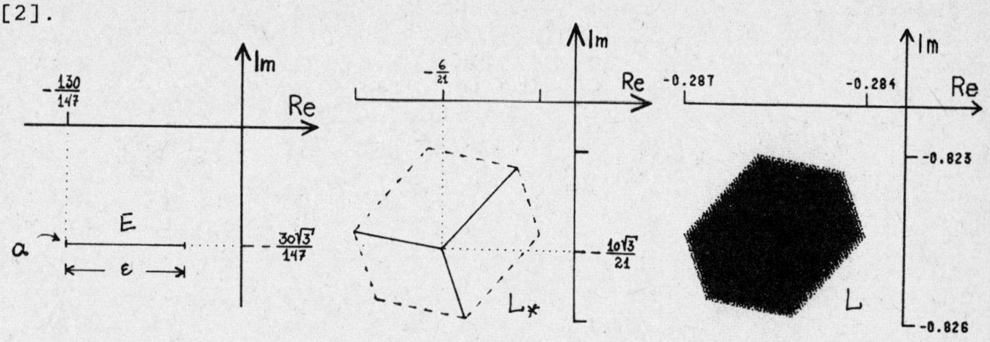

4. Application for convergence acceleration

Following standard notation we shall for a given continued fraction $K(z_n/1)$ define

$$S_n(w) := \frac{z_1}{1} + \frac{z_2}{1} + \ldots + \frac{z_n}{1+w} \quad . \tag{4.1}$$

Convergence of the continued fraction to f means convergence of $\{S_n(0)\}$ to f. Often $\{S_n(w_n)\}$ for some known (or easy to determine) sequence $\{w_n\}$ converges to f <u>faster</u> than $\{S_n(0)\}$, see for instance [10], [4]. The simplest case is when $z_n \to a \notin (-\infty, -\frac{1}{4}]$. Then it is well known that for $f \neq \infty$ and $a \neq 0$ we have

$$\frac{f-S_n(\Gamma)}{f-S_n(0)} \to 0 \quad , \tag{4.2}$$

Γ defined in (2.2).

Let $\{\varrho_N\}$ be a monotone sequence of positive numbers with $\varrho_N \to 0$, and $\{\delta_N\}$ a sequence of complex numbers with $|\delta_N| \leq \varrho_N$. Let ϱ_0 be sufficiently small to make the C-condition hold for (a, ε, R) for some $\varepsilon > \varrho_0$. Then it follows from Theorem 2 that for some $K > 0$ the tail values $f^{(N)}$ of $K((a+\delta_n)/1)$ satisfy the set of inequalities:

$$|f^{(N)} - \Gamma - \frac{1}{1+\Gamma} \cdot \sum_{n=0}^{\infty} (\frac{-\Gamma}{1+\Gamma})^n \cdot \delta_{N+n+1}| \leq K\varrho_N^2 \quad . \tag{4.3}$$

Put

$$\tilde{f}^{(N)} = \Gamma + \frac{1}{1+\Gamma} \sum_{n=0}^{\infty} \left(\frac{-\Gamma}{1+\Gamma}\right)^n \delta_{N+n+1} . \tag{4.4}$$

From a well known formula for $f - S_N(w_N)$, see for instance [10, p. 156], we find

$$\frac{f-S_N(\tilde{f}^{(N)})}{f-S_N(\Gamma)} = \frac{h_N+\Gamma}{h_N+\tilde{f}^{(N)}} \cdot \frac{f^{(N)}-\tilde{f}^{(N)}}{f^{(N)}-\Gamma} , \tag{4.5}$$

where

$$h_N = 1 + \frac{a_N}{1} + \frac{a_{N-1}}{1} + \ldots + \frac{a_2}{1} .$$

A complete discussion is beyond the scope of the present paper. We shall briefly indicate what is "normally" to be expected: Under the present conditions we have for sufficiently small ε that $h_N \to 1 + \Gamma$, see e.g. [11, Thm. 2.3]. Since $\Gamma \neq -\frac{1}{2}$ the factor $(h_N+\Gamma)/(h_N+f^{(N)}) \to 1$. For the rightmost factor we find

$$\frac{f^{(N)}-\tilde{f}^{(N)}}{f^{(N)}-\Gamma} = \frac{\frac{R_N}{\varrho_N}}{\frac{1}{1+\Gamma} \sum_{n=0}^{\infty} \left(\frac{-\Gamma}{1+\Gamma}\right)^n \frac{\delta_{N+n+1}}{\varrho_N} + \frac{R_N}{\varrho_N}} .$$

Here $\left|\frac{R_N}{\varrho_N}\right| \leq K\varrho_N$, hence $\frac{R_N}{\varrho_N} \to 0$. Under mild conditions the first sum in the denominator is bounded away from zero, in which case the ratio (4.5) tends to 0 at the same rate as ϱ_N. From [10] we know that the ratio in (4.2) tends to zero at the same rate. Thus the transition $S_N(\Gamma) \rightsquigarrow S_N(\tilde{f}^{(N)})$ represents roughly the same order of improvement of convergence as the transition $S_N(0) \rightsquigarrow S_N(\Gamma)$. We conclude this section by presenting three numerical examples, with ordinary approximants, Γ-modified approximants and $\tilde{f}^{(N)}$-modified approximants.

<u>Example 3</u> The continued fraction

$$\frac{6+2^{-1}}{1} + \frac{6+2^{-2}}{1} + \frac{6+2^{-3}}{1} + \ldots$$

has the value f = 2.12283, correctly rounded in the 5th decimal place. The N-value indicated in the table is the smallest for which the nth approximant, rounded in the 5th place, takes this value for all $n \geq N$:

	For ordinary approx. $S_N(0)$	For Γ-modified approx. $S_n(\Gamma) = S_N(2)$	For $\tilde{f}_N$-modified approx. $S_N(\tilde{f}) = S_N(2+1/2^{n+3})$
N	34	10	4

Example 4

$$\frac{56+2^{-1}}{1} + \frac{56+2^{-2}}{1} + \frac{56+2^{-3}}{1} + \ldots = 7.04338 \quad ,$$

rounded as above. The N-values for the three types of approximants are here

114 12 3 .

Example 5

$$\mathop{K}_{n=1}^{\infty} \frac{6+n^{-1}}{1} = 2.2474, \text{ rounded in the 4th place.}$$

The three N-values are here

28 15 5 .

5. Final remarks

In the present paper we have restricted the discussion to the very simplest case, in order to present the idea essentially undisturbed by technical details. There are several natural extensions, the most natural one being the study of c. f. $K(z_n/1)$ "near" a given continued fraction $\mathop{K}_{n=1}^{\infty}(a_n/1)$ whose value and tail values are known, in particular cases when $a_n \to \infty$. Since derivatives are established also for c. f. $K(z_n/u_n)$, extensions to continued fractions of this type are also possible. Perhaps some of the ideas are extendable to n-fractions or to branched continued fractions.

Acknowledgement The author is indebted to Stephan Ruscheweyh for calling his attention to the paper [1] , and to Lisa Jacobsen for encouragement and constructive criticism.

References

1. Atkinson, F. V., A value-region problem occuring in the theory of continued fractions, MRC Technical Summary Report # 419, December 1963, Madison, Wisconsin.

2. Istad, R. M., Om limitområder og strategier for numerisk bestemmelse av dem. Thesis in preparation (Norwegian).

3. Jacobsen, L., Some periodic sequences of circular convergence regions, Lecture Notes in Mathematics, Vol. 932, Springer Verlag, Berlin, Heidelberg, New York 1982, pp. 87-98.

4. Jacobsen, L., Convergence acceleration for continued fractions $K(a_n/1)$, Transactions of the American Mathematical Society, Vol. 275, Number 1, January 1983.

5. Jacobsen, L., Further results on convergence acceleration for continued fractions $K(a_n/1)$. Transactions of the American Mathematical Society, Vol, 281, Number 1, January 1984.

6. Jacobsen, L., Nearness of continued fractions, Math. Scand. To appear.

7. Jacobsen, L. and Thron, W. J., Element regions belonging to circular limit regions. In preparation.

8. Jones, W. B. and Thron, W. J., Continued Fractions: Analytic Theory and Applications. Encyclopedia of Mathematics and its Applications, Vol. 11, Addison-Wesley, Reading, Mass., 1980.

9. Rye, E. and Waadeland, H., Reflections on value regions, limit regions and truncation errors for continued fractions, Numerische Mathematik. To appear.

10. Thron, W. J. and Waadeland, H., Accelerating convergence of limit periodic continued fractions $K(a_n/1)$, Numerische Mathematik 34, 155-170 (1980).

11. Thron, W. J. and Waadeland, H., Truncation error bounds for limit periodic continued fractions, Mathematics of Computation, Volume 40, Number 162, April 1983, pp. 589-597.

12. Waadeland, H., A note on partial derivatives of continued fractions, Lecture Notes in Mathematics, Springer-Verlag. To appear.

A STIELTJES ANALYSIS OF THE $K^{\pm}p$ FORWARD ELASTIC AMPLITUDE

J. Antolin and A. Cruz

Departamento de Física Teórica. Facultad de Ciencias.
Universidad de Zaragoza. 50009 Zaragoza (SPAIN)

ABSTRACT

The positivity hypothesis on an unknown function $\chi^*(x)$, related to the imaginary part of the $K^{\pm}p$ scattering amplitude on the unphysical region, allows the construction of a Stieltjes function $H(z)$, known in a discrete set of real points and affected by errors owing to experimental measurements.

The Stieltjes character of $H(z)$ imposes constraints on the coefficients of its formal expansion which limit the universe of approximant functions, so acting as stabilizers of the analytic extrapolation.

The Pade approximants (P.A.) to $H(z)$, built with the coefficients of the formal expansion, provide rigorous bounds on the function in the cut complex plane.

These bounds on $H(z)$ can be translated to the $K^{\pm}$ amplitude, $F_{\pm}(\omega)$, obtaining bounds on the coupling constants $g^2_{KN\Lambda}$ and $g^2_{KN\Sigma}$.

Taking advantage of the fact that P.A. are valid for complex values of z, the position of the complex conjugate zeros of the amplitude has also been calculated.

The consistency of the calculated real part has been successfully checked by taking different absorption points with the latter values of real parts.

The stability of the method of extrapolation has been confirmed using a model function, whose analytical structure is perfectly known, perturbed randomly according to the experimental errors.

The addition of the hypothesis of unimodality of $\chi^*(x)$ provides tighter rigorous bounds on H(z) on the cut complex plane and the obtention of upper and lower moment sequences of $\chi^*(x)$ allowed by our two general hypotheses.

The inversion of these moment sequences using a Stieltjes-Tchebycheff technique allows the calculation of the scattering amplitude $F_{\pm}(\omega)$ even on the unphysical cut, so achieving the rational parametrization of the amplitude in the whole ω complex plane.

1. INTRODUCTION

The analytic continuation has been widely used in high energy physics: once experimental data have been measured in a certain domain of the complex plane, a fitting procedure may be used to interpolate the data, and the resulting parametrization is extrapolated to regions where data have not been obtained yet, or, in some cases, where data are not physically accessible.

However, for direct analytic continuation to be used properly, the function concerned must be exactly known in some continuum and this is never the case in experimental physics where data have some statistical errors and are measured at discrete points.

These facts make the task of analytic continuation, using experimental data,impossible, because there is an arbitrary number of parametrizations which, agreeing in the experimental region, give absolutely different values when extrapolated to other regions.

This instability in analytic extrapolation forces the search of other properties, besides analycity, which act as stabilizers of the analytic continuation in such a way that small perturbations in the data region do not give rise to very different predictions outside the experimental region.

In other words we have to limit the number of admissible parametrizations by adding some information or constraints on the type of functions we can use.

Several methods exist of stabilization of the analytic extrapolation [1] . We shall use, besides analycity, two properties of the functions ,positivity and unimodality, which allows the use of bounding and convergence properties of P.A. and some properties of the Hausdorff moment sequences, to produce a stable extrapolation.

This analytic extrapolation method is applied to the analysis of the $K^{\pm}p$ amplitude, which has become the testing ground for analytic extrapolation methods owing to its long and checkered history [2] .

2. THE $K^{\pm}p$ FORWARD ELASTIC SCATTERING AMPLITUDE.

The $K^{\pm}p$ forward elastic scattering amplitude can be represented by an analytic function of the complex laboratory kaon energy, ω ,$F_{\pm}(\omega)$, satisfying the Schwartz reality condition and supposed to be asymptotically polynomically bounded.

Unitarity predicts the analytic structure of $F_-(\omega)$ consisting of two poles at the (unphysical) values ω_Λ, ω_Σ, corresponding to the K^-p system having the mass of the hyperons Λ and Σ respectively, a left-hand cut going from $-m_K$ to $-\infty$ (K^+p scattering), and a right-hand cut from $\omega_{\Lambda\pi}$ to ∞(K^-p scattering). The latter cut has an unphysical region from $\omega_{\Lambda\pi}$ (the unphysical energy corresponding to the K^-p system having the invariant mass $m_\Lambda + m_\pi$) to $\omega = m_K$.

By applying Cauchy's integral theorem one can write a once subtracted dispersion relation at $\omega = \omega_o$

$$\frac{\mathrm{Re}\, F_-(\omega)}{\omega - \omega_o} = \frac{\mathrm{Re}\, F_-(\omega_o)}{\omega - \omega_o} + \sum_Y \frac{X_Y}{(\omega_Y - \omega)(\omega_Y - \omega_o)} - \frac{P}{\pi}\int_{m_K}^{\infty} \frac{\mathrm{Im}F_+(\omega')\,d\omega'}{(\omega'+\omega)(\omega'+\omega_o)} + \frac{P}{\pi}\int_{\omega_{\Lambda\pi}}^{\infty} \frac{\mathrm{Im}F_-(\omega')\,d\omega'}{(\omega'-\omega)(\omega'-\omega_o)}, Y=\Lambda,\Sigma \tag{2.1}$$

Then we define a discrepancy function [3]

$$\Delta_-(\omega) = \frac{1}{\pi}\int_{\omega_{\Lambda\pi}}^{m_K} \frac{\mathrm{Im}F_-(\omega')\,d\omega'}{(\omega'-\omega_o)(\omega'-\omega)} + \sum_Y \frac{X_Y}{(\omega_Y-\omega)(\omega_Y-\omega_o)}, Y = \Lambda,\Sigma \tag{2.2}$$

which has the analytic structure of fig. 1, and the dispersion relation can be written in the form

$$\Delta_-(\omega) = \frac{\mathrm{Re}F_-(\omega) - \mathrm{Re}F_-(\omega_o)}{\omega - \omega_o} + \frac{P}{\pi}\int_{m_K}^{\infty} \frac{\mathrm{Im}F_+(\omega')\,d\omega'}{(\omega'+\omega)(\omega'+\omega_o)} - \frac{P}{\pi}\int_{m_K}^{\infty} \frac{\mathrm{Im}F_-(\omega')\,d\omega'}{(\omega'-\omega)(\omega'-\omega_o)} \tag{2.3}$$

which allows the evaluation of $\Delta_-(\omega)$ in those points where $\mathrm{Re}F_-(\omega)$ has been measured. Δ_- is known, with errors, in 218 points, 119 on the K^+p cut and 99 on the K^-p cut (fig. 1).

The integrals in (2.3) are evaluated using the optical theorem, which relates $\mathrm{Im}F_-$ to the experimentally measured total $K^{\pm}p$ cross-section. Conventionally, the residues X_Y in (2.2) are parametrized as

$$X_Y = G_Y^2 \frac{(m_Y - m_p)^2 - m_K^2}{4m_p^2}, \quad G_Y^2 = g^2_{KNY} \qquad (2.4)$$

where G_Y is the so called coupling constant.

Our problem is to extrapolate the known values of $\Delta_-(\omega): \Delta(\omega_i) \pm \Delta^e(\omega_i), i = 1,218$, to the following regions :

a) The positions of the Λ and Σ poles in order to calculate their residues easily related to G^2_Λ and G^2_Σ

b) The point $\omega = 0$, to calculate $F_-(0)$ which is an important parameter related to the chiral symmetry breaking term [4]

c) We want to find the positions of the real and complex zeros of the $F_-(\omega)$ function

d) We are also interested in the values of the amplitude on the unphysical cut: $[\omega_{\Lambda\pi}, m_K]$

The main difficulties in the calculation of the coupling constants and other parameters are the instability of the extrapolation methods [1,3,5], the unphysical cut, experimentally inaccesible to the K^-p channel, and dominated by the Y^*_{1405} resonance and the proximity of the Λ and Σ poles.

Then we have two possibilities: use an approximation and consider only one reduced pole at ω_Λ with only one reduced coupling constant $G^2 = G^2_\Lambda + 0.9\, G^2_\Sigma$ to account for both physical poles, or try and separate both contributions and calculate simultaneously G^2_Λ and G^2_Σ

The values for G^2 extrapolated by several authors range from 6 to 22 [5] giving a measure of the instability in the analytic extrapolation referred to above, and there is no (model independent) simulta-

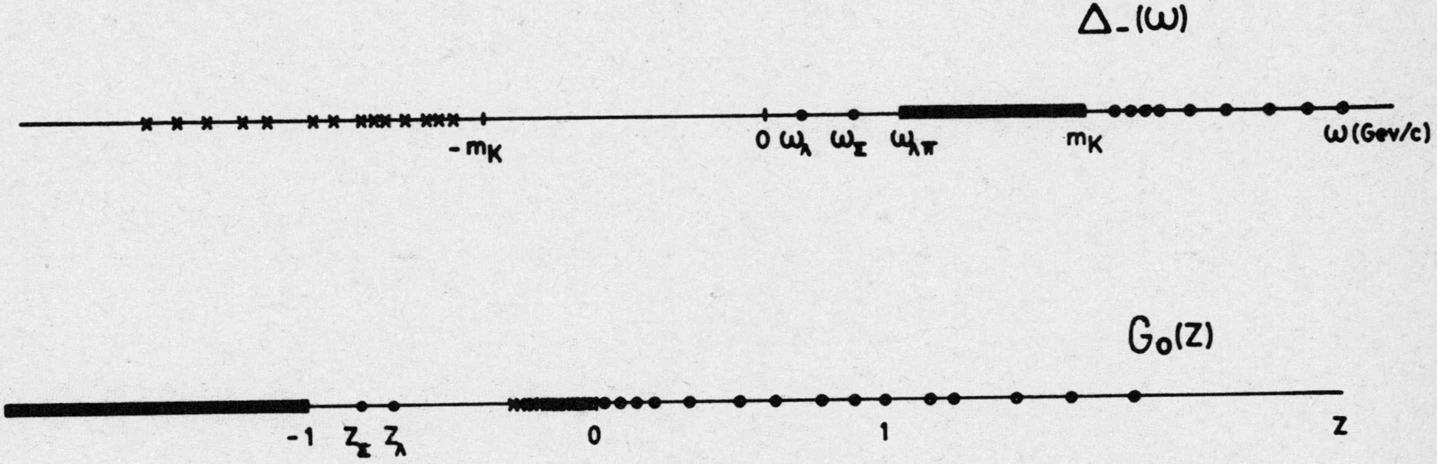

Figure 1 Analytic structure of the Δ_- discrepancy function in the kaon laboratory energy plane, ω , and of the function G_0 in the transformed z plane. Crosses and blank points indicate the experimental zone in the K^+p and K^-p regions respectively and their location in the z plane.

neous calculation of G_{Λ}^{2} and G_{Σ}^{2} .

In order to stabilize the analytic extrapolation, to reduce these discrepancies in the G^2 calculations and to separate the contributions of both poles we are going to use, besides analycity, two general hypotheses on the imaginary part of the amplitude on the unphysical cut: positivity and unimodality ($\mathrm{Im}F_{-}(\omega')$ has a unique maximum on the physical cut). Both hypotheses are supported by all the low energy models, [6] so our calculations are only based on experimental data, the analytic properties of the amplitude and these two general hypotheses.

3. DISCREPANCY FUNCTIONS AND STIELTJES FUNCTIONS. THE GRONWAL TRANSFORMATIONS.

The basic idea is to transform Δ_{-} into a Stieltjes function and then construct the P.A. by using the experimental data.

By means of the transformations

$$x(\omega') = \frac{\omega' - m_K}{\omega_{\Lambda\pi} - m_K}, \qquad z(\omega) = \frac{m_K - \omega_{\Lambda\pi}}{\omega - m_K} \tag{3.1}$$

the discrepancy function is turned into

$$G_o(z) = -\frac{\Delta_{-}(z)}{z} = \int_o^1 \frac{\chi(x)\,dx}{1 + xz} + \frac{R_1}{z + \varepsilon_1} + \frac{R_2}{z + \varepsilon_2} \tag{3.2}$$

$$\chi(x) = \frac{\mathrm{Im}F_{-}(\omega')}{\pi(\omega' + m_K)} \geq o\,,\; x \in [o,1]\,,\;\; \omega_o = -m_K \tag{3.3}$$

$$R_1 = -0.222\; G_{\Lambda}^2 \quad \text{and } R_2 = -0.196\; G_{\Sigma}^2 \qquad (\mathrm{Gev/c})^{-2} \tag{3.4}$$

Where $-\varepsilon_1 = z_{\Lambda} = (-x_{\Lambda})^{-1} = -0.6544$ and $-\varepsilon_2 = z_{\Sigma} = (-x_{\Sigma})^{-1} =$ $= -0.80729$ are the pole positions in the z-plane (See fig. 1)

Using now the expression

$$G_{K+1}(z) = \frac{G_K(z)\;(z+\varepsilon_{K+1}) - G_K(z_{K+1})(z_{K+1} + \varepsilon_{K+1})}{z - z_{K+1}}\,, K = 0,1 \tag{3.5}$$

we can absorb the two pole terms by choosing a couple of absorption points, z_1 and z_2. These are points where we know the value of $G_o(z)$.

As one can see the method is independent of the number of poles one has, then in the reduced coupling constant case we have to use only $K = 0$ to absorb the unique reduced pole at $z = -\varepsilon_1 = z_\Lambda$. In this case we have

$$G_1(z) = \int_o^1 \frac{\chi_1(x)}{1+xz} \qquad \chi_1(x) = \chi(x)\,\frac{1-\varepsilon_1 x}{1+z_1 x} > o \tag{3.6}$$

and in this case of two poles

$$G_2(z) = \int_o^1 \frac{\chi_2(x)}{1+xz} \qquad \chi_2(x) = \chi(x)\,\frac{1-\varepsilon_1 x}{1+z_1 x}\,\frac{1-\varepsilon_2 x}{1+z_2 x} \tag{3.7}$$

$$\text{where} \quad 0 < \varepsilon_1 < \varepsilon_2 < 1 \quad \text{and} \quad z_1,\ z_2 > -1 \tag{3.8}$$

G_1 and G_2 are now pure Stieltjes functions known with errors in 218-1 and 218-2 points respectively. We call in general $H(z)$ these Stieltjes functions and $\chi^*(x)$ the corresponding weight function which has the same positivity and unimodality properties as $\mathrm{Im}F_-(\omega)$ on the unphysical cut.

The formal series expansion of the Stieltjes function

$$H(z) = \int_o^1 \frac{\chi^*(x)\,dx}{1+xz} = \sum_{n=o}^{\infty} h_n(-z)^n \tag{3.9}$$

$$h_n = \int_o^1 \chi^*(x)\, x^n dx$$

is divergent in most of the points where H is known, therefore one cannot determine the coefficients by a fit of the series to the known values.

Instead, we transform the cut z-plane into the unit circle $|s| \leq 1$ by means of the conformal transformation

$$s = \frac{\sqrt{1+z}-1}{\sqrt{1+z}+1} \tag{3.10}$$

which unfolds the cut in the z-plane into the unit circumference, and the z_i values that range from -0.3 to 11.73 map into a real interval close to the origin from $s = -0.076$ to 0.56.

The function $G(z) = H(z)\sqrt{1+z}$, having the same analytical properties as $H(z)$, can be expanded in series in s,

$$G(z(s)) = \sum_{i=0}^{\infty} g_i\, s^i \qquad (3.11)$$

and its coefficients g_i can be determined by fitting the experimental values $G(z_i)$ with errors $G^e(z_i)$ by the least squares method. The importance of these coefficientes g_i lies in that they are related to the coefficients h_i of the series expansion of $H(z)$ by the remarkable formula, obtained by Gronwall [7]

$$h_p = 4^{-p} \sum_{r=o}^{p} \binom{2p}{p-r}\, g_r (-1)^r \qquad (3.12)$$

which allows us, once the p first coefficients are known, to calculate the p first coefficients of the expansion (3.9).

The advantage the present procedure has over any other possible conformal transformation is that the h_i´s so determined are the coefficients of a Stieltjes series or, in other words, the moments of a positive function.

Conversely, in the next section we shall introduce the constraints the coefficients h_i must fulfill due to their being the moments of a positive function. [8,9]

4. CONSTRAINTS IMPOSED BY POSITIVITY AND UNIMODALITY

a) Positivity

Given the coefficients h_i we construct the table of differences

$$\Delta_p^o = h_p \qquad p = 0,1... \qquad (4.1)$$

$$\Delta_p^k = \Delta_p^{k-1} - \Delta_{p-1}^{k-1} \qquad k = 1,... \quad p = o,1,... \qquad (4.2)$$

and due to h_i being the moments of a positive function, the following inequalities must be satisfied

$$H^m_n \; (\Delta^k) > 0 \qquad (4.3)$$

H^m_n being the Hankel determinants [8] .

In our case, h_i being determined from the experimental data with errors, h^e_i, we shall use (4.3) to obtain bounds that the h_i should satisfy if they are the moments of a positive function. In this way, using the Hankel determinants of the first line, $\Delta^o_p = h_p$, we get lower bounds for the coefficient h_p given the previous $h_o \ldots h_{p-1}$.

The same procedure as with row 1 can be followed with the rest of table Δ^k_m , yelding lower bounds for Δ^k_m with increasing m, which translate into lower bounds for h_m for even k, and upper bounds for h_m for odd k.

A subroutine has been designed which, given the m first coefficients $\Delta^o_o \ldots \Delta^o_{m+1}$, calculates the tightest possible lower and upper bounds for the next coefficient to be one of a Stieltjes series with non null convergence radius.

b) Unimodality

Let $\beta \in [0,1]$ be the position of the maximum of $\chi^*(x)$, then , the function $\phi'(x) = (\beta - x) \dfrac{d\chi^*(x)}{dx}$ is positive in $[0,1]$ and its moments are related with those of $\chi^*(x)$ in the following way

$$\mu_\nu = \int_o^1 x^\nu \phi'(x)\,dx = (\nu + 1) h_\nu - \beta\nu h_{\nu-1} \qquad (4.4)$$

where μ_ν , being the moments of a positive function, form a totally monotonical sequence, for which the positivity property of the

Hankel determinants $H_n^m(\Delta^k)$ applies, Δ^k being the table constructed with the μ_ν .

The inequalities so obtained for the μ_ν turn into inequalities in the h_ν which are more restrictive than those obtained previously with positivity alone.

These conditions are stabilizers of the analytic extrapolation because they limit the number of possible functions fitting the data.

Now we construct with the first p coefficients the P.A. and we apply their bounding and convergence properties to Stieltjes functions in the cut complex plane [8,9,10]

By undoing the transformations and absorptions these bounds turn into bounds for the residues of the G_o function so we have upper and lower bounds on G^2 or G^2_Λ and G^2_Σ , zero positions, bounds on $F_-(0)$ etc. (see fig 2)

5. RESULTS IN THE CUT COMPLEX PLANE

Fig 3 shows the bounds obtained on G^2 using positivity and unimodality with the results for the simultaneous determination of G^2_Λ and G^2_Σ . The bounds on G^2_Λ and G^2_Σ are:

$$11.7 < G^2_\Lambda < 13.6 \qquad \text{and} \qquad 1.0 < G^2_\Sigma < 4.3 \qquad (5.1)$$

The amplitude has three zeros, a real zero between the poles at ω_R

$$0.0641 = \omega_\Lambda < 0.147 < \omega_R < 0.150 < \omega_\Sigma = 0.159 \qquad (5.2)$$

and a pair of complex conjugate zeros at $\omega_C = (0.245 \pm 0.035, \pm 0.345 \pm 0.025)$ and the value for $F_-(0)$ is: $F_-(0) = -2.8 \pm {}^{0.12}_{0.06}$ fm.

We have found a set of points which are incompatible with the positivity hypothesis and the stability of the method of analytic extrapolation has been studied using a model function perturbed randomly according to the experimental errors.

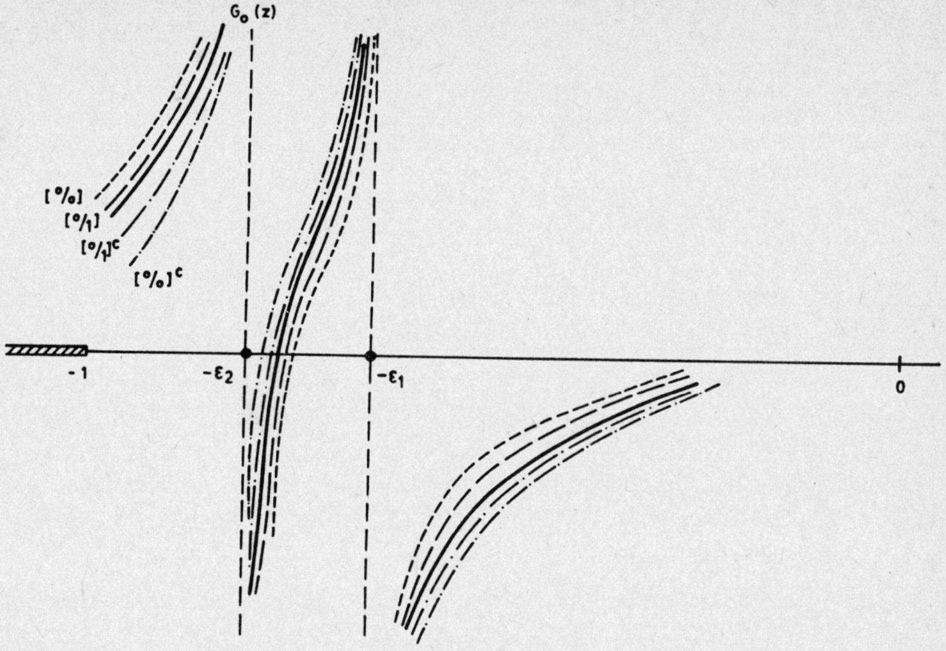

<u>Figure 2</u> Scheme of the bounding properties of [0/0] , [0/1] P.A. and their complementaries to the G_0 function in the pole region.

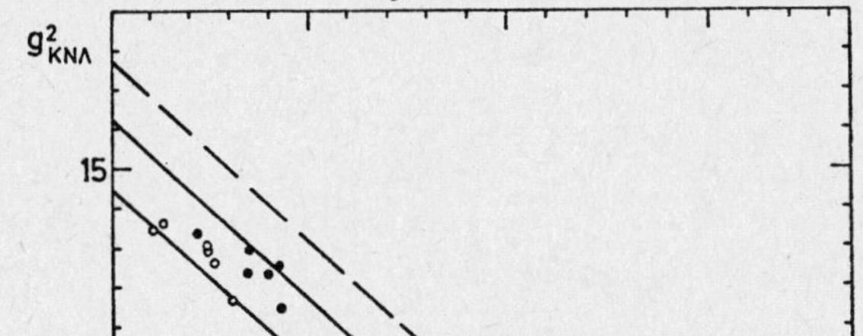

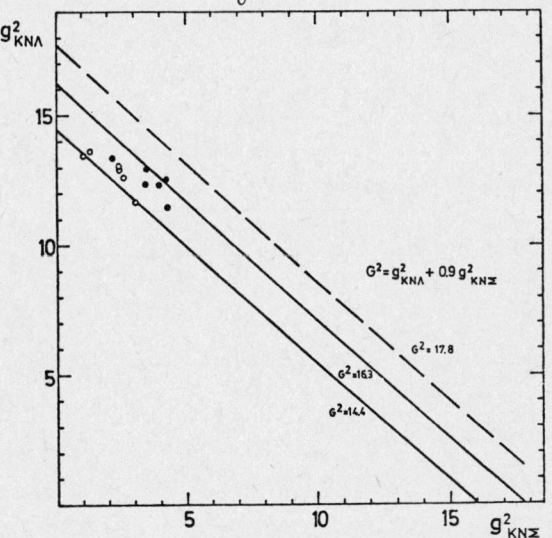

<u>Figure 3</u> Values obtained for $g^2_{KN\Lambda}$ and $g^2_{KN\Sigma}$ (6 absorption cases) and bounds obtained for G^2 with positivity (dashed line) and unimodality (continous line). The lower bound is common to both. The blank points correspond to the [1/1] P.A., for each absorption case, and the dots to the complementary P.A.

Other alternative rational approximations have been analyzed: Type II P.A. (multipoint or interpolation P.A.) and Type III P.A.(fitting P.A.) but we think that the type I P.A. have some advantage over these alternative parametrizations.

The bounding and convergence properties of P.A. fail on the real cut (the Stieltjes cut) but the orthogonality properties of Pade denominators [8,9,10] are used in the last section to get approximations to the amplitude on the unphysical cut.

6. CUT EXTRAPOLATION. STIELTJES AND TCHEBYCHEF DISTRIBUTIONS AND WEIGHTS.

We need to invert the first p calculated moments in the fitting procedure to obtain an approximation to the weight function $\chi^*(x)$.

Even if the number of calculated moments is insufficient to obtain a faithful approximation to $\chi^*(x)$ we can again use the Hankel constraints (4.3) to obtain upper and lower moment sequences allowed by our hypotheses

$$h_o, h_1 \ldots h_{p-1} \begin{array}{l} \nearrow h_p^u \longrightarrow h_{p+1}^u \longrightarrow \\ \searrow h_p^l \longrightarrow h_{p+1}^l \longrightarrow \end{array} \tag{6.1}$$

In this case we have to check if the results with both sequences are compatible.

In view of the Pade approximation: $[L/M] = \dfrac{P_L^{(N)}}{Q_M^{(N)}}$, $N = L - M$

$$\left[n-1/n\right] = \frac{P_{n-1}^{(-1)}(z)}{Q_n^{(-1)}(z)} = \sum_{i=1}^{n} \frac{f_i^{(n)}}{1+z\,\varepsilon_i^{(n)}} \sim H(z) = \int_o^1 \frac{d\psi(x)}{1+xz} = \int_o^1 \frac{\chi^*(x)\,dx}{1+xz} \tag{6.2}$$

we can obtain an order n approximation to $d\psi(x) = \chi^*(x)\,dx$:

$$d\psi_s^{(n)}(x) = \sum_{i=1}^{n} f_i^{(n)}\,\delta\,(x-\varepsilon_i^{(n)})\,dx \tag{6.3}$$

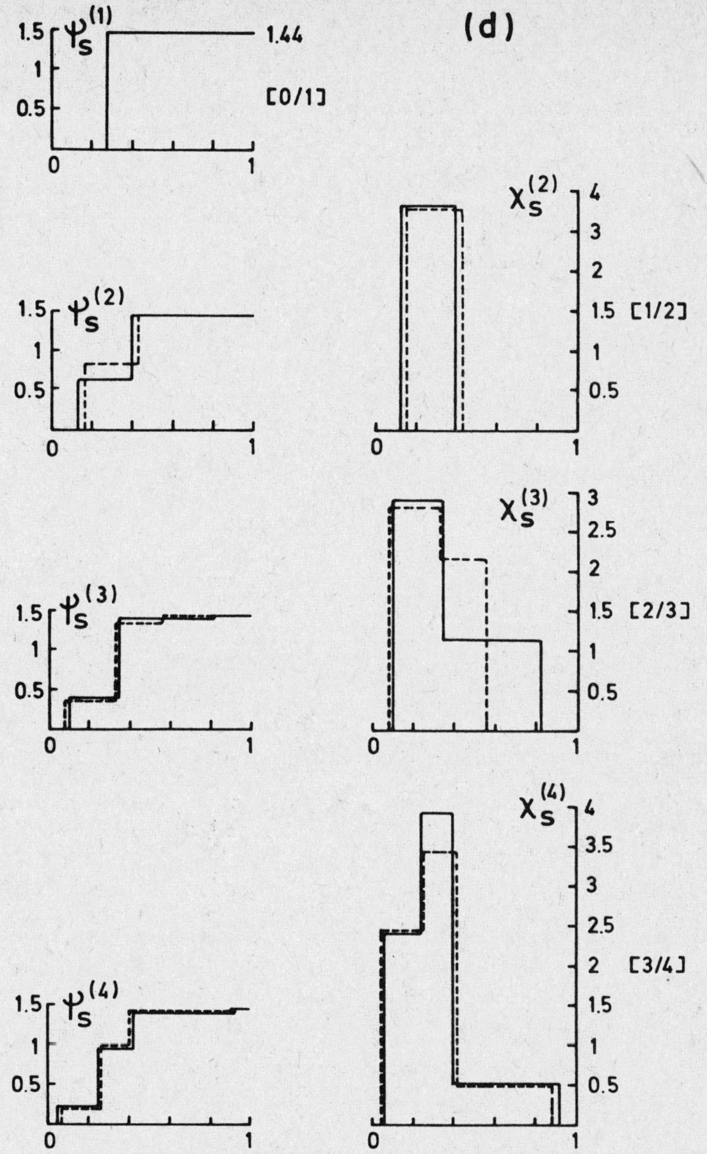

Figure 4 The succesive approximations $\Psi_s^{(n)}$ and $\chi_s^{(n)}$ (for one of the absorption cases, d) to the distribution $\Psi(x)$ and function $\chi^*(x)$ generated by the approximants $[n-1/n]$ are shown. The continous line corresponds to the upper moment sequence approximation, and the dashed line to the lower one.

which yields a step approximation $\psi_s^{(n)}(x)$ to $\psi(x)$, providing bounds for the distribution function at the points of increasing $\varepsilon_i^{(n)}$ $i = 1,\dots n$ in the form of Tchebycheff inequalities [11, 12] (See fig 4).

An approximation to $\chi^*(x)$, $\chi_s^{(n)}(x)$, is obtained from the slopes of the segments joining the mid point of the discontinuities of $\psi_s^{(n)}(x)$.

As can be seen in fig. 4, the Stieltjes distribution and weight histograms: $\psi_s^{(n)}(x)$ and $\chi_s^{(n)}(x)$ $n = 1, 2, 3, 4$ calculated with 2, 4, 6, 8 moments respectively, are strikingly similar despite the fact of having been built taking the farthest moment sequences allowed by unimodality, providing a first estimate of the shape of $\chi^*(x)$.

Satisfactory as the Stieltjes distributions and weights are in many aspects, it is desirable to have continuous approximations to $\psi(x)$ and $\chi^*(x)$ in any point of the interval $[0,1]$, not only in the points related to the poles of the approximants.

Consequently, it is convenient to consider approximations to the formal series expansion of $H(z)$ having a pole in an arbitrarilly prespecified position on the cut $]-\infty, -1]$ [12, 13]. The denominators of these new approximants are related to the system of quasi-orthogonal polynomials associated with the distribution function $\psi(x)$.

The interpolation of the mid-points of the discontinuities of the histogram approximations which now have an arbitrary spectral point in $[0,1]$ provides a continuous approximation to $\psi(x)$. Its derivative is a continuous approximation to $\chi^*(x)$.

Figure 5 shows the Tchebycheff approximations: $\psi_T^{(4)}(x)$ and $\chi_T^{(4)}(x)$ obtained by varying the position of one of the spectral points of the distribution histogram, using the upper sequences.

Taking advantage of our approximation for $\chi^*(x)$ and therefore to $\mathrm{Im}\, F_-(\omega)$ on the unphysical cut and the values for the coupling constants obtained previously, we have calculated the real parts of the K^-p amplitude on the unphysical cut obtaining in particular a

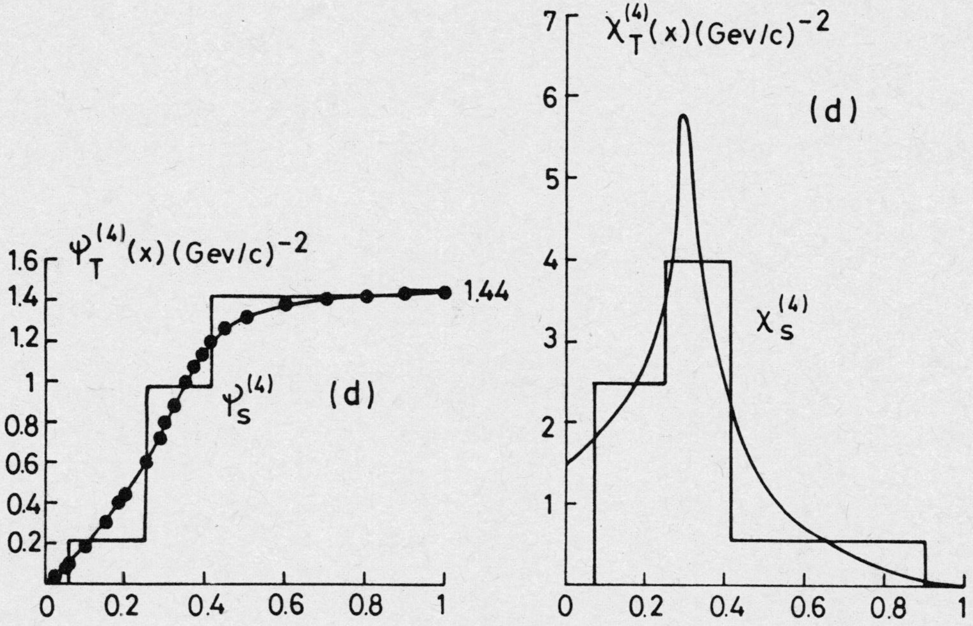

Figure 5 The left hand figure shows the interpolation of the Tchebycheff values obtained when varying the pole position in the interval [0,1] .The histogran is the [3/4] approximation. The right hand figure shows the approximation $\chi_T^{(4)}$ obtained differentiating the previous interpolation, together with the histogram approximation.

range from -1.075 to 0.865 fm for the real part of the amplitude in the elastic K^-p threshold (Fig 6) and achieving the rational parametrizatior of the $K^{\pm}p$ amplitude in the whole complex plane

7. COMMENTS AND CONCLUSIONS

We have parametrized the $K^{\pm}p$ amplitude in the whole complex plane using only:

A) - Experimental data on the $K^{\pm}p$ total cross-sections: $\sigma^{\pm}(\omega)$ [15].
- A Regge parametrization for $\sigma_{\pm}(\omega)$ above 300 GeV/c [16].
- The 218 values of the real parts of the amplitude in the physical region compiled by Queen [17].
- The well established value of the scattering length $a_s^{I=1}$, in order to use $\omega_o = -m_K$ as a subtraction point. [18].

B) - Two general hypotheses on the imaginary part of the amplitude in the unphysical region: positivity and unimodality.

The two hypotheses allow one to apply rigorously the bounding properties of P.A. in the whole complex plane, except on the unphysical cut $[\omega_{\Lambda\pi}, m_K]$, which is most important, since regions of the ω-plane which are physically inaccesible (complex ω, Λ and Σ poles, $-m_K < \omega < \omega_{\Lambda\pi}$ and particularly the important point $\omega = 0$ [19]) are dealt with by P.A.'s on the same footing as the experimental region.

The application of the bounds on the experimental region has led to the rejection of a set of experimental measurements of the real parts. We think that the rejection of that set of points is crucial if one uses a parametrization based on the positivity hypothesis, since the inconsistence of those points with the hypothesis reflects itself in smaller coupling constants than allowed by positivity [20, 21].

The bounds obtained for the reduced coupling constant, $14.4 < G^2 < 16.3$, are compatible with recent determinations using low energy models [14] and with other ones based on discrepancy functions [22]. On the other hand, our extrapolation method allows us to treat the two poles Λ and Σ separately in a natural way, which is an improvement on other model independent extrapolation methods.

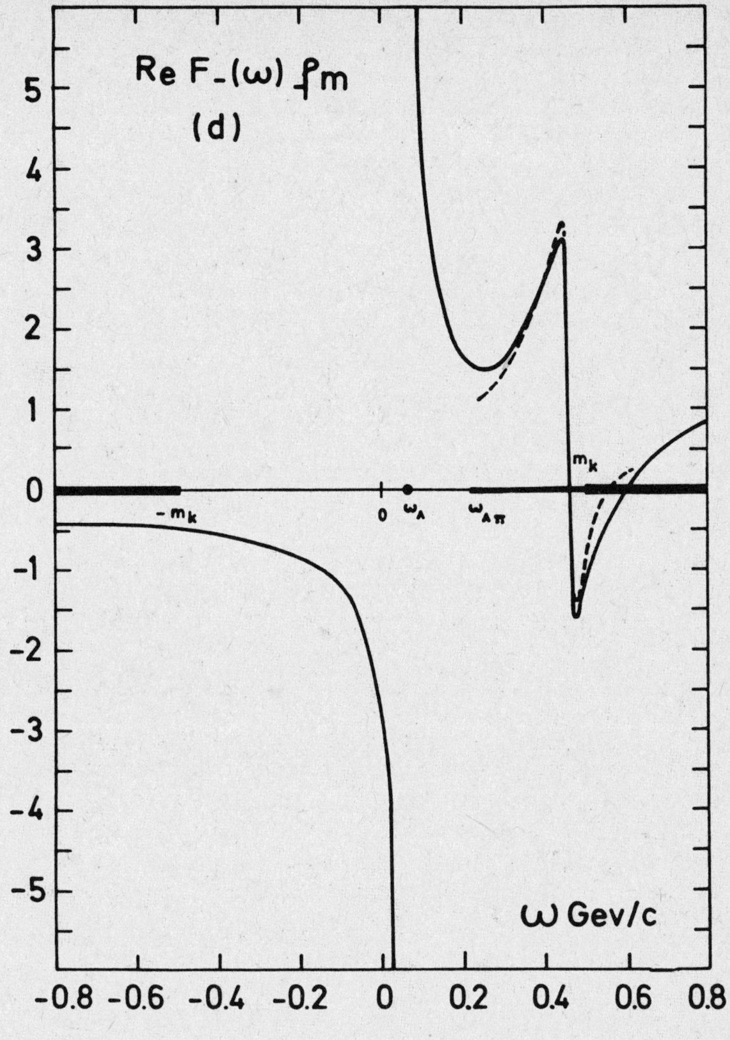

Figure 6 Real part of the $F_-(\omega)$ amplitude on the unphysical cut. The continous line is our result and the dashed line is the real part obtained from a low energy parametrization, |14|. We have taken into account the reduced coupling constant approximation, then one has only a reduced pole at ω_Λ and no real zero.

Regarding the position of the complex zeros, our results are well inside the allowed region of Atkin et al. [23] .

Finally, our method allows us too to give information on the amplitude on the unphysical cut, something which, up to now, had been achieved only using more sofisticated methods and all the available low energy data, obtaining results compatible with the latter [24] .

ACKNOWLEDGEMENTS

We thank Drs. J. Gilewicz, M. Pindor and W. Siemaszko for their interest in this work and for their kind invitation to participate in the conference.

REFERENCES

1 "Analytic Extrapolations Techniques and Stability Problems in dispersion Theory". S. Ciulli, C. Pomponiu and I. Sabba-Stefanescu, Phys. Rep. 17 (1975) 133.

2 "Low and intermediate Energy Kaon-Nucleon Physics". E. Ferrari and G.Violini. Proceedings of the Workshop at the Institute of Physics of the University of Rome, March 1980. (D.Reidel P.Co.1980)

3 "Dispersion Theory in high energy physics". N.M. Queen and G. Violini. MacMillan Press Limited 1974.

4 "Magnitude of the σ-Term, Chiral Symmetry and Scale Invariance". G. Altarelli et al. Phys. Lett. B 35 (1971) 415.

5 "Compilation of Coupling Constants and Low-Energy Parameters". N.M. Nagels et al. Nucl. Phys. B 147 (1979) 189.

6 "Phenomenological Dispersion Theory of KN Scattering". N.M. Queen, M. Restignoli and G. Violini, Fortschr. Phys. 21 (1973) 569.

7 "Summation of Series and Conformal Mapping". T.H. Gronwall, Annals of Math. (2), 33 (1932) 101.

8 "Approximants de Padé" J. Gilewicz. Lecture Notes in Mathematics 667 (1978).

9 "Padé type approximation and General Orthogonal Polynomials". C. Brezinski. Birkhauser Verlag ISNM 50 1980.

10 "Essentials of Padé Approximants". G.A. Baker. Academic Press 1975.

11 "Moment-theory Approximations for Non-negative Spectral Densities". C.T. Corcoran and P.W. Langhoff, J. Math. Phys. 18 (1977) 651.

12 "The problem of moments". J.A. Shoat and J.D. Tamarkin. American Mathematical Society. Mathematical Surveys I (1943).

13 "An introduction to orthogonal polynomials". T.S. Chihara, Gordon and Breack 1978.

14 "Dispersion Relations Constraints on Low Energy $\bar{K}N$ Scattering". A.D. Martin, Phys. Lett. 65 B (1976) 346.

15 "Compilation of $K^{\pm}$ Cross Sections". CERN-HERA 79-02 (1979).

16 "Phenomenology of Total Cross Sections and Forward Scattering at High Energy". V. Barger and R.J.N. Phillips, Nucl. Phys. B 32 (1971) 93.

17 "Compilation of Real Parts of the $K^{\pm}p$ Forward Elastic Amplitude, Report UB-Kp-1-78 (1978). Birmingham.

18 "$K^{\pm}p$ Elastic Scattering from 130 to 755 Mev/c". W. Cameron et al. Nucl. Phys. B 78 (1974) 93.

19 "The Zero-Energy Kp Scattering Amplitude and the Evaluation of the Kaon-Nucleon Sigma-Terms". A.D. Martin and G. Violini. Lett. Nuovo Cim. 30 (1981) 105.

20 "On Determinations of the KN Effectiv Coupling Constant from Forward Amplitudes". N. Sznalder-Hald et al., Nucl. Phys. B 59 (1973) 93.

21 "A Use of Zeros in Evaluating the $K^{\pm}p$ Forward Scattering Amplitudes". G.K. Atkin, J.E. B owcock and N.M. Queen, J. Phys. G: Nucl. Phys. 7 (1981) 613.

22 "Dispersion Analysis of the $K^{\pm}p$ Forward Scattering Amplitudes" B. di Claudio, G. Violini and N.M. Queen, Nucl. Phys. B 161 (1979) 238.

23 "Positivity as a Constraint on the Positions of the Zeros of the $K^{-}p$ Forward Scattering Amplitude". G.K. Atkin, J.E. B owcock and N.M. Queen, Z. Phys. C 15 (1982) 129.

24 "Kaon Nucleon Parameters". A.D. Martin. Nucl. Phys. B 179 (1981) 33.

SMOOTHNESS CONDITIONS FOR STIELTJES MEASURES
FROM PADE APPROXIMANTS

D. Bessis, Georgia Institute of Technology, Atlanta Georgia, USA
G. Turchetti, Dipartimento di Fisica, Università di Bologna
Istituto Nazionale di Fisica Nucleare, Sezione di Bologna, ITALY
W. Van Assche, University of Leuven, BELGIUM : Research assistant
Belgian National Fund for Scientific Research

Introduction and definitions

In many physical problems we are faced with solutions which can be related to Stieltjes functions and we would like to determine their relevant features. The Padé Approximants (PA) do not only provide approximations uniformly convergent in compact sets of the cut plane and bounds on the real line but can also give some information about the structure of the measure.

In this note we analyze some smoothness criteria by means of two known results from the theory of orthogonal polynomials (Theorems 1 and 2) and we quote a new theorem (Theorem 3) on the behavior of PA at an end point where the measure has a singularity. Applications to the Lee-Yang measure for the Ising model are briefly discussed.

Let $d\sigma(t)$ be a positive measure in $[0,1]$ and

$$f(z) = \int_0^1 \frac{d\sigma(t)}{1-tz} = 1 + \sum_{n=1}^{\infty} f_n z^n \tag{1}$$

be the associated Stieltjes function regular at the origin (if the support of the measure is $[0,A]$ the representation (1) still holds after the scaling $t \to At$ and $z \to A^{-1}z$). We introduce another function $R(z) = \frac{1}{z} f(\frac{1}{z})$ regular at infinity and denote by

$$[n-1/n]_f(z) = \frac{\overline{Q}_{n-1}(z)}{\overline{P}_n(z)} = f(z) + \mathcal{O}(z^{2n}) \tag{2}$$

the PA to $f(z)$. The polynomials $Q_n(z) = z^n\overline{Q}_n(\frac{1}{z})$ and $P_n(z) = z^n\overline{P}_n(\frac{1}{z})$ are orthogonal with respect to $d\sigma$ and satisfy the recursion

$$\begin{aligned} P_{n+1}(z) &= (z-\alpha_n)\, P_n(z) - \beta_n P_{n-1}(z) \\ Q_{n+1}(z) &= (z-\alpha_{n+1})Q_n(z) - \beta_{n+1}Q_{n-1}(z) \end{aligned} \tag{3}$$

initialized by $P_{-1} = Q_{-1} = 0$ and $P_0 = Q_0 = 1$, where α_n, β_n are the coefficients of the continued fraction $R(z)$:

$$R(z) = \cfrac{1}{z-\alpha_0-\cfrac{\beta_1}{z-\alpha_1-\cfrac{\beta_2}{z-\alpha_2-\ddots}}} \tag{4}$$

The truncations of R(z) at order n (that is setting $\beta_n = 0$) are given by

$$Q_{n-1}(z)/P_n(z) = \frac{1}{z}\,[n-1/n]_f(\tfrac{1}{z}).$$

We denote by $x_{j,n}^{-1}$ and $\lambda_{j,n}$ $(j=1,2,\ldots,n)$, where $0 < x_{1,n} < x_{2,n} < \ldots < x_{n,n} < 1$ and $\lambda_{j,n} > 0$, the poles and residues of $[n-1/n]_f(z)$ and by $d\sigma_{[n-1/n]}$ the associated measure

$$[n-1/n]_f(z) = \int_0^1 \frac{d\sigma_{[n-1/n]}(t)}{1-tz} \quad ; \quad d\sigma_{[n-1/n]}(t) = \sum_{j=1}^{n} \lambda_{j,n}\,\delta(t-x_{j,n}) \tag{5}$$

Smoothness conditions

A necessary condition for the smoothness of $d\sigma(t)$ in the neighborhood of an end point is given by the law of convergence of the zeros of $P_n(z)$ to it. Given two sequences a_n, b_n we shall write $a_n \approx b_n$ if two constants c_1 and c_2 exist such that

$$0 < c_1 < a_n/b_n < c_2 < \infty$$

So if

$$1 - x_{n,n} \approx n^{-s} \tag{6}$$

then the smoothness condition is s = 2, which follows from the next theorem, which is an immediate corollary of a result found by Nevai [1] :

THEOREM 1. If $d\sigma$ is a measure on $[-1,1]$ and absolutely continuous near 1 with $\sigma'(t) \approx (1-t)^{-\gamma}$ $(\gamma < 1)$ in the interval $1-\varepsilon \leqslant t \leqslant 1$ for some $\varepsilon > 0$, then for every zero in $]1-\varepsilon,1]$

$$1 - x_{n+1-j,n} \approx \frac{j^2}{n^2} \tag{7}$$

Proof: define $x_{n+1-j,n} = \cos\theta_{j,n}$ $(j=1,2,\ldots,n)$ and $\theta_{0,n} = 0$, then from Theorem 21 ([1], p. 165) we obtain

$$\theta_{k+1,n} - \theta_{k,n} \approx \frac{1}{n}$$

so that

$$\theta_{j,n} = \sum_{k=0}^{j-1} (\theta_{k+1,n} - \theta_{k,n}) \approx \frac{j}{n}$$

and the result follows immediately. □

Another smoothness condition is obtained if the coefficients α_n, β_n converge for

$n \to \infty$ to limits α, β. In such a case it has been proved that the measure is continuous in $]\alpha - 2\sqrt{\beta}, \alpha + 2\sqrt{\beta}[$ and can have point masses outside this interval. Smooth approximations $d\hat{\sigma}_n$ to the measure $d\sigma$ have been obtained for this case [1],[2] . The idea is to replace the continued fraction $R(z)$ no longer by a rational fraction obtained by truncation, but rather by another continued fraction $R_n(z)$ whose coefficients $\hat{\alpha}_n, \hat{\beta}_n$ are chosen to be constant after the order n

$$\begin{aligned} \hat{\alpha}_j &= \alpha_j \\ \hat{\beta}_{j+1} &= \beta_{j+1} \end{aligned} \quad j < n \qquad\qquad \begin{aligned} \hat{\alpha}_j &= \alpha \\ \hat{\beta}_{j+1} &= \beta \end{aligned} \quad j \geqslant n \tag{8}$$

The measure $d\hat{\sigma}_n$ explicitely reads

$$d\hat{\sigma}_n(t) = w_n(t)\, dt + \sum_j c_j\, \delta(t-t_j) \tag{9}$$

where

$$w_n(t) = \frac{\sqrt{4\beta - (t-\alpha)^2}}{2\pi\, S_n(t)} \qquad \alpha - 2\sqrt{\beta} < t < \alpha + 2\sqrt{\beta} \tag{10}$$

and S_n is a polynomial whose zeros do not belong to $]\alpha - 2\sqrt{\beta}, \alpha + 2\sqrt{\beta}[$. Denote by $p_n(t)$ the normalized orthogonal polynomials with respect to $d\sigma$ (the $P_n(t)$ are not normalized) then we can write

$$S_n(t) = \beta\, p_n^2(t) - \beta_n\, p_{n-1}^2(t) - (t-\alpha)\sqrt{\beta_n}\, p_n(t) p_{n-1}(t) \tag{11}$$

and the mass points t_j in (9) are the zeros of S_n such that $|p_{n+1}(t_j)/p_n(t_j)| < 1$.

The theorem about the smoothness of $d\sigma$ and the convergence of $d\hat{\sigma}_n$ is

<u>THEOREM 2</u>. If α_n, β_n converge to α, β in such a way that

$$\sum_{n=1}^{\infty} n^q \left\{ \frac{|\beta_n - \beta|}{\beta} + \frac{|\alpha_{n-1} - \alpha|}{\sqrt{\beta}} \right\} < \infty \tag{12}$$

with $q = 1$, then $d\hat{\sigma}_n$ converges weakly to $d\sigma$ which is absolutely continuous in $[\alpha - 2\sqrt{\beta}, \alpha + 2\sqrt{\beta}]$ with at most a finite number of point masses outside this interval. If the convergence in (12) occurs only for $q = 0$ then the number of point masses is countable and they may accumulate at $\alpha \pm 2\sqrt{\beta}$.

The proof can be found in [1] or [2].

End point singularities

If the measure is absolutely continuous near the endpoint 1 of its support with singularities such as $\sigma'(t) \approx (1-t)^{-\gamma}$ with $0 < \gamma < 1$, then the behavior of the PA at $z = 1$ determines the index γ according to the following result

<u>THEOREM 3</u>. If $d\sigma$ is absolutely continuous in $[1-\varepsilon, 1]$ for some $\varepsilon > 0$ and if $\sigma'(t) \approx (1-t)^{-\gamma}$ with $\gamma < 1$, then

$$[n-1/n]_f(1) \approx \begin{cases} n^{2\gamma} & 0 < \gamma < 1 , \\ \log n & \gamma = 0, \\ 1 & \gamma < 0. \end{cases} \tag{13}$$

Proof: according to (5) we can write

$$[n-1/n]_f(1) = \sum_{|1-x_{j,n}|<\varepsilon} \frac{\lambda_{j,n}}{1-x_{j,n}} + \sum_{|1-x_{j,n}|\geqslant\varepsilon} \frac{\lambda_{j,n}}{1-x_{j,n}} = S_1 + S_2 .$$

The second sum can be bounded according to

$$0 \leqslant S_2 \leqslant \frac{1}{\varepsilon} \sum_{j=1}^{n} \lambda_{j,n} = \frac{1}{\varepsilon} .$$

Using Theorem 27 ([1], p. 119) the first sum behaves as

$$S_1 \approx \frac{1}{n} \sum_{|1-x_{j,n}|<\varepsilon} \frac{(\sqrt{1-x_{j,n}} + \frac{1}{n})^{-2\gamma+1}}{1 - x_{j,n}}$$

and by (7)

$$S_1 \approx n^{2\gamma} \sum_{|1-x_{j,n}|<\varepsilon} j^{-2\gamma-1} .$$

By Theorem 12.7.2 ([3], p. 310) the number of terms in this sum is approximately

$$\frac{n}{\pi} \int_{1-\varepsilon}^{1} \frac{dt}{\sqrt{1-t^2}} = n\Delta$$

(this provided $\sigma' > 0$ almost everywhere in $[-1,1]$)so that

$$S_1 \approx n^{2\gamma} \sum_{j=1}^{n\Delta} j^{-2\gamma-1} .$$

Since ε is kept fixed the last sum can easily be estimated for large n and leads to the result (13). □

Under essentially the same hypothesis as in the last theorem but without the restriction $\gamma > 0$ one can relate the index γ to the behavior at $z = 1$ of the normalized orthogonal polynomials p_n :

$$p_n(1) \approx n^{-\gamma+1/2} \quad ; \quad p_n(x_{n+1,n+1}) \approx n^{-\gamma-1/2} \tag{14}$$

(see corollary 34, p. 171 and Theorem 31, p. 134 of the reference [1]).

Algorithms for the indices

The smoothness and end point singularity indices s and γ defined by (6) and (13) are the exponents of the leading terms in n of the sequences $x_{n,n}$ and $[n-1/n]_f(1)$ and can be computed with the following algorithms.

Given a sequence

$$r_n = r + c_o n^{\omega} \left(1 + \frac{c_1}{n} + \frac{c_2}{n^2} + \dots\right) \tag{15}$$

then we determine ω as the limit of the new sequences

$$\omega_n^{(1)} = n \log \frac{r_n - r}{r_{n+1} - r} = -\omega + O\left(\frac{1}{n}\right) \tag{16}$$

and

$$\omega_n^{(2)} = n \log \frac{r_n - r_{n-1}}{r_{n+1} - r_n} = 1 - \omega + O\left(\frac{1}{n}\right), \tag{17}$$

the latter to be used if r is not known.

Accordingly s and γ are computed as limits of the sequences $s_n^{(1)}$ and $\gamma_n^{(1)}$ obtained by replacing in (16) r_n, r by $x_{n,n}$, 1 and $[n-1/n]_f(1)$, 0 respectively. The slow convergence (linear in $1/n$) is substantially improved by using rational extrapolation algorithms derived from the Thiele continued fraction [4] .

Several measures for which α_n, β_n are exactly known were considered to test the above criteria. For the Jacobi measure $d\sigma(t) = (1-t)^a(1+t)^b dt$ with support in $[-1,1]$ the extrapolations of the sequence $s_n^{(1)}$ converge to 2 exponentially fast. The coefficients

$$\alpha_n = \frac{b^2 - a^2}{(2n+a+b)(2n+2+a+b)}$$

$$\beta_n = \frac{4n(n+a+b)(n+a)(n+b)}{(2n-1+a+b)(2n+a+b)^2(2n+1+a+b)}$$

satisfy the conditions of Theorem 2 with $q = 1$ if $|a| = |b| = \frac{1}{2}$ and with $q = 0$ otherwise, as one should expect. The extrapolations of $\gamma_n^{(1)}$ also converge to $-a$ if $-1 < a < 0$ but the convergence rate is maximum when $a \cong -1/2$ and decreases when a approaches -1 or 0.

Conversely for singular continuous measures, such as the balanced measures on Julia sets [5] of the mapping $T(z) = (z-\lambda)^2$ with $\lambda > 2$ (proved to be Cantor sets) for which

$$\alpha_n = \lambda \quad \cdot \quad \beta_1 = \lambda \quad ; \quad \beta_2 = 1 .$$

$$\beta_{2n+1} = \lambda - \beta_{2n} \quad , \quad \beta_{2n+2} = \beta_{n+1}/\beta_{2n+1}$$

the sequence $\gamma_n^{(1)}$ does not converge, nor α_n, β_n do.

The Lee-Yang measure

The smoothness analysis was succesfully carried out for the Lee-Yang measure [6] of the Ising model. We recall that for a system of spins in a lattice, interacting with a magnetic field according to the Hamiltonian

$$\mathcal{H} = - J \sum^{*} \sigma_i \sigma_j - H \sum \sigma_i \tag{18}$$

where $\sum^{*}$ denotes the sum over nearest neighbors, the magnetization M at temperature T can be written as [7]

$$M(H,T) = \overline{M}(w,u) \tanh(\frac{H}{kT})$$

where

$$u = e^{-2p \frac{J}{kT}} \quad ; \quad w = \frac{e^{-2 \frac{H}{kT}}}{(1 + e^{-2 \frac{H}{kT}})^2} (1-u) \tag{19}$$

and $\overline{M}$ is a Stieltjes function in w; p is 1 cr 2 when the coordination number is odd or even. The representation of $\overline{M}$ is

$$\overline{M}(w,u) = \int_0^{A(u)} \frac{d\sigma_u(t)}{1-wt} = 1 + \sum_{j=0}^{\infty} w^j P_j(u) \tag{20}$$

where

$$A(u) = \frac{4}{1-u} \cos^2 \frac{\theta_o(u)}{2}$$

and $\theta_o(u)$ is the Lee-Yang angle which vanishes for $T \leqslant T_c$, if T_c denotes the critical temperature.

The Lee-Yang measure is expected to behave as $\sigma_u'(t) \approx (A(u) - t)^{-\gamma(u)}$ for t close to A(u), which implies $\overline{M}(w,u) \approx (1 - wA(u))^{-\gamma(u)}$ for w close to A(u).

Having determined the Taylor expansion of $\overline{M}$ up to order 15 for various models (the coefficients $P_j(u)$ are polynomials in u [8]) the computation of s and γ was carried out. The result gave $s = 2$ for all models at any temperature with an accuracy better than 10^{-3}. The coefficients α_n, β_n converge to finite limits in such a way that (12) of Theorem 2 is satisfied with $q = 1$ for $T < T_c$ and with $q = 0$ for $T \geqslant T_c$. These results are good evidence for the absence of a singular continuous or a discrete part in the measure near its endpoint A(u).

The end point singularity γ was determined within 10^{-6} for $T < T_c$ and 10^{-3} for $T = T_c$ in agreement with the result $\gamma = 1/2$ for $T < T_c$ and $\gamma = (1 - 1/\delta)/2$ for $T = T_c$ (to be expected from scaling arguments) where δ is the magnetization critical index for which $M \approx H^{1/\delta}$, when $H \to 0$ at $T = T_c$. In the table the results obtained for the square lattice are reported; the exact value of γ is $\frac{7}{15} = 0.4666\ldots$.

Estimates of γ for $T > T_c$ and smooth approximations to the measure $d\sigma_u$ were also computed with 15 Taylor coefficients (for the square lattice) and the results agree with previous results obtained by different methods [9],[10].

n	α_n	β_n	n	$x_{n,n}$	extrapolated
0	2.3431458		1	2.4556630	2.4556630
1	2.4365942	2.8629150	2	4.3340575	6.2124519
2	2.4186407	1.4691123	3	4.7828793	5.5135212
3	2.4162785	1.4609487	4	4.9511531	5.0976430
4	2.4153710	1.4590917	5	5.0315480	5.1827451
5	2.4149786	1.4582799	6	5.0760265	5.1803692
6	2.4147502	1.4578737	7	5.1031697	5.1802921
7	2.4146053	1.4576594			

n	$s_n^{(1)}$	extrapolated	n	$\gamma_n^{(1)}$	extrapolated
1	1.2035535	1.2035535	1	0.3285207	0.3285207
2	1.5488992	1.8942448	2	0.3824543	0.4363878
3	1.6846040	1.9959982	3	0.4060066	0.4660757
4	1.7575563	2.0017012	4	0.4192738	0.4670162
5	1.8031112	2.0001684	5	0.4277866	0.4667896
6	1.8342546	1.9996412	6	0.4337116	0.4667312
7	1.8568959	2.0000889	7	0.4380746	0.4667877

Table : results for the square lattice $(T = T_c)$

REFERENCES

[1] P. Nevai, *Orthogonal Polynomials*, Mem. Amer. Math. Soc. 213 (1979), Providence, Rhode Island.

[2] J.S. Geronimo, W. Van Assche, *Orthogonal polynomials with asymptotically periodic recurrence coefficients*, J. Approx. Theory 46 (1986)

[3] G. Szegö, *Orthogonal polynomials*, Amer. Math. Soc. Colloq. Publ. 23, 4th edition (1975), Providence, Rhode Island.

[4] J. Stoer, S. Burlisch, *Einfuhrüng in die Numerische Mathematik*, Springer, Berlin (1973).

[5] D. Bessis, M.L. Mehta, P. Moussa, *Orthogonal polynomials on a family of Cantor sets and the problem of iterations of quadratic mappings*, Letters Math. Phys. 6 (1982), 123-140.

[6] C.N. Yang, T.D. Lee, *Statistical theory of equations of state and phase transitions I, theory of condensation*, Phys. Rev. 87 (1952), 404-409.
T.D. Lee, C.N. Yang, *Statistical theory of equations of state and phase transitions II, lattice gas and Ising model*, Phys. Rev. 87 (1952), 410-419.

[7] M. Barnsley, D. Bessis, P. Moussa, *The diophantine moment problem and the analytic structure in the activity plane of the ferromagnetic Ising model*, J. Math. Phys. 20 (1979), 535-546.

[8] D. Bessis, J.M. Drouffe, P. Moussa, *Positivity constraints for the Ising ferromagnetic model*, J. Phys. A 9 (1976), 2105-2124.

[9] P.J. Kortman, R.B. Griffiths, *Density of zeros on the Lee-Yang circle for two Ising ferromagnets*, Phys. Rev. Letters 27 (1971), 1439-1442.
D.A. Kurtze, M.E. Fisher, *Yang-Lee edge singularities at high temperature*, Phys. Rev. B 20 (1979), 2785-2796.

[10] R. Jullien, K. Uzelac, P. Pfeuty, P. Moussa, *The Yang-Lee edge singularity studied by a four-level quantum renormalization-group blocking method*, J. Physique 42 (1981), 1075-1080.
J.A. Baker Jr., L.P. Benofi, I.G. Enting, *Yang-Lee edge for the two dimensional Ising model*, Los Alamos preprint (1985).

Exact multisoliton properties of rational approximants to the iterated solution of nonlinear evolution equations.

F. Lambert and M. Musette
Theoretische Natuurkunde
Vrije Universiteit Brussel
Pleinlaan, 2, B - 1050 Brussel, Belgium.

1. Introduction

Recent progress in the description of nonlinear phenomena has been achieved by recognizing that nonlinear partial differential equations or integro-differential equations may have regular solitary wave solutions of permanent profile that can be expressed in a closed form. Whereas the occurence of such solitary waves is rather common - it is often viewed as reflecting a balance between some kind of nonlinearity and the dispersive properties of the linearized equation - few of these waves will preserve their identity upon interaction with one another. Solitons are solitary waves with this exceptional property.

Though powerful techniques, such as the IST-method [1], have led to an impressive list of integrable nonlinear equations which possess soliton solutions, it is as yet not known how to ascertain definitely whether a given nonlinear system has solitons.

Here we consider a particular kind of kink-shaped solitary waves in 1+1 dimensions (padeons) and we examine the circumstances in which their x-derivatives, which have the typical sech-squared form of a KdV-like soliton, possess the "particle-like" interaction properties of a true soliton.

These "padeons" can be associated with an exponential solution of the linearized equation :

$$V_0 = \exp \theta \quad , \quad \theta = -kx + \omega(k)t \quad , \quad k > 0 \quad , \tag{1}$$

through the sum of a geometrical iteration series - the ROSALES series [2] - in powers of the nonlinearity parameter $\varepsilon > 0$. The ones we consider are [1/1] fractions in ε of the particular form :

$$V^{(1)} = -2\partial_x \ln (1 + \varepsilon z) \quad , \quad z = \frac{1}{2k} \exp \theta \tag{2}$$

Examples of padeon equations are the BURGERS equation, the potential KdV equation, the potential BOUSSINESQ equation [3], the potential KOTERA-SAWADA equation and various (potential) model equations for shallow water waves [4] ...

The "particle-like" interaction properties of the soliton refer to the fact that a soliton equation sould also possess N-soliton solutions, $N = 2, 3, \ldots$, which as $t \to \pm \infty$ tend asymptotically to a sum of N separated sech-squared solitary waves (solitons) with parameters $0 < k_1 < \ldots < k_N$. Thus, in order to be accepted as a potential soliton equation a padeon equation should also possess "N-padeon" solutions, the x-derivative of which would have the asymptotic properties of an N-soliton solution.

At $N = 2$ a reasonable ansatz for the "dipadeon" suggests that one should look for "regular" N-padeons in the form of a sum of N partial [1/1] fractions in ε, generalizing the functional form of the padeon (sect. 2) :

$$V^{(N)} = -2\partial_x \ln \prod_{i=1}^{N} (1+\varepsilon z_i) \quad \text{with} \quad \sum_{i_1<\ldots<i_n=1}^{N} z_{i_1}\ldots z_{i_n} > 0 \;,\; n = 1,2,\ldots N, \tag{3}$$

which are the sum of a particular iteration series (the Nth ROSALES series) generated by a sum of N exponential solutions of the linearized equation :

$$V_0^{(N)} = \sum_{i=1}^{N} \exp \theta_i \quad , \quad \theta_i = -k_i x + \omega(k_i)t + \tau_i \quad , \tag{4}$$

with $0 < k_1 < \ldots < k_N$, $v_i \equiv \dfrac{\omega(k_i)}{k_i} \neq v_j$ for $i \neq j$, and τ_i : constant.

In sect. 3 we show that if a padeon equation possesses a sequence of regular N-padeons, $N = 2, 3, \ldots \bar{N}$, the x-derivative of any such solution tends, for $t \to \pm \infty$, to a sum of N sech-squared solitary waves of the form :

$$V_x = -\frac{k_i^2}{2} \operatorname{sech}^2 \left(\frac{\theta_i + \tau_i^{\pm}}{2}\right) \quad , \tag{5}$$

and that : $\sum_{i=1}^{N} \tau_i^+ - \tau_i^- = 0$.

This result opens a direct way to decide whether a given padeon equation possesses exact N-soliton solutions for $N = 2, 3, \ldots$ Explicit conditions on a set of mixing coefficients in the second and the third ROSALES series, which are necessary and sufficient for the existence of regular dipadeons ($N = 2$) and tripadeons ($N = 3$), can be checked right away by calculating particular contributions to some lower order iteration terms (sect. 4). Though the above regular ($N \geqslant 3$)-padeons behave as N-soliton potentials, the phaseshifts $\delta_i = \tau_i^+ - \tau_i^-$ which result from the interaction between the N solitary waves do not always obey the "two-wave" collision law which characterized the KdV-like multisoliton interaction. This additional property is found to coincide with a factorization property of the interaction parameters $K_{i_1\ldots i_n}$ which enter into the explicit form of the N-padeons (sect. 5)

It follows that the N-padeons which for $N \geqslant 3$ possess the two-wave collision property are expressible in terms of only one interaction parameter K_{12} which can be obtained from the first order mixing coefficient α_{12}.

Another consequence of the particular structure of the ROSALES series associated with a padeon equation is that the [2/2] fraction of the form :

$$V^{(2)} = -2\partial_x \ln \prod_{i=1}^{2} (1 + \varepsilon z_i) \quad ,$$

with $z_1 + z_2 > 0$ and $z_1 z_2 > 0$, which matches the first two terms $V_0^{(2)}$ and $V_1^{(2)}$ of the second ROSALES series, behaves as an exact two-soliton potential even when it fails to solve the padeon equation (sect. 3). In the same way it turns out that if a padeon equation possesses a sequence of (regular) n-padeons, from $n = 2$ up to $N-1$, the [N/N] fraction of the form (3) which matches the first N terms $V_0^{(N)}, \ldots$ $\ldots V_{N-1}^{(N)}$ of the Nth ROSALES series behaves as an N-soliton potential even when this approximant fails to be an N-padeon. The latter results could be used to explain the quasi-soliton behaviour of the solitary wave solutions of some (generalized) padeon equations, such as the regularized long wave equation [5], which are not soliton equations but which may be equally important model equations in a given physical context.

2. Padeons and (generalized) N-padeons.

To introduce the padeon-ansatz we consider the example of BURGERS' nonlinear diffusion equation :

$$V_t - V_{xx} + VV_x = 0 \quad , \tag{6}$$

which is the simplest nonlinear partial differential evolution equation which possesses solitary wave solutions $V(\xi = x-vt)$ with a kink-shaped profile :

$$\lim_{\xi\to-\infty} V(\xi) = c > 0 \quad \text{and} \quad \lim_{\xi\to+\infty} V(\xi) = 0 \ . \tag{7}$$

These solitary waves are easily obtained by scaling the field variable $V = \varepsilon\overline{V}$, $\varepsilon > 0$, and by looking for solutions $\overline{V}(x,t;\varepsilon)$ of the scaled equation :

$$\overline{V}_t - \overline{V}_{xx} + \varepsilon\, \overline{V}\, \overline{V}_x = 0 \tag{8}$$

displaying particular properties with respect to ε.
Starting with a formal power series expansion $\overline{V} = \sum_{n=0}^{\infty} \varepsilon^n V_n$ and collecting equal powers of ε we find that the (scaled) BURGERS equation produces the following

hierarchy of equations (iteration hierarchy) :

$$V_{0,t} - V_{0,xx} = 0 \tag{9}$$

$$V_{n,t} - V_{n,xx} = - \sum_{j=0}^{n-1} V_j \, V_{n-j-1,x} \quad , \quad n > 1 \quad , \tag{10}$$

from which the successive terms V_n can be obtained by selecting a particular solution at each step of the iteration.

At zero order we observe that the linearized BURGERS equation possesses positive solutions :

$$V_0^{(1)} = \exp \theta \quad , \quad \theta = -kx + k^2 t \quad , \quad k > 0 \tag{11}$$

As we are looking for kink-shaped solitary wave solutions of the full equation, subject to the conditions (7), it is natural to start the iteration with $V_0^{(1)}$ and to try to end up with a fraction in exp θ, the simplest possible form of which is :

$$\overline{V} = \frac{\exp \theta}{1 + \frac{\varepsilon}{c} \exp \theta} \tag{12}$$

This suggests that we should aim at a geometrical iteration series, i.e. that we should solve the further iteration equations (10) subject to the condition :

$$V_n \div \exp (n + 1) \theta \quad .$$

We thus obtain the first order term : $V_1^{(1)} = - \frac{1}{2k} \exp 2\theta$, which fixes the height of the kink in terms of its velocity : $c = 2k$.

Hence, a candidate solitary wave solution of the scaled BURGERS equation has the form :

$$\overline{V}^{(1)} = \frac{\exp \theta}{1 + \frac{\varepsilon}{2k} \exp \theta} \tag{13}$$

Insertion of this fraction into the l.h. side of equ. (8) produces an identity of the form :

$$\partial_t \overline{V}^{(1)} - \partial_{xx}^2 \overline{V}^{(1)} + \varepsilon \overline{V}^{(1)} (\partial_x \overline{V}^{(1)}) = \frac{A + \varepsilon B}{(1 + \frac{\varepsilon}{2k} \exp \theta)^3} \tag{14}$$

By expanding $\overline{V}^{(1)}$ and the fraction at the r.h. side of equ. (14) in powers of ε, and by collecting equal powers, it is clear that since $V_0^{(1)}$ and $V_1^{(1)}$ solve, respectively, equ. (9) and the first equ. (10), the coefficients A and B must vanish identically. It follows that the fraction $\overline{V}^{(1)}$ is an actual solution of the scaled BURGERS equation. The corresponding two-parameter family ($k > 0$, $\varepsilon > 0$) of regular solitary wave solutions of the BURGERS equation takes the form :

$$V^{(1)} = \varepsilon\, \overline{V}^{(1)} = -2\partial_x \ln\left[1 + \frac{\varepsilon}{2k}\exp(-kx + k^2 t)\right]$$

It is remarkable that quite different nonlinear dispersive evolution equations (including the potential KdV equation) have two-parameter families of solitary wave solutions [3,4] of the same functional form :

$$V^{(1)} = -\,2\partial_x \ln\left(1 + \frac{\varepsilon}{2k}\exp\theta\right) \quad , \tag{15}$$

which are equally generated by exponential solutions of the linearized equation :

$$V_0^{(1)} = \exp\theta \quad , \quad \theta = -\,kx + \omega(k)t \quad , \quad k > 0 \tag{16}$$

where $\omega(k)$ satisfies the linear dispersion law. These solitary waves have been called "padeons". The x-derivative of a padeon is well-shaped and has the functional form of the KdV-soliton :

$$V_x^{(1)} = -\,\frac{k^2}{2}\,\mathrm{sech}^2\left(\frac{\theta + \tau}{2}\right) \quad , \quad \tau = \ln(\varepsilon/2k) \tag{17}$$

In the following we call "padeon equation" any nonlinear partial differential evolution equation in one space-dimension with constant coefficients and with a polynomial nonlinearity, which possesses solitary wave solutions of the form (15).

Given a padeon equation we wish to examine in which circumstances the x-derivative of its padeon can be regarded as a true soliton. This means that we should look for N-soliton potentials, N = 2, 3, ..., the x-derivative of which would consist essentially (except for some finite period of interaction) of N separated well-shaped solitary waves of type (17), up to some phaseshift resulting from their mutual interaction.

Let us consider a padeon equation for which the linear dispersion law produces a phasevelocity $v(k) = \frac{\omega(k)}{k}$. We assume that $\frac{dv}{dk} \neq 0$, as it is the case for a linear dispersive equation. As one considers the iteration hierarchy associated with the scaled padeon equation, it is clear that the zero-order equation (linearized padeon equation) possesses solutions of the form (4). The full padeon equation may, as a counterpart, possess families of real N-pole solutions in ε which generalize the one-pole form (15) :

$$V^{(N)} = -\,2\partial_x \ln\left[\prod_{i=1}^{N}(1 + \varepsilon z_i)\right] \tag{18}$$

in terms of N functions $z_i(x,t)$, taking values which are either real and positive or complex conjugate in pairs.

At N = 2 we may particularize the ansatz by considering an explicit 2-pole form which is almost equal to the sum of two padeons (with parameters $k_1 < k_2$), except for a real phase τ_{12} $(k_1,k_2) \neq 0$ which could account for the nonlinear interaction :

$$V^{(2)} = -2\partial_x \ln \left[1+\varepsilon\left(\frac{\exp\theta_1}{2k_1} + \frac{\exp\theta_2}{2k_2}\right) + \frac{\varepsilon^2}{4k_1k_2}\exp(\theta_1+\theta_2+\tau_{12})\right] \tag{19}$$

It is easy to verify that this form possesses all the asymptotic properties of a two-soliton potential. It suffices to follow each of the solitary waves as $t \to \pm\infty$. As $x = v_i t + \xi$, $i = 1$ or 2, and $t \to \pm\infty$, it is clear that θ_i remains fixed while the other θ goes either to $+\infty$ or to $-\infty$ $(v_1 \neq v_2)$, so that :

$$V_x^{(2)} \underset{t\to\pm\infty,\ \theta_i \text{ fixed}}{\longrightarrow} -\frac{k_i^2}{2}\operatorname{sech}^2\left(\frac{\theta_i+\tau_i^{\pm}}{2}\right) , \tag{20}$$

with $\tau_1^- = \ln(\varepsilon/2k_1)$, $\tau_2^+ = \ln(\varepsilon/2k_2)$, $\tau_1^+ = \tau_1^- + \tau_{12}$ and $\tau_2^- = \tau_2^+ + \tau_{12}$ (21)

if $v_1 < v_2$, and the time-reversed analogue if $v_2 < v_1$.

Furthermore, it follows from these asymptotic results that, as t goes from $-\infty$ to $+\infty$, the sum of the two phaseshifts $\delta_i = \tau_i^+ - \tau_i^-$, experienced by each of the two solitary waves, equals zero :

$$\delta_1 + \delta_2 = 0 \tag{22}$$

Let us now examine under which conditions a fraction of the form (19) will solve the given padeon equation.

Expansion of the r.h. side in powers of ε produces the "perturbation terms" :

$$V_0^{(2)} = \exp\theta_1 + \exp\theta_2 \tag{23}$$

$$-V_1^{(2)} = \frac{\exp 2\theta_1}{2k_1} + \frac{(k_1+k_2)}{2k_1k_2}(1-\exp\tau_{12})\exp(\theta_1+\theta_2) + \frac{\exp 2\theta_2}{2k_2} \tag{24}$$

and further terms $V_n^{(2)}$ of the form :

$$(-)^n V_n^{(2)} = \frac{\exp(n+1)\theta_1}{(2k_1)^n} + \alpha_{1\ldots12}\exp(n\theta_1+\theta_2) + \alpha_{1\ldots122}\exp[(n-1)\theta_1+2\theta_2] + \ldots$$

$$\ldots + \alpha_{12\ldots2}\exp(\theta_1+n\theta_2) + \frac{\exp(n+1)\theta_2}{(2k_2)^n} . \tag{25}$$

We remark that the insertion of the [2/2] fraction (19) into the padeon equation produces several higher-order fractions which can all be summed by a fraction of the form :

$$\frac{N(\varepsilon)}{D_2^m(\varepsilon)} , \tag{26}$$

where $D_2(\varepsilon)$ stands for the denominator of the fraction (19), where $N(\varepsilon)$ is a polynomial in ε of degree 2m, and where the integer m depends on the order of the padeon equation.

At first sight one might expect that for a padeon equation of order r : $m = r + 1$. However, it is easy to check that when the highest order derivatives in a padeon equation have the form :

$$V^{(p+q)}_{x..x\ t..t} = \frac{\partial^{p+q}}{\partial x^p\ \partial t^q} V \quad , \quad p + q = r \geqslant 2 \ ,$$

the fraction (26) can actually be reduced to a $[2\ell/2\ell]$ fraction of type :

$$\frac{\varepsilon^2\ P_{2\ell-2}(\varepsilon)}{D_2^{\ell}(\varepsilon)} \qquad \text{with} \quad \ell \leqslant r \tag{27}$$

Indeed, it suffices to remark that the insertion of any fraction of the form $V = -2\partial_x \ln D$ into a padeon equation of order $r \geqslant 2$ produces several fractions, the sum of which can be written in the form :

$$\frac{-2}{D^{r+1}} \{D^r\ [B_1] + D^{r-1}\ [B_2] + \ldots + [B_{r+1}]\} \quad , \tag{28}$$

where each bracket $[B_n]$, $1 \leqslant n \leqslant r+1$, is homogeneous of degree n in the derivatives $D^{(i+j)}_{x..x\ t..t}$ with $0 < i+j \leqslant r+2-n$. The first bracket $[B_1]$ is obtained by setting $V = D_x$ in the linearized padeon equation. The last bracket $[B_{r+1}]$ contains only terms of the form a $D_x^{p+1}.D_t^q$, p+q=r. When $D = 1 + \frac{\varepsilon}{2k} \exp\theta$, the existence of a padeon implies that each bracket $[B_n]$, which in that case is of order ε^n, should vanish separately. The first bracket vanishes in this case as a result of the linear dispersion law. The last bracket, on the other hand, vanishes identically for whatever form of D. If follows that when $D=D_2$, $[B_{r+1}]$ will still vanish identically, whereas $[B_1]$ will be of order ε^2.

By expanding the fraction (27) as well as the fraction (19) in powers of ε, and by collecting equal powers at both hand sides of the identity obtained by inserting $V^{(2)}$ into the padeon equation, one finds that the polynomial $P_{2\ell-2}(\varepsilon)$ will vanish identically if, and only if, the $2\ell-1$ first perturbation terms in the expansion of $V^{(2)}$: $V_1^{(2)}$, ... $V_{2\ell-1}^{(2)}$, solve, respectively, the $2\ell-1$ corresponding equations of the iteration hierarchy. This means that the fraction $V^{(2)} = -2\partial_x \ln D_2$ will solve the given padeon equation iff it coincides with the $[\ell/\ell]$ Padé approximant to the particular iteration series $\varepsilon \sum_{n=0}^{\infty} \varepsilon^n V_n$ which arises from $V_0 = \exp\ \theta_1 + \exp\theta_2$ by solving the hierarchy under the condition that V_n should be a linear combination of the various exponentials : $\exp(n_1\theta_1 + n_2\theta_2)$, $n_1 + n_2 = n+1$, $n_{1,2} = 1, 2, \ldots n+1$. A similar iteration series $\sum_{n=0}^{\infty} \varepsilon^{n+1} V_n^{(N)}$ can be obtained from the hierarchy at arbitrary values of $N > 2$, by starting with $V_0^{(N)} = \sum_{i=1}^{N} \exp\theta_i$ and by applying the prescription that at each stage of the iteration $V_n^{(N)}$ should be a linear combination of the various exponentials which appear as non-homogeneous terms at the r.h. side of the nth equation. This series coincides with that considered by ROSALES [3]. We therefore call it the "N^{th} ROSALES series".

For a padeon equation the first ROSALES series (N = 1) is geometrical and one has :

$$V_n^{(1)} = (-1/2k_1)^n \exp (n+1)\, \theta_1 \quad ,$$

whereas the terms of the Nth ROSALES series generalize the former expressions (23, 24, 25) :

$$\begin{aligned}
V_0^{(N)} &= \sum_{i=1}^{N} \exp \theta_i \\
-V_1^{(N)} &= \sum_{i=1}^{N} \frac{\exp 2\theta_i}{2k_i} + \sum_{i<j=1}^{N} \alpha_{ij} \exp (\theta_i + \theta_j) \\
V_2^{(N)} &= \sum_{i=1}^{N} \frac{\exp 3\theta_i}{4k_i^2} + \sum_{i\neq j=1}^{N} \alpha_{iij} \exp (2\theta_i+\theta_j) + \sum_{i<j<\ell=1}^{N} \alpha_{ij\ell} \exp (\theta_i+\theta_j+\theta_\ell) \\
&\vdots \\
(-)^n V_n^{(N)} &= \sum_{i=1}^{N} \frac{\exp (n+1)\,\theta_i}{(2k_i)^n} + \sum_{i\neq j=1}^{N} \alpha_{i..ij} \exp (n\theta_i+\theta_j) + \dots \\
&\quad \dots + \sum_{i_1<..<i_{n+1}}^{N} \alpha_{i_1..i_{n+1}} \exp (\theta_{i_1} + \dots + \theta_{i_{n+1}}) \;, \\
&\vdots
\end{aligned} \qquad (29)$$

with mixing coefficients $\alpha_{ij} = \alpha(k_i,k_j)$, $\alpha_{iij} = \beta(k_i,k_j)$, $\alpha_{ij\ell} = \gamma(k_i,k_j,k_\ell)$, ... the structure of which does not depend on the particular values of the indices, but on the form of the padeon equation.

The results obtained at N=2 suggest that one should look for possible families of real N-pole solutions to the padeon equation, of the form (18), which would sum the Nth **ROSALES** series. If such solutions exist we call them "generalized N-padeons". The above structure (29) of the ROSALES terms leads to the following :

Lemma. If a padeon equation possesses a sequence of generalized N-padeons $V^{(N)}$ for N = 2, 3, ..., then these solutions take the form : $V^{(N)} = -2\partial_x \ln D_N$, with :

$$\begin{aligned}
D_N = 1 &+ \varepsilon\, (e_1 + \dots + e_N) + \varepsilon^2 \sum_{i<j=1}^{N} K_{ij}\, e_i e_j + \dots \\
&+ \varepsilon^r \sum_{i_1<..<i_r=1}^{N} K_{i_1..i_r}\, e_{i_1} \dots e_{i_r} + \dots + \varepsilon^N K_{12..N}\, e_1 \dots e_N \;,
\end{aligned} \qquad (30)$$

where $e_i = \dfrac{\exp \theta_i}{2k_i}$, and $K_{i_1 \dots i_r} = K^{(r)}(k_{i_1}, \dots k_{i_r})$.

<u>Proof</u>. Setting $D_N = \prod_{i=1}^{N} (1 + \varepsilon z_i) = 1 + \sum_{j=1}^{N} \varepsilon^j d_j^{(N)}(x,t)$, the N-pole solution (18) reads $V^{(N)} = \varepsilon D_N^{-1}.[\sum_{j=0}^{N-1} \varepsilon^j c_j^{(N)}(x,t)]$, with $c_j^{(N)}(x,t) = -2\partial_x d_{j+1}^{(N)}(x,t)$.

As this N-padeon should sum the Nth ROSALES series, its coefficients $\{c_j, d_j\}$ must obey the following relations :

$$c_0^{(N)} = V_0^{(N)} \tag{31}$$

$$c_1^{(N)} = V_0^{(N)} d_1^{(N)} + V_1^{(N)} \tag{32}$$

$$c_2^{(N)} = V_0^{(N)} d_2^{(N)} + V_1^{(N)} d_1^{(N)} + V_2^{(N)} \tag{33}$$

$$\vdots$$

$$c_{N-1}^{(N)} = V_0^{(N)} d_{N-1}^{(N)} + V_1^{(N)} d_{N-2}^{(N)} + \ldots + V_{N-2}^{(N)} d_1^{(N)} + V_{N-1}^{(N)} \tag{34}$$

$$0 = V_q^{(N)} d_N^{(N)} + V_{q+1}^{(N)} d_{N-1}^{(N)} + \ldots + V_{N+q-1}^{(N)} d_1^{(N)} + V_{N+q}^{(N)} , \quad q = 0,1,2,\ldots \tag{35}$$

Together with the relation $d_j^{(N)} = \frac{1}{2} \int_x^\infty c_{j-1}^{(N)} dx'$, the first N relations (31-34) determine the N-padeon from the only knowledge of the first N ROSALES terms : $V_0^{(N)}, \ldots V_{N-1}^{(N)}$. The further relations (35) constitute a sequence of consistency relations, relating the further terms $V_{N+q}^{(N)}$, $q = 0,1,2,\ldots$, to the former ones. The relation (31) yields $c_0^{(N)} = \sum_{i=1}^{N} \exp \theta_i$, so that : $d_1^{(N)} = \frac{1}{2} \int_x^\infty c_0^{(N)} dx' = \sum_{i=1}^{N} e_i$

$$\tag{36}$$

To verify the lemma for the other coefficients $d_{j>1}^{(N)}$ we proceed by introduction. At N = 2, we get from rel. (32) : $c_1^{(2)} = V_0^{(2)} d_1^{(2)} + V_1^{(2)}$, from which we conclude that $c_1^{(2)}$ contains only exponential terms proportional to $\exp(\theta_i + \theta_j)$, i, j = 1, 2. Yet, the terms which in $V_0^{(2)} d_1^{(2)} + V_1^{(2)}$ are proportional to $\exp(2\theta_i)$ cancel each other on account of the first consistency relation (35) for the padeon (N = 1) : $V_0^{(1)} d_1^{(1)} + V_1^{(1)} = 0$. This proves the lemma at N = 2, and one has :

$$c_1^{(2)} = \left(\frac{k_1+k_2}{2k_1k_2} - \alpha_{12}\right) \exp(\theta_1+\theta_2), \quad d_2 = \frac{1}{2} \int_x^\infty c_1^{(2)} dx' = K_{12}\, e_1 e_2$$

$$\text{with} \quad K_{12} = \left[1 - \frac{2k_1k_2}{k_1+k_2}\, \alpha_{12}\right] . \tag{37}$$

Let now the lemma be true for some N > 2, and let us consider an (N+1)-padeon solution. Comparison of the nth ROSALES terms $V_n^{(N)}$ and $V_n^{(N+1)}$ in two successive series shows that for $n \leqslant N-1$ these corresponding terms are quite similar since they contain exponentials involving up to $n + 1 \leq N$ different indices, whereas $V_N^{(N+1)}$ contains just one "new" term involving N+1 different indices :

$$\alpha_{12\ldots N+1} \exp (\theta_1 + \ldots + \theta_{N+1}) \quad ,$$

which did not appear in $V_N^{(N)}$. It follows that since the structure of the coefficients $d_j^{(N)}$ agrees, by hypothesis, with fomula (30), this will also be true for the first N coefficients $d_{j\leq N}^{(N+1)}$. As to the last coefficient $d_{N+1}^{(N+1)}$, we remark that all the exponential terms of $c_N^{(N+1)} = V_0^{(N+1)} d_N^{(N+1)} + V_1^{(N+1)} d_{N-1}^{(N+1)} + \ldots + V_{N-1}^{(N+1)} d_1^{(N+1)} + V_N^{(N+1)}$, which do not involve N+1 different indices, must cancel each other on account of the first consistency relation (35) which results from the existence of the N-padeon. It follows that both $c_N^{(N+1)}$ and $d_{N+1}^{(N+1)}$ must be proportional to $\exp (\theta_1 + \ldots + \theta_{N+1})$.

3) Regular N-padeons.

At this stage it is useful to distinguish among the generalized N-padeons of the form (30) a class of "regular" N-padeons of type (3), for which :

$$K_{i_1\ldots i_r} = K^{(r)}(k_{i_1},\ldots k_{i_r}) > 0 \quad , \quad r = 2, \ldots N \tag{38}$$

By using the structure (30) of successive regular N-padeons one can show that these solutions behave asymptotically as N-soliton potentials.

Theorem. If a padeon equation possesses a sequence of regular N-padeons, N=2,3,..., and if $V^{(N)} = -2\partial_x \ln D_N(\theta_1,\ldots\theta_N)$ is any such solution with $\theta_i = -k_i(x-v_i t)+\tau_i$, $0 < k_1 < \ldots < k_N$, $v_i = \omega(k_i)/k_i$ and $v_i \neq v_j$, for $i \neq j$, then :

$$\lim_{t\to\pm\infty} \partial_x V^{(N)} = - \sum_{i=1}^{N} \frac{k_i^2}{2} \operatorname{sech}^2 \left(\frac{\theta_i + \tau_i^{\pm}}{2}\right) \quad , \tag{39}$$

with

$$\sum_{i=1}^{N} (\tau_i^+ - \tau_i^-) = 0 \quad . \tag{40}$$

Proof. Let the N phasevelocities v_i be ordered as follows : $v_{i_1} < v_{i_2} < \ldots < v_{i_N}$. Given a reference frame $x = vt + \xi$ we remark that the particular structure of D_N, displayed by formula (30), together with the linearity of the functions θ_i with respect to x and t, leads to the following property : when $t \to \pm\infty$, $\partial_x \ln D_N$ tends to a constant (independent of ξ) unless the asymptotic behaviour of D_N is dominated by at least two "competing" terms which are either finite or which grow to infinity at the same exponential rate, but which differ by a factor $\exp(k_0\xi)$, $k_0 \neq 0$.

Under the condition (38) it is easy to verify that in any reference frame which is not the rest-frame of one of the solitary waves with parameters $k_i, \ldots k_N$ (a frame in which one of the θ's remains constant) the asymptotic behaviour of D_N is dominated by one single term. As $x = vt + \xi$ with $v \neq v_i$, $i = 1, 2, \ldots N$, one has :

$$\theta_i = k_i\,(v_i - v)\,t + \tau_i - k_i\,\xi \quad ,$$

with the following possibilities :

i) $v < v_{i_1}$ or $v > v_{i_N}$: as $t \to \pm\infty$ one has either $\theta_i \to +\infty$, $\forall$ i or $\theta_i \to -\infty$, $\forall$ i, so that either $D_N \sim \varepsilon^N K_{12\ldots N}\, e_1 \ldots e_N$ or $D_N \to 1$.

ii) $v_{i_r} < v < v_{i_{r+1}}$: as $t \to \pm\infty$ one has $\theta_{i_{r+1}}, \ldots \theta_{i_N} \to \pm\infty$ and $\theta_{i_1}, \ldots \theta_{i_r} \to \mp\infty$, so that, as $t \to +\infty$: $D_N \sim \varepsilon^{N-r} K_{i_{r+1}\ldots i_N}\, e_{i_{r+1}} \ldots e_{i_N}$, whereas for $t \to -\infty$: $D_N \sim \varepsilon^r K_{i_1\ldots i_r}\, e_{i_1} \ldots e_{i_r}$.

If, on the other hand, one follows the pth solitary wave by taking $v = v_p \equiv v_{i_r}$, and if $t \to \pm\infty$ one has : $\theta_{i_{r+1}}, \ldots \theta_{i_N} \to \pm\infty$ and $\theta_{i_1}, \ldots \theta_{i_{r-1}} \to \mp\infty$, whereas $\theta_{i_r} = \theta_p$ remains fixed.

As $t \to +\infty$ one has : $D_N \sim \varepsilon^{N-r} K_{i_{r+1}\ldots i_N}\, e_{i_{r+1}} \ldots e_{i_N} + \varepsilon^{N-r+1} K_{i_r\ldots i_N}\, e_{i_r} \ldots e_{i_N}$,

$$\text{so that :} \quad -2\partial_x \ln D_N \to \sum_{j=r+1}^{N} 2k_{i_j} - 2\partial_x \ln\left[1 + \varepsilon\,\frac{K_{i_r\ldots i_N}}{K_{i_{r+1}\ldots i_N}}\, e_{i_r}\right]$$

As $t \to -\infty$ one has : $D_N \sim \varepsilon^{r-1} K_{i_1\ldots i_{r-1}}\, e_{i_1} \ldots e_{i_{r-1}} + \varepsilon^r K_{i_1\ldots i_r}\, e_{i_1} \ldots e_{i_r}$,

$$\text{so that :} \quad -2\partial_x \ln D_N \to \sum_{j=1}^{r-1} 2k_{i_j} - 2\partial_x \ln\left[1 + \varepsilon\,\frac{K_{i_1\ldots i_r}}{K_{i_1\ldots i_{r-1}}}\, e_{i_r}\right] \quad ,$$

with $K_{i_{r+1}\ldots i_N} = 1 = K_{i_r\ldots i_N}$ if $r = N$ and $K_{i_1\ldots i_{r-1}} = 1 = K_{i_1\ldots i_r}$ if $r = 1$.

We thus get :

$$\lim_{t\to\pm\infty;\,\theta_p \text{ fixed}} \{\partial_x V^{(N)}\} = -\frac{k_p^2}{2}\,\text{sech}^2\left(\frac{\theta_p + \tau_p^{\pm}}{2}\right) \quad ,$$

$$\text{with} \quad \tau_p^+ = \ln\left[\frac{\varepsilon}{2k_p}\,\frac{K_{i_r\ldots i_N}}{K_{i_{r+1}\ldots i_N}}\right] \quad , \quad \tau_p^- = \ln\left[\frac{\varepsilon}{2k_p}\,\frac{K_{i_r\ldots i_r}}{K_{i_1\ldots i_{r-1}}}\right] \quad . \tag{41}$$

In any other reference frame $\partial_x V^{(N)}$ vanishes asymptotically for whatever value of ξ.

We also remark that if $\delta_p = \tau_p^+ - \tau_p^-$ denotes the phaseshift experienced by the pth solitary wave as t goes from $-\infty$ to $+\infty$, one has :

$$\sum_{p=1}^{N} \delta_p = \sum_{r=1}^{N} \ln \left[\frac{K_{i_1 \ldots i_{r-1}} \cdot K_{i_r \ldots i_N}}{K_{i_1 \ldots i_r} \cdot K_{i_{r+1} \ldots i_N}}\right] = 0 \quad . \tag{42}$$

With this proof and that of the preceeding lemma we are led to the following :

<u>Corollary 1</u>. If a padeon equation possesses a sequence of (regular) n-padeons, from n=2 up to N-1, then each [N/N] fraction of the form (3) which matches the first N terms of the Nth ROSALES series, generated by an expression $-V_0^{(N)}$ of the form (4), has an x-derivative which behaves as an N-soliton solution.

<u>Corollary 2</u>. If the linear dispersion relation associated with a padeon equation produces a phasevelocity $v(k) = \omega(k)/k$ which is a continuous, strictly monotonic function of k, and if this padeon equation possesses a sequence of N-padeons, regular for a set of parameters $0 < k_1 < \ldots < k_N$, N = 2, 3, ..., then the x-derivative of any such regular N-padeon behaves asymptotically as an N-soliton solution.

The last assumptions on v(k) guarantee that the phasevelocities v_i are ordered for any choice of parameters $0 < k_1 < \ldots < k_N$. According to whether v(k) increases or decreases with k one has : $v_1 < v_2 < \ldots < v_N$ or $v_N < v_{N-1} < \ldots < v_1$. In the former case the phaseshift of the pth solitary wave becomes :

$$\delta_p = \ln \left[\frac{K_{1\ldots p-1} \cdot K_{p\ldots N}}{K_{1\ldots p} \cdot K_{p+1\ldots N}}\right] \quad , \tag{43}$$

whereas in the latter case δ_p is given by formula (43) with an over-all change of sign. In view of the standard soliton equations these assumptions on v(k) do not seem to constitute an important restriction on padeon equations. If the linear dispersion law of a padeon equation produces two phasevelocities $v_{\pm}(k) = \pm \frac{|\omega(k)|}{k}$, and if $|v(k)|$ is continuous and strictly monotonic on some interval, it is clear that the velocities $v_i = \varepsilon_i \, |v(k_i)|$, $\varepsilon_i = \pm 1$, remain ordered for whatever choice of ε_i and of the parameters k_i on that interval. It follows that the N-padeons will still behave as N-soliton potentials. This is the case for the potential BOUSSINESQ equation [3] for which $v_{\pm}(k) = \pm \sqrt{1 + k^2}$.

4) <u>Necessary and sufficient conditions for the existence of regular N-padeons</u>.

Given a padeon equation the existence of lower regular N-padeons (N = 2, 3, ...) can be checked straight away by inspection of some particular mixing coefficients which appear in the lower-order terms of the Nth ROSALES series.

To simplify the form of the explicit conditions it is convenient to consider the "reduced" mixing coefficients :

$$\bar{\alpha}_{i_1 \dots i_n} = \frac{\alpha_{i_1 \dots i_n}}{k_{i_1} + \dots + k_{i_n}} .$$

a) Dipadeons (N = 2)

The relations (36, 37) determine an explicit [2/2] fraction (19) in terms of the first-order mixing coefficient α_{12}, which appears in $V_1^{(2)}$. This fraction will solve a padeon equation of order r iff it matches the 2(ℓ-1) further ROSALES terms : $V_2^{(2)}, \dots V_{2\ell-1}^{(2)}$, as required by the first 2(ℓ-1) consistency relations (35) :

$$V_q^{(2)} d_2^{(2)} + V_{q+1}^{(2)} d_1^{(2)} + V_{q+2}^{(2)} = 0 \quad , \quad q = 0, 1, \dots 2\ell-3 \quad , \tag{44}$$

where the integer $\ell \leqslant r$ equals r + 1 minus the number of brackets in the expression (28) which vanish identically for $D = D_2$.

These 2(ℓ-1) conditions, necessary and sufficient for the existence of a dipadeon, are equivalent with $n_\ell = \sum_{m=2}^{\ell} (2m-1)$ explicit consistency conditions relating the mixing coefficients of order j, $1 \leqslant j \leqslant 2\ell-1$, to the first-order one : α_{12}.

Indeed, by using the expressions (36, 37) for $d_1^{(2)}$ and $d_2^{(2)}$ one finds that the first relation (44) amounts to the single condition :

$$\bar{\alpha}_{112} = \frac{1}{2k_1} \bar{\alpha}_{12} \tag{45}$$

The two first relations (44) are equivalent to the condition (45) plus the two conditions :

$$\bar{\alpha}_{112} = \frac{1}{4k_1^2} \bar{\alpha}_{12} \quad , \quad \bar{\alpha}_{1122} = \frac{1}{4k_1k_2} \bar{\alpha}_{12} + \frac{1}{4} \bar{\alpha}_{12}^{\ 2} \quad , \tag{46}$$

whereas the first four relations (44) are equivalent with the conditions (45, 46) plus the five conditions :

$$\bar{\alpha}_{11112} = \frac{1}{8k_1^3} \bar{\alpha}_{12} \quad , \quad \bar{\alpha}_{11122} = \frac{1}{8k_1^2k_2} \bar{\alpha}_{12} + \frac{1}{4k_1} \bar{\alpha}_{12}^2 \tag{47}$$

$$\bar{\alpha}_{111112} = \frac{1}{16k_1^4} \bar{\alpha}_{12} \quad , \quad \bar{\alpha}_{111122} = \frac{1}{16k_1^3k_2} \bar{\alpha}_{12} + \frac{3}{16k_1^2} \bar{\alpha}_{12}^2$$

$$\bar{\alpha}_{111222} = \frac{1}{16k_1^2k_2} \bar{\alpha}_{12} + \frac{1}{24k_1k_2} \bar{\alpha}_{12}^2 + \frac{1}{12} \bar{\alpha}_{12}^3 \quad , \tag{48}$$

and so on ...

If ℓ = 2, as it is the case for candidate soliton equations of the third order for which the two last brackets in the expression (28) vanish identically for whatever form of D, the 3 first conditions (45, 46) are necessary and sufficient for the existence of dipadeons. In this case the search for a two-soliton solution requires that the iteration be carried up to the third order. If $\ell = 3$, one must also check the further conditions (47, 48) by going up to the fifth order ...

If dipadeons have been found, their regularity is subject to the following "positivity" condition on $\bar{\alpha}_{12}$ which results from the rel. (37) and the condition (38) :

$$\bar{\alpha}_{12} < \frac{1}{2\,k_1\,k_2} \tag{49}$$

b) Tripadeons (N = 3)

Let there be dipadeons. The relations (31-33), together with the relation $d_j^{(3)} = \frac{1}{2}\int_x^\infty c_{j-1}^{(3)}\,dx'$ and the explicit form (29) of $V_0^{(3)}$, $V_1^{(3)}$ and $V_2^{(3)}$, determine a [3/3] fraction of the form $V^{(3)} = -2\partial_x \ln D_3$, where D_3 is given by an expression of type (30) with :

$$K_{ij} = 1 - 2k_i k_j\,\bar{\alpha}_{ij} \quad , \quad \text{and} \quad K_{123} = 1 + 4k_1k_2k_3\,\bar{\alpha}_{123} - 2\sum_{i<j=1}^{3} k_i k_j\,\bar{\alpha}_{ij} \tag{50}$$

Insertion of this fraction into the padeon equation shows that $V^{(3)}$ will actually solve the padeon equation iff it matches the $3(\ell-1)$ further perturbation terms : $V_3^{(3)}, \ldots V_{3-1}^{(3)}$, as required by the first $3(\ell-1)$ consistency relations (35) at N = 3 :

$$V_q^{(3)} d_3^{(3)} + V_{q+1}^{(3)} d_2^{(3)} + V_{q+2}^{(3)} d_1^{(3)} + V_{q+3}^{(3)} = 0 \quad , \quad q = 0, 1, \ldots 3\ell-4 \tag{51}$$

As the assumed existence of dipadeons fixes, through the former consistency relations (44), the form of any mixing coefficient which at N = 3 involves but two different indices,it turns out that the first relation (51) amounts to just one new condition relating α_{1123} to α_{123} :

$$\bar{\alpha}_{1123} = \frac{\bar{\alpha}_{123}}{2k_1} + \frac{1}{2}\,\bar{\alpha}_{12}\,\bar{\alpha}_{13} \tag{52}$$

The first 3 consistency relations (51) are equivalent with the condition (52) plus the further conditions :

$$\bar{\alpha}_{11123} = \frac{1}{4k_1^2}\,\bar{\alpha}_{123} + \frac{1}{2k_1}\,\bar{\alpha}_{12}\,\bar{\alpha}_{13}, \quad \bar{\alpha}_{11223} = \frac{1}{4k_1k_2}\,\bar{\alpha}_{123} + \frac{1}{2}\,\bar{\alpha}_{12}\,\bar{\alpha}_{123} + \bar{\alpha}_{12}\left(\frac{\bar{\alpha}_{13}}{4k_2} + \frac{\bar{\alpha}_{23}}{4k_1}\right) \tag{53}$$

$$\bar{\alpha}_{111123} = \frac{1}{8k_1^3}\,\alpha_{123} + \frac{3}{8k_1^2}\,\bar{\alpha}_{12}\,\bar{\alpha}_{13}$$

$$\bar{\alpha}_{111223} = \frac{1}{8k_1^2k_2}\bar{\alpha}_{123} + \frac{1}{2k_1}\bar{\alpha}_{12}\bar{\alpha}_{123} + \frac{1}{4k_1k_2}\bar{\alpha}_{12}\bar{\alpha}_{13} + \frac{1}{8k_1^2}\bar{\alpha}_{12}\bar{\alpha}_{23} + \frac{1}{4}\bar{\alpha}_{12}^2\bar{\alpha}_{13}$$

$$\bar{\alpha}_{112233} = \frac{1}{8k_1k_2k_3}\bar{\alpha}_{123} + \frac{1}{4}\bar{\alpha}_{123}^2 + \bar{\alpha}_{123}\left(\frac{\bar{\alpha}_{12}}{4k_3} + \frac{\bar{\alpha}_{13}}{4k_2} + \frac{\bar{\alpha}_{23}}{4k_1}\right) + \frac{1}{8k_2k_3}\bar{\alpha}_{12}\bar{\alpha}_{13}$$
$$+ \frac{1}{8k_1k_3}\bar{\alpha}_{12}\bar{\alpha}_{23} + \frac{1}{8k_1k_2}\bar{\alpha}_{13}\bar{\alpha}_{23} + \frac{1}{2}\bar{\alpha}_{12}\bar{\alpha}_{13}\bar{\alpha}_{23} \qquad (54)$$

If $\ell = 2$, the conditions (52-53) are necessary and sufficient for the existence of tripadeons. In this case, the search for tri-soliton solutions requires that the iteration be carried up to the fifth order.
If it exists, the solution $V^{(3)}$ will be a regular tripadeon provided that the following "positivity" conditions are satisfied :

$$K_{ij} > 0 \quad \text{or} \quad \bar{\alpha}_{ij} < \frac{1}{2k_ik_j} \qquad \text{for} \qquad 1 \leqslant i < j \leqslant 3 \qquad (55)$$

$$K_{123} > 0 \quad \text{or} \quad 2\sum_{i<j=1}^{3} k_ik_j\,\bar{\alpha}_{ij} < 1 + 4k_1k_2k_3\,\bar{\alpha}_{123} \qquad (56)$$

As one proceeds step by step to higher values of N, one must verify at each step $N(\ell-1)$ consistency relations of the form (35), involving the perturbation terms $V_N^{(N)}$, $V_{N+1}^{(N)}$, ... up to $V_{N\ell-1}^{(N)}$. Yet, we remark that at each step these consistency relations should only be checked for those exponential terms which involve N different θ's.

5) <u>Two-wave collision properties at $N \geqslant 3$</u>

A striking feature of the multisoliton interaction which is described by the standard ($N \geqslant 3$)-soliton solutions, is that the collisions occur in pairs [6] : the phaseshift experienced by the pth soliton is the sum of two-soliton phaseshifts, each of which corresponds to the interaction of that soliton with the N-1 other ones.
If such a "two-wave" collision property is regarded as being a characteristic of genuine multisoliton solutions, we must examine which additional properties it imposes on regular N-padeons at $N \geqslant 3$.
Without any loss of generality we assume in the following that the N phasevelocities v_i which enter into the N-padeon are ordered as the N k_i's : $v_1 < v_2 < \dots < v_N$, so that the phaseshift of the pth solitary wave is given by formula (43).
At N = 2 we have : $\delta_1^{(2)} = \ln K_{12} = -\,\delta_2^{(2)}$, whereas at N = 3 :

$$\delta_1^{(3)} = \ln K_{123} - \ln K_{23}\ ,\ \delta_2^{(3)} = \ln K_{23} - \ln K_{12}\ ,\ \delta_3^{(3)} = \ln K_{12} - \ln K_{123} \qquad (57)$$

Each of the phaseshifts $\delta_i^{(3)}$ will be the sum of two phaseshifts $\delta_j^{(2)}$, as it is already the case for $\delta_2^{(3)}$, provided that K_{123} satisfies the factorization property :

$$K_{123} = K_{12}\, K_{13}\, K_{23} \tag{58}$$

By using the above expression (50) for K_{123} one finds that this factorization property determines the form of α_{123} in terms of that of α_{ij} :

$$\bar{\alpha}_{123} = k_1\, \bar{\alpha}_{12}\, \bar{\alpha}_{13} + k_2\, \bar{\alpha}_{12}\, \bar{\alpha}_{23} + k_3\, \bar{\alpha}_{13}\, \bar{\alpha}_{23} - 2k_1k_2k_3\, \bar{\alpha}_{12}\, \bar{\alpha}_{13}\, \bar{\alpha}_{23} \tag{59}$$

As one repeats the same argument at higher values of N one finds that the regular N-padeons will possess the two-wave collision property iff all the coefficients $K_{i_1 \ldots i_r}$ which appear in the expression (30) are linked to K_{ij} by the factorization condition :

$$K_{i_1 \ldots i_r} = \prod_{s<t=1}^{r} K_{i_s i_t} \tag{60}$$

It follows that if a padeon equation possesses a sequence of regular n-padeons, $n \leqslant N$, whose x-derivatives behave as n-soliton solutions with the two-wave collision property at $n \geqslant 3$, this will be reflected by the fact that the form of each mixing coefficient which involves more than 2 indices but less than N+1 different indices is determined by the form of the first order mixing coefficient α_{12}, either through the factorization condition (60), or through the consistency relations (35).

It also follows that the regularity of an $(n \geqslant 3)$-padeon, with the factorization property (60), is only subject to the conditions :

$$K_{ij} > 0 \quad , \quad 1 \leqslant i < j \leqslant n \tag{61}$$

As a contrast to HIROTA's families of nonlinear equations which may be cast in the s.c. "bilinear form" [7], the regular $(N \geqslant 3)$-padeons do not always possess the two-wave collision property which characterized the KdV-solitons. This means that the padeon ansatz may produce multisoliton solutions which fall beyond the standard IST-scheme and also beyond HIROTA's multisoliton formulas. Due to the symmetry between the x and t variables, displayed by the basic solutions (2) of the linearized padeon equations, we finally remark that similar results could also be obtained by starting with "t-padeons" of the form : $\tilde{V} = \partial_t \ln (1+\varepsilon z)$, $z = \dfrac{\exp \theta}{\omega}$, instead of "x-padeons" of the form (2). An example of a t-padeon is the solitary wave solution of the potential regularized long wave equation [5].

References

[1] : M.J. Ablowitz, H. Segur. *Solitons and the inverse scattering transform.* Siam, Philadelphia (1981).

[2] : R. ROSALES. *Exact solutions of some nonlinear evolution equations.* Stud. Appl. Math. 59, 117 (1978).

[3] : F. lambert, M. Musette. *Solitary waves, padeons and solitons.* Lect. Notes Math. 1071, Springer-Verlag, 197 (1983).

[4] : R. Hirota, M. Ito. *Resonance of solitons in one dimension.* J. Physic. Soc. Japan 52, 744 (1983).

[5] : J. Eilbeck. *Numerical studies of solitons.* Springer series in Solid-state Sciences 8, Springer-Verlag, 28 (1978).

[6] : V.E. Zakharov. *Kinetic equation for solitons.* Soviet Phys. JETP, 33, 538 (1971).

[7] : R. Hirota. *Direct Methods in Soliton Theory.* Solitons, Springer-Verlag, 157 (1980).

APPLICATION OF RATIONAL APPROXIMATIONS TO SOME FUNCTIONAL EQUATIONS

Pierre Moussa
Service de Physique Théorique
CEN - Saclay
91191 Gif-sur-Yvette cedex France

1) Introduction

Let $T(z)$ and $W(z)$ be monic complex polynomials with respective degrees d and $(d-1)$:

$$T(z) = z^d + \sum_{i=1}^{d} t_i z^{d-i} \quad , \qquad W(z) = z^{d-1} + \sum_{i=1}^{d-1} w_i z^{d-1-i} \quad , \tag{1}$$

and let F be the set of formal power series with complex coefficients, expanded around infinity, such that $g \in F$ when :

$$g(z) = \sum_{k=0}^{\infty} \mu_k / z^{k+1} \qquad , \qquad \mu_0 = 1. \tag{2}$$

Now we define the transformation τ by :

$$(\tau g)(z) = W(z)\, g(T(z)) \ . \tag{3}$$

It is easily checked that $g \in F$ implies $\tau g \in F$. We are in particular interested in the properties of the fixed points of τ in F, which satisfy the functional equation :

$$g(z) = W(z)\, g(T(z)) \qquad . \tag{4}$$

Similar functional equations appear in many areas of mathematics such as the theory of iteration of polynomials [1], the analysis of the geometrical properties of the invariant sets [2], the orthogonality properties of iterated polynomials [3, 4], and in theoretical physics in the study of the almost periodic discrete Schrödinger equation [5, 6], in the renormalisation group approach of critical behaviour in some statistical mechanics models [7,8,9], and in the vibration spectrum of fractal structures [10]. We shall first analyse the properties of (3) and (4) in the sense of formal power series, then describe the analyticity properties of g, and finally discuss how (3) is related to a polynomial change of variable in a measure, and (4) to an invariant measure under the polynomial transformation T of the complex plane.

2) Formal power series analysis

We shall denote $T^{(n)}(z)$ the iterates of T, defined recursively by $T^{(o)}(z) = z$ and $T^{(n)}(z) = T(T^{(n-1)}(z))$ for $n > 0$. The Padé approximants $[n-1/n]_g(z)$ to the series g given by (2) is defined as usual by:

$$g(z) - [m-1/m]_g(z) = O(z^{-(2m+1)}) , \quad (5)$$

$$[m-1/m]_g(z) = Q^g_m(z)/P^g_m(z) , \quad (6)$$

where Q^g_m and P^g_m are monic complex polynomials with respective degrees $(m-1)$ and m. We have the following results :

Theorem 1 : The equation (4) has a unique solution g_∞ in F, satisfying $\tau g_\infty = g_\infty$. The iteration scheme $g_{n+1} = \tau g_n$ starting from any g_o in F fulfils : for $k < d^n$, $\mu_k^{(n)} = \mu_k^{(\infty)}$, where $g_n(z) = \sum_k \mu_k^{(n)}/z^{k+1}$, and $g_\infty(z) = \sum_k \mu_k^{(\infty)}/z^{k+1}$.

Proof : we equate the coefficient of $z^{-(k+1)}$ in both sides of (4), which express μ_k as a linear combination of μ_l, with $d^l \leqslant k$. Therefore we get unambiguously the μ_k in successive order starting from $\mu_o = 1$. Consider now the iterated series g_n, we get from (4), $g_n(z) = R_n(z)(1+O(z^{-d^n}))$, where :

$$R_n(z) = W(z)W(T(z))\ldots W(T^{(n-1)}z))/T^{(n)}(z) \sim \frac{1}{z} . \quad (7)$$

Therefore the $\mu_n^{(k)}$ are independent on g_o for $k < d^n$. Choosing $g_o = g_\infty$ gives the result.

Theorem 2 : When they exist, the Padé approximants to g and τg are related by:

$$[dm-1/dm]_{\tau g}(z) = W(z)[m-1/m]_g(T(z)) \quad . \tag{8}$$

When they exist, the Padé approximant to g_∞ fulfil :

$$[dm-1/dm]_{g_\infty}(z) = W(z)[m-1/m]_{g_\infty}(T(z)) \quad . \tag{9}$$

Proof: we substitute T(z) to z in (5) and multiply both sides by W(z). Thus we get in the right hand side $O(z^{-d(2m+1)+d-1}) = O(z^{-(2dm+1)})$. Then (3) and the uniqueness of the Padé approximant give (8). Since $\tau g_\infty = g_\infty$, we get (9). Using (6) we get the following corollary :

Corollary 3 : The numerators and denominators of Padé approximants fulfil :

$$Q^{\tau g}_{dm}(z) = W(z)\, Q^{g}_{m}(T(z)) \quad , \qquad P^{\tau g}_{dm}(z) = P^{g}_{m}(T(z)) \quad , \tag{10}$$

$$Q^{g_\infty}_{dm}(z) = W(z)\, Q^{g_\infty}_{m}(T(z)) \quad , \qquad P^{g_\infty}_{dm}(z) = P^{g_\infty}_{m}(T(z)) \quad , \tag{11}$$

The previous corollary permits a recursion argument to compute the continued fraction expansion of $g_\infty(z)$. More precisely, the numerator and denominator polynomials in (6) fulfil the classical three terms relation :

$$P^{g}_{m+1}(z) = (z-A^{g}_{m})P^{g}_{m}(z) - R^{g}_{m}\, P^{g}_{m-1}(z) \quad , \tag{12}$$

associated with the (formal) continued fraction expansion of g(z) :

$$g(z) = 1/(z-A^{g}_{0}-R^{g}_{1}/(z-A^{g}_{1}-R^{g}_{2}/ \ \dots \ /(z-A^{g}_{n}-R^{g}_{n}/(z-A^{g}_{n+1} \dots))\dots)) \quad . \tag{13}$$

It is convenient to define the three diagonal semi infinite Jacobi matrix associated to g by its only non vanishing elements :

$$H^{g}_{ii} = A^{g}_{i} \ , \ H^{g}_{i+1,i} = R^{g}_{i+1} \quad , \quad H^{g}_{i,i+1} = 1 \ , \quad i = 0,1,2, \ \dots \tag{14}$$

We also define what is called in theoretical physics a decimation matrix D by its matrix elements : For i, j = 0, 1, 2, ..., we have $D_{ij} = 1$ if $j = di$, 0 otherwise. Then we have :

Theorem 4 : When g and τg admit a continued fraction expansion, their coefficients defined in (13) fulfil a relation which is easily displayed in matrix form :

$$H^{g}D = DT(H^{\tau g}) \quad . \tag{15}$$

The coefficients of the continued fraction expansion of g_{∞} fulfil :

$$H^{g_{\infty}} D = DT(H^{g_{\infty}}) \quad . \tag{16}$$

Proof : first one easily checks that (15) holds when applied to the column vector $\psi^{\tau g}(z)$ with components $\psi_n^{\tau g}(z) = P_n^{\tau g}(z)$. For this check, one only needs equation (10), the definition of D and equation (12) written in matrix form : $H^g\psi^g(z) = z\,\psi^g(z)$. Due to the particular structure of the matrix H^g and D, all matrix products involved contain only sums with finite number of terms, therefore no specification on the growth of the components of vector is needed at this stage. Now, defining $M = H^gD - DT(H^{\tau g})$, we have $M\psi^{\tau g}(z) = 0$. But in each line of M, there is only a finite number of matrix elements. Therefore each component of $M\psi^{\tau g}(z)$ is an identically vanishing polynomial. Since we have assumed the existence of the continued fraction expansion each $P_n^{\tau g}$ is monic with degree n. From this one deduces that each matrix element of M vanishes, which gives (15). Choosing $g = g^{\infty}$ gives (16), which achieves the proof.

The Theorem 4 permits a recursive computation of the coefficients of the continued fraction expansion of τg knowing g, and also of the coefficients of the continued fraction expansion of g_{∞}. More precisely, the matrix element $(n,(n-1)d)$ of (15) gives for $n \geqslant 1$ the relation :

$$R_{dn}^{\tau g}\, R_{dn-1}^{\tau g} \cdots R_{dn-d+1}^{\tau g} = R_n^{g} \quad , \tag{17}$$

which express $R_{dn}^{\tau g}$ knowing the coefficients of lower order. From the matrix element $(n,(n-1)d+2k)$ one gets $R_{dn+k}^{\tau g}$ for $k = 1$ to $d-1$, and from the element $(n,(n-1)d+2k+1)$ one gets $A_{dn+k}^{\tau g}$ for $k = 0, 1, \ldots d-1$. For $n = 0$ one needs the knowledge of $A_0^{\tau g}, A_1^{\tau g}, \ldots A_{p-1}^{\tau g}, R_2^{\tau g}, \ldots, R_{p-1}^{\tau g}$ for $d = 2p$ even, and the knowledge of $A_0^{\tau g}, A_1^{\tau g}, \ldots A_{p-1}^{\tau g}, R_1^{\tau g}, R_2^{\tau g}, \ldots R_p^{\tau g}$ for $d = (2p+1)$ odd, to start the recursion. We therefore need $(d-1)$

initial conditions. The required initial parameters do not depend on the particular g considered, and can be obtained by a direct computation as the (d-1) first coefficients of the continued fraction expansion of the rational fraction W(z)/T(z).

As an example [11], we consider the case d = 2, $T(z) = z^2-\lambda$ and W(z) = z. That is for g_∞ the equation :

$$g_\infty(z) = z\ g_\infty(z^2-\lambda) \quad . \tag{18}$$

The starting condition is $A_o = 0$, and we first get $R_1 = \lambda$. Then the recursion relations give $A_n = 0$, for any n, and for n > 0:

$$R_{2n}\ R_{2n-1} = R_n \quad , \quad R_{2n} + R_{2n+1} = \lambda \quad , \tag{19}$$

which gives all R_n as rational fractions in the parameter λ. In some cases [5], for instance |λ| large enough, the R_n are limit periodic functions of n, which shows that H^{g_∞} is a limit periodic discrete operator [5,12].

Remarks : The present results generalise previous works where the particular case W(z) = (1/d) T'(z) was considered [3,4,11] . We have also here deliberately used the formal series point of view. In fact the existence of the continued fraction expansion of g_∞ depends only on the conditions $R_n^{g_\infty} \neq 0$. One can check that in general, one gets $R_n^{g_\infty}$ and $A_n^{g_\infty}$ as rational fractions of the coefficients of T and W given in equations (1) and (2). No $R_n^{g_\infty}$ does identically vanish. So g_∞ always admits an expansion of the type (13), where the coefficients are rational fraction of w_i and t_j.

3) Analyticity properties

The point at infinity is an attractive fixed point for the polynomial transformation T. Let A(∞) be the immediate basin of attraction of infinity, that is the open connected part containing infinity of the set of points z such that $|T^{(n)}(z)|$ goes to infinity with n.

Theorem 5 : The power series g_∞ obtained in Theorem 1 represents a holomorphic function in $A(\infty)$.

Proof : We first show that $g_\infty(z)$ is analytic around infinity : one first easily finds R such that for $|z| > R$, one has $2|z|^d > |T(z)| > (1/2)|z|^d > R$, from which one gets for $|z| > R$: $2^j > |T^{(j)}(z)|/|z|^{d^j} > (1/2)^j$. Now let a be an arbitrary point not in $A(\infty)$, for instance a fixed point of T, and let $g_o(z) = (z-a)^{-1}$. To study the convergence of the sequence $g_\ell = \tau^\ell g_o$ we consider the associated infinite product :

$$\pi(z) = g_o(z) \times \prod_{j=o}^{\infty} \frac{g_{j+1}(z)}{g_j(z)} = g_o(z) \prod_{j=o}^{\infty} \rho(u_j) \quad , \tag{20}$$

where $\rho(u) = (u-a)W(u)/(T(u)-a)$, and $u_j = T^{(j)}(z)$. Since $a \notin A(\infty)$, the general term $\rho(u_j)$ has no pole for $z \in A(\infty)$, and by enlarging R if necessary to avoid possible zeroes of W, one gets a bound for $\log|\rho|$ of the following form: for $|u|>R$, $|\ell n|\rho(u)||<C/|u|$, where C is a constant. Joining these estimates shows that for $|z| > 2R$ the infinite product is uniformly convergent to an analytic function, the expansion of which is nothing else than $g_\infty(z)$. Now given any z in $A(\infty)$ there exists N such that $|T^{(N)}(z)| > 2R$, and using the functional equation, we have $g_\infty(z) = W(z)W(T(z)) \ldots W(T^{(N-1)}(z))g_\infty(T^{(N)}(z))$ which shows that $g_\infty(z)$ is analytic in the vicinity of z, which achieves the proof.

Remark : the boundary of $A(\infty)$ is nothing else than the Julia set J of the polynomial T [1]. J can be connected or made of an infinite number of connected parts. So $A(\infty)$ is not always simply connected. All rational fractions g_n considered in the proof of theorem 5 have all their poles outside $A(\infty)$. If a is in J, all their poles are also in J. These rational functions g_n are uniformly convergent rational approximations in any compact set included in $A(\infty)$, to the function $g_\infty(z)$.

4) Polynomial transformation on a measure in the complex plane.

We shall show in this section how equation (3) is related to a polynomial change of variable in a measure on the complex plane : let μ be a bounded complex measure on the complex plane, with bounded support. Let $T_i^{(-1)}(z)$, $i = 1, \ldots d$ be a complete assignment of branches of the inverse of the polynomial function $T(z)$ given in (1). In addition let $\gamma_i(x)$, $i = 1, \ldots d$ be measurable complex functions on the complex plane, such that :

$$\sum_{i=1}^{d} \gamma_i(x) = 1 \tag{21}$$

To the measure μ, we associate the transformed measure μ^T by :

$$\int f(x)\, d\mu^T(x) = \sum_{i=1}^{d} \int \gamma_i(x)\; f(T_i^{(-1)}(x))\, d\mu(x) \quad , \tag{22}$$

for any measurable f. For any Borel set E, $T_i^{(-1)} E \cap T_j^{(-1)} E$, $i \neq j$, is contained in the finite set C of critical points of T, that is points x such that $T'(x) = 0$. We shall assume that $\mu(C) = 0$. Using (22) for $f(x) = \chi_{T_i^{(-1)}E}(x)$, the characteristic function of the set $T_i^{(-1)} E$, we get :

$$\mu^T(T_i^{(-1)} E) = \int_E \gamma_i(x)\, d\mu(x) = \int \gamma_i(x)\, \chi_E(x)\, d\mu(x) \quad , \tag{23}$$

from which we get :

$$\mu^T(T^{(-1)}E) = \mu(E) \quad . \tag{24}$$

Therefore (22) is a particular choice of a non one to one change of variable in the measure μ. If μ satisfies $\mu = \mu^T$, it will be an invariant measure under the transformation T. Let μ_n (resp μ_n^T) be the moments of the measure μ (resp μ_n^T) :

$$\mu_n = \int x^n\, d\mu(x) \quad , \quad \mu_n^T = \int x^n\, d\mu^T(x) \quad , \quad n \geqslant 0 \; . \tag{25}$$

and let g (resp g^T) be the associated generating series :

$$g(z) = \sum_{n=0}^{\infty} \mu_n / z^{n+1} \quad , \quad g^T(z) = \sum_{n=0}^{\infty} \mu_n^T / z^{n+1} \; . \tag{26}$$

For z outside the support of the corresponding measure these series converge to

$$g(z) = \int \frac{d\mu(x)}{z-x} \quad , \qquad g^T(z) = \int \frac{d\mu^T(x)}{z-x} \quad . \tag{27}$$

Using (22), we get :

$$g^T(z) = \int \frac{W(z,x)}{T(z)-x} \, d\mu(x) \quad , \tag{28}$$

with:

$$\frac{W(z,x)}{T(z)-x} = \sum_{i=1}^{d} \frac{\gamma_i(x)}{z - T_i^{(-1)}(x)} \quad . \tag{29}$$

$W(z,x)$ is a monic polynomial with degree $(d-1)$ in z. If we further assume W to be independant on x, we get the transformation (3) :

$$g^T(z) = W(z)\, g(T(z)) = \tau g(z) \tag{30}$$

which corresponds to the choice :

$$\gamma_i(x) = W(T_i^{(-1)}(x))/T'(T_i^{(-1)}(x))$$

The particular choice $W(z) = \frac{1}{d} T'(z)$ leads to the so called balanced choice [4] of the transformed measure and has already been analyzed [4,3]. Here we have shown that the algebraic results valid in this particular case extend to the more general situation considered here. Now the following question arises, wether $g_\infty(z)$ is the generating function of an invariant measure μ_∞. In fact, as we have seen in the concluding remark of the previous section, $g_k(z) = \tau^k g_o(z)$ has poles on the Julia set associated to T, provided we choose $g_o(z) = (z-a)^{-1}$ with a in J. In addition if we choose a outside the orbit of iterates of the critical point of J (when they belong to J) all poles of $g_k(z)$ are simple. Therefore $g_k(z)$ is the generating function of a discrete complex measure μ_k, with d^k masspoints on the preimages of a. If this sequence of measures has accumulation points in the set of measures, these limiting measures have their support on J. So we have to discuss existence and uniqueness of these limiting measures. We list here the results :

Theorem 6 :

i) when $W(z) = T'(z)/d$, $g_\infty(z)$ is the generating function of an invariant measure supported by J.

ii) when $W(z)/T'(z)$ is real and positive on J, then there exists limiting measures.

iii) when the complement of $A(\infty)$ has an empty interior, then if the limiting measures exist, they all coincide with an invariant measure supported by J.

iv) when the Julia set is real, and when $W(z)/T'(z)$ is positive on J, then conditions ii) and iii) are both fulfilled, and $g_\infty(z)$ is the generating function of an invariant measure on J.

Proof : the statement in i) is due to Brolin [13]. In case ii) one easily sees that all measure μ_k are probability measures (positive and normalized) on J. The set of probability measures on the bounded closed set J being compact, one gets the statement. Statement iii) is an immediate consequence of Mergelyan's theorem [14]. Statement iv) is a consequence of ii) and iii). A connected result is given in [15].

Remarks : no general result is known yet. The main difficulty comes from the lack of positivity of the discrete measures μ_k. Condition iv) in the previous theorem is easy to fulfil when the Julia set of T is a real Cantor set [1]. In this case, the zeroes of T' belong to the open complement of J. There is no difficulty to find W having each of is zeroes belonging to the same open connected interval of the complement of J, which achieves condition ii in Theorem 6.

More generally, when the measure μ_∞ associated to g_∞ exists, the polynomials $P_m^{g_\infty}$ defined in (11) are orthogonal with respect to μ_∞ in the sense of the scalar product $\langle f, g\rangle = \int f(z)g(z)d\mu(z)$, which is not hermitian, unless T, W and the Julia set J are real.

This work has benefited from the permanent collaboration with J. Geronimo and D. Bessis.

References

1. For a review including classical references see : P. Blanchard, *Complex analytical dynamics on the Riemann sphere*, Bull. Amer. Math. Soc. 11 (1984) 85-141

2. M. Barnsley, J. Geronimo, A. Harrington, *On the invariant sets of a family of quadratic maps*, Commun. Math. Phys., 88 (1983) 479-501

3. D. Bessis, P. Moussa, *Orthogonality properties of iterated polynomial mappings*, Commun. Math. Phys. 88 (1983) 503-529

4. M. Barnsley, J. Geronimo, A. Harrington, Bull. Amer. Math. Soc., 7 (1982) 381-384.

5. J. Bellissard, D. Bessis, P. Moussa, *Chaotic states of almost periodic Schrödinger operators*. Phys. Rev. Lett. 49 (1982) 701-704.

6. P. Moussa, D. Bessis, *A solvable almost periodic Schrödinger operator*, *Stochastic aspects of Classical and Quantum systems*, 136-147, Lectures Notes in Math., 1109 Springer 1985.

7. B. Derrida, J.P. Eckmann, A. Erzan, *Renormalisation groups with periodic and aperiodic orbits*, J. Phys. A: Math. Gen. 16 (1983) 893-906

8. Th. Niemeijer, J.M. Van Leeuwen, *Renormalisation theory for Ising like spin systems*, *Phase transition and Critical Phenomena*, Vol. 6, 425-505, C. Domb and M.S. Green editors, Academic Press, N.Y. 1976.

9. D. Bessis, J. Geronimo, P. Moussa, *Mellin transforms associated with Julia sets and physical applications*. J. Stat. Phys., 34 (1984) 75-110.

10. R. Rammal, *Spectrum of harmonic excitations on fractals*, J. Physique 45 (1984) 191-206.

11. D. Bessis, M.L. Mehta, P. Moussa, *Orthogonal polynomials on a family of Cantor sets, and the problem of iteration of quadratic mappings*, Letter Math. Phys. 6 (1982) 123-140

12. G. Baker, D. Bessis, P. Moussa, *A family of almost periodic Schrödinger operators*, Physica, 124A (1984) 61-78

13. H. Brolin, *Invariant sets under iteration of rational functions*, Ark. Mat. 6 (1965) 103-144

14. See for instance, W. Rudin, *Real and complex analysis*, p. 386, Mc Graw Hill 1970

15. D. Bessis, J. Geronimo, P. Moussa. *Function weighted measure and orthogonal polynomials on Julia sets* (preprint)

OPERATOR RATIONAL FUNCTIONS AND VARIATIONAL METHODS FOR THE MODEL OPERATOR

Maciej Pindor
Institute of Theoretical Physics
Warsaw University
00-681 Warszawa, ul. Hoza 69, Poland .

1. Introduction

Operator Continued Fractions (OCF) and Operator Padé Approximants (OPA) have been earlier applied to sum perturbation series for scattering amplitudes [1,2] and for the model operator [3] with promising results. It is the purpose of this work to reanalyse the way in which the model operator is expressed as an OCF.

In section 2 we define our notation and recall results of [3] . In section 3 we derive a compact expression for approximants of the OCF for the model operator and show to what solution of the equation for the model operator it converges. A connection with the Rayleigh-Ritz method is pointed out and simultaneously the convergence region for the OCF for the model operator is greatly enlarged. Other solutions of the original operator equation are also expressed as OCF's.

In section 4 we concentrate on the simplest case of one dimensional model space and show that then the model operator can be expressed as an OPA. An equivalence between this expression and the variational method is also established.

2. Operator Continued Fraction for the model operator

Let us consider a system with a hamiltonian:

$$H = H_0 + gV \qquad (1)$$

acting on a Hilbert space $\mathcal{H}$ and having eigenvectors and eigenstates defined through the following equations:

$$H\,\Psi_\beta = E_\beta\,\Psi_\beta$$

$$H_0\,\Phi_\beta = E^0_\beta\,\Phi_\beta$$

$$\sum_\beta |\Phi_\beta\rangle\langle\Phi_\beta| = 1 \qquad (2)$$

$$\langle\Phi_\sigma|\Phi_\beta\rangle = \delta_{\sigma\beta}$$

We consider now $\mathcal{H}_0^D$ spanned by a group of Φ_β , $\beta \in D$:

$$\mathcal{H}_0^D = [\Phi_\beta]_{\beta\in D} \quad ; \quad \dim \mathcal{H}_0^D = d$$

which is called the model space, and we introduce:

$$P = \sum_{\beta\in D} |\Phi_\beta\rangle\langle\Phi_\beta|$$

$$P\,\mathcal{H} = \mathcal{H}_0^D \qquad (3)$$

Now we consider d vectors Ψ_β such that:

$$P\,\Psi_\beta = \Psi_\beta^0 \neq 0 \qquad (4)$$

and the Ψ_β^0's are linearly independent. We call the latter " model functions ". Finally, we can define the " model operator " Ω through the equations:

$$\Omega\,\Psi_\beta^0 = \Psi_\beta \qquad (\,\Omega\,\mathcal{H}_0^D = \mathcal{H}^D\,)$$

$$\Omega(\,I-P)\mathcal{H} = \Omega Q\mathcal{H} = 0 \qquad (5)$$

Ω satisfies two obvious identities frequently used below:

$$\Omega P = \Omega$$

$$P\Omega = P \qquad (6)$$

The effective hamiltonian can be expressed in terms of Ω :

$$H_{eff} = PH\Omega = PH_0P + gPV\Omega \qquad (7)$$

The name " effective hamiltonian " comes from the fact that:

$$H_{eff}\,\Psi_\beta^0 = PH\Omega\,\Psi_\beta^0 = PH\,\Psi_\beta = PE_\beta\,\Psi_\beta = E_\beta\,\Psi_\beta^0$$

Therefore it is a finite dimensional operator which has, in $\mathcal{H}_0^D$, the same eigenvalues as H in $\mathcal{H}^D$.
The above considerations are interesting only if we can find Ω . Lindgren [4] has shown that:

$$[\Omega, H_0] = gV\Omega - g\Omega V\Omega \tag{8}$$

and this equation is our starting point.
We concentrate now on the degenerate case i.e. the one when:

$$H_0 \Phi_\beta = E^0 \Phi_\beta \qquad \beta \in D$$

If so then (8) leads to:

$$(E^0 - H_0)\Omega = gV\Omega - g\Omega V\Omega \tag{9}$$

and after introducing:

$$R = \frac{Q}{E^0 - H_0} \tag{10}$$

finally to:

$$gR\Omega V\Omega + (I - gRV)\Omega - P = 0 \tag{11}$$

This is a quadratic operator equation and, in general, it has infinitely many solutions. In fact, we expect this to be the case: Ω is defined by a choice of d Ψ's (4) and, generally, there will be infinitely many such choices. However, we are mainly interested in a solution which converges to P when $g \to 0$. This solution can be expressed as an operator (periodic) continued fraction. Namely, putting

$$\Omega = \frac{1}{I - gRV + gR\Omega V} P \tag{12}$$

and iterating this equation with $\Omega^{(0)} = P$ we get:

$$\Omega = \cfrac{1}{I - gRV + gR \cfrac{1}{I - gRV + gR \cfrac{1}{I - \dots} PV} PV} P \tag{13}$$

By expanding $\Omega^{(n)}$ in powers of g one can check that this expansion agrees up to the nth order with that obtained directly from (11).
Sufficient conditions for the convergence of (13), written as in ref. [5] are:

$$\| gR(1-gRV)^{-1} \| . \| PV(1-gRV)^{-1} \| < 1/4$$

or

$$\| g(1-gRV)^{-1}R \| . \| (1-gRV)^{-1}PV \| < 1/4 \tag{14}$$

3. Operator Continued Fractions and the Rayleigh-Ritz method

The conditions (14) are presumably much too restrictive for periodic OCF and

we shall indeed show that in our special case the OCF (13) converges almost everywhere.
In fact, the equ.(12) can be written as:

$$\Omega = P + g \frac{1}{Q(I-gRV+gR\Omega V)Q}(RV-R\Omega V)P \tag{15}$$

(by $\frac{1}{QAQ}$ we understand $Q\frac{1}{\beta P+QAQ}$ for any β) and it is the same as:

$$\Omega = P + g \frac{1}{Q(E^0-H_0-gV+\Omega V)Q} Q(V-\Omega V)P \tag{16}$$

and finally (remembering that $(E^0-H_0)P = 0$):

$$\Omega = P - \frac{1}{Q[E^0-H-\Omega(E^0-H)]Q} Q\{E^0-H-\Omega(E^0-H)]P \tag{17}$$

Now we can easily check that:

$$\Omega^{(n)} = P - \frac{1}{Q(E^0-H)^n Q} Q(E^0-H)^n P \tag{18}$$

It follows from the theorem discussed in the Appendix that (18) converges to an operator producing out of $P\mathcal{H}$ a subspace spanned by d eigenvectors Ψ_{β_k}, $\beta_k \in D$ of H such that:

a) their projections onto $P\mathcal{H}$ are linearly independent
b) $\left| E^0-E_{\beta_k} \right| < \left| E^0-E_{\beta_\ell} \right| \qquad \ell=1,\dots d\ ,\ \ell \neq k$
$\qquad \beta_k \in D \qquad\qquad \beta_\ell \in D$

To see more clearly what it means let us take $d = 1$. Then (18) converges for $n \to \infty$ to an operator which produces out of Φ_0 an eigenvector of H with the eigenvalue nearest to E^0 and a nonzero projection on Φ_0. The method converges also in a degenerate case, though there will, in general, be a discontinuity for this value of g which produces the degeneracy (see the final remark in the Appendix).
In other words (18) can be seen as nothing more but the iterative method for finding d eigenvectors of H belonging to d dominant eigenvalues of $\frac{1}{E^0-H}$.
At this stage we can observe that E^0 is no more necessarily an eigenvalue of H_0. Indeed, we can always make a transformation:

$$H_0 \to H_0 + \varepsilon I \ ; \ E^0 \to E^0 + \varepsilon \ ; \ V \to V - \varepsilon I \tag{19}$$

if so, then:

$$E^0 - H \to E^0 + \varepsilon - H$$

This transformation does not change the original equation for Ω (11) because $\Omega\Omega = \Omega$. However, it changes the continued fraction (13). On the other hand, the nth approximant of this new continued fraction is represented by (18) with $E^0 \rightarrow E^0 + \varepsilon$, and therefore it converges, in general, to a different operator Ω. Thus, using the transformation (19) we can generate OCF representations for different solutions of (11).

Of course, in any application of the method we must face a problem of finding the non-trivial operator inverses appearing in (13), a problem which could be more difficult than the original one of finding the spectrum of H. A practical approach to this problem could be (without discussing here all the subtleties involved) to project a finite dimensional subspace $Q_L\mathcal{H}$ out of $Q\mathcal{H}$, to calculate the operator inverses there, and to look for the limit in which this finite dimensional subspace becomes dense in $Q\mathcal{H}$. If so , then Q in (17) must be changed to Q_L:

$$\Omega_L = P - \frac{1}{Q_L[E^0-H-\Omega(E^0-H)]Q_L} Q_L[E^0-H-\Omega(E^0-H)]P \tag{20}$$

Iterating this equation with $\Omega_L^{(0)} = P$ we get:

$$\Omega_L^{(n)} = P - \frac{1}{Q_L[(E^0-H)_L]^n Q_L} Q_L[(E^0-H)_L^n]P \tag{21}$$

where

$$(E^0-H)_L = (P+Q_L)(E^0-H)(P+Q_L) \tag{22}$$

It is visible that with this additional approximation the OCF is entirely equivalent to the Rayleigh-Ritz method [6] with d+L trial vectors spanning $(P+Q_L)\mathcal{H}$.

This equivalence becomes even more evident if we remark that the original equation (11) is invariant under the more general transformation than (19) i.e. under:

$$H_0 \rightarrow \beta P + \sigma Q \; ; \; gV \rightarrow H - \beta P - \sigma Q \; ; \; E^0 \rightarrow \beta \tag{23}$$

Now the Φ_β^0's can be chosen almost at will and their interpretation as " trial functions " is fully justified.

Closing this section we want to point out that the equivalence we have shown holds only in the case of a degenerate model space. If the model space is non-degenerate then (17) is replaced by:

$$\Omega = P - g \sum_\beta \frac{1}{Q[E_\beta^0-H-\Omega(E_\beta^0-H)]Q} Q[E_\beta^0-H-\Omega(E_\beta^0-H)]P_\beta \tag{24}$$

This formula leads to a branched continued fraction [7] and its approximants cannot be transformed into a form analogous to (18). As a consequence, even if the operator inverses appearing in the calculation of the approximants are approximated by inverses in finite dimensional subspaces of QH , the results do not correspond to the Rayleigh-Ritz method with any choice of trial vectors.

4. Operator Padé Approximants

We shall now introduce another operator ω related to Ω through:

$$\Omega = \omega + P \tag{25}$$

When this relation is introduced into (11) we get:

$$gR\,\omega V\omega + gR\,\omega VP + (I-gRV)\omega - gRVP = 0 \tag{26}$$

We can find a continuous fraction solution of this equation only when $d = 1$. In this case $PVP = \langle \Phi_0 | V | \Phi_0 \rangle P = V_0 P$ (let us observe that $Q\omega P=\omega$), and therefore (26) takes the form:

$$gR\,\omega\, V\omega + (Q-gRV+gRV_0 Q)\,\omega = gRVP \tag{27}$$

which leads to:

$$\omega = g \frac{1}{Q(I-gRV+gRV_0+gR\,\omega\, V)Q} RVP \tag{28}$$

Now it is very interesting to observe that this equation is invariant with respect to the transformation (19), which was not the case for (12). Therefore, an OCF being a solution of (28) must correspond to a model operator different from the one found from (12). Let us now notice that the OCF following from (28) is actually an Operator Padé Approximant.

Indeed, the nth approximant of this OCF, $\omega^{(n)}$ is given by:

$$\omega^{(n)} = g \frac{1}{Q\,[\, I-gR(V-V_0)+gR\,\omega^{(n-1)}V\,]\,Q} RVP \tag{29}$$

i.e.

$$gR\,\omega^{(n)}V\,\omega^{(n-1)} - gR(V-V_0)\,\omega^{(n)} + \omega^{(n)} = gRVP \tag{30}$$

and it is evident that if the power expansion of $\omega^{(n-1)}$ is exact (i.e. if it coincides with the power expansion following directly from (27)) up to k^{th} order , that of $\omega^{(n)}$ is exact up to order $k+2$ (the expansion of ω starts with the first order).It then follows from this last observation that the nth

iteration of (28) is exact up to the $(2n+1)^{th}$ order, i.e. that it is the $[n-1/n]$ OPA to ω and, consequently, that $P + \omega^{(n)}$ is the $[n/n]$ OPA to Ω.

When we now transform (28) in the same way as we have done with (16) we get:

$$\omega = - \frac{1}{Q\,[E^0+gV_0-H-\omega\,(E^0+gV_0-H)]\,Q}\; Q(E^0+gV_0-H)P \tag{31}$$

and using the fact that $P(E^0+gV_0-H)P = 0$ we immediately get:

$$\omega^{(n)} = - \frac{1}{Q\,[(E^0+gV_0-H)^{(n)}]\,Q}\; Q\,[(E^0+gV_0-H)^{(n)}]\,P \tag{32}$$

Therefore $(P+\lim_{n\to\infty}\omega^{(n)})\,\Phi_0$ is an eigenstate of H which corresponds to the eigenvalue nearest to E^0+gV_0.

Again, when operators in denominators are inverted in a finite dimensional subspace of $\mathcal{H}$, the limit for $n\to\infty$ gives the operator producing an eigenstate of $(P+Q_L)H(P+Q_L)$ corresponding to the eigenvalue nearest to E^0+gV_0. This is equivalent again to the Rayleigh-Ritz method for $H-gV_0$.

5. Conclusions

We have demonstrated that OCF and OPA for the model operator have much wider regions of convergence than those following from [5], though the conditions we have found are expressed in terms of an unknown spectrum of the operator H. We have also shown that the OCF for the model operator corresponds to eigenstates of the hamiltonian having eigenvalues closest to a selected eigenvalue of the unperturbed hamiltonian. This result strongly supports a similar conjecture expressed in [8], though in that paper one considered an OCF for the effective hamiltonian rather than for the model operator.
However, we have also shown for the degenerate model space that when the OCF or OPA are calculated in an approximate way using the projection technique, the results obtained are reproducible by the Rayleigh-Ritz method. In particular, for finite dimensional operators the results correspond exactly to those obtained with the iterative method of calculating eigenvectors.

Appendix

Lemma. Let A be a linear operator in a Hilbert space $\mathcal{H}$ and let P be a projection operator and $Q = I-P$. Assume that $P\mathcal{H} \subset D(A^n)$ and that QA^nQ

has an inverse for all integer n. Then:

$$P - \frac{1}{QA^nQ} QA^nQP = A^{-n}P \frac{1}{PA^{-n}P} P \qquad \text{for all } n \tag{A.1}$$

Proof:

Let $$\Phi^{(n)} = \frac{1}{QA^nQ} QA^nPu \qquad \text{for some } u \in H \tag{A.2}$$

then: $$QA^nQ\,\Phi^{(n)} - QA^nPu = 0 \tag{A.3}$$

i.e. $$A^nQ\,\Phi^{(n)} - A^nPu = p \in PH \tag{A.4}$$

Thus: $$Q\,\Phi^{(n)} = Pu + A^{-n}p = \Phi^{(n)} \tag{A.5}$$

and $$Pu + PA^{-n}p = 0 \tag{A.6}$$

or $$p = -\frac{1}{PA^{-n}P} Pu \tag{A.7}$$

Inserting this into (A.5) we get:

$$\Phi^{(n)} = Pu - A^{-n}\frac{1}{PA^{-n}P} Pu \tag{A.8}$$

As this is true for any $u \in H$, (A.1) follows.

Theorem

Let B be a selfadjoint, bounded, linear operator in a Hilbert space H. Let P be a projection operator on a d ($< \infty$) dimensional subspace of H. Assume that $PH \subset D(A^n)$ and that PA^nP has an inverse for all integer n. Assume moreover that among the eigenvectors of B one can choose d vectors v_i (i = 1,...d) such that:

a) the eigenvalues corresponding to these eigenvectors are farther from the origin than the eigenvalues corresponding to other eigenvectors and than the continuous spectrum,

b) the Pv_i are linearly independent.

Then:

$$\lim_{n\to\infty} B^n \frac{1}{PB^nP} Pu = \sum_{i=1}^{d} \beta_i v_i \tag{A.9}$$

for any u such that $Pu = 0$, and some β_i (i=1,...d)

We present here a sketch of the proof for the case of a purely discreet spectrum. There are no substantial modifications necessary if the spectrum has a continuous component.

We introduce a basis in PH : $\{u_k\}_{k=1}^{d}$ and a set of eigenvectors of B: $\{v_k\}_{k=1}^{d}$.

We have: $B\, v_k = \mu_k v_k \qquad |\mu_1| > |\mu_2| > \dots |\mu_d| \quad \dots$ (A.10)

Then:

$$B = \sum_{k=1} \mu_k\, |v_k\rangle\langle v_k| = \sum_{k=1} \mu_k P_k \tag{A.11}$$

and for any $u \in H$:

$$B^n \frac{1}{PB^nP} Pu = \sum_{k=1} \sum_{m,\ell=1}^{d} \mu_k^n\, |v_k\rangle\langle v_k\,|\,u_m\rangle\, a_{m\ell}\, \langle u_\ell\,|\,u\rangle \tag{A.12}$$

where $a_{m\ell}$ are matrix elements of a matrix inverse of:

$$b_{ij} = \sum_{k=1} \langle u_i | v_k\rangle\, \mu_k^n\, \langle v_k | u_j\rangle \tag{A.13}$$

$$a_{m\ell} = \beta_{m\ell} \,/\, \det(b_{ij})$$

where $\beta_{m\ell}$ is the algebraic complement of the $(m,\ell)^{th}$ element of b_{ij}.

$\mathrm{Det}(b_{ij})$ contains terms of the form:

$$\mu_{i_1}^n * \dots * \mu_{i_d}^n * \text{ something}$$

We see immediately that " something " multiplying μ_ℓ^{nd} is zero for any ℓ because it is $\det(\langle u_i | v_\ell\rangle\langle v_\ell | u_j\rangle)$. Similarly, all coefficients standing at products of μ_k's with repeated indices vanish because they are proportional to vanishing determinants. The largest nonvanishing term is therefore proportional to:

$$\mu_1^n * \mu_2^n * \dots * \mu_{d-1}^n$$

However, in (A.12) $\beta_{m\ell}$ is summed with $|v_k\rangle\, \mu_k^n\, \langle v_k | u_m\rangle$ and therefore the coefficients at corresponding powers of μ_k^n 's appearing in this expression are proportional either to the same determinants as those appearing in the denominator or to sums of products of algebraic complements of b_{ij} with matrix elements from a different row, such sums vanishing identically.

As a result, only terms with:

$$|v_k\rangle * \mu_{i_1}^n * \dots * \mu_k^n * \dots * \mu_{i_d}^n$$

remain in the numerator.

Finally, only coefficients at $|v_k\rangle$ $k = 1,\dots d$ in (A.12) will not vanish in the limit $n \to \infty$.

It is also evident, from the above considerations that if any μ_k $k = 1,\dots d$ is degenerate then

$$\lim_{n \to \infty} A^n \frac{1}{PA^nP} P$$

produces out of PH a d-dimensional subspace spanned by eigenvectors belonging to other largest eigenvalues and a specific linear combination of eigenvectors belonging to this degenerate eigenvalue.

References.

[1] J. Fleisher, M. Pindor : "Evaluation of operator Padé approximants for perturbation expansions in scattering theory". Phys.Rev.D24,1978(1981)

[2] M. Pindor, G. Turchetti :"Padé approximants and variational methods for operator series". Nuovo Cimento A71, 171 (1982)

[3] M. Pindor : "Operator continued fractions and boud states" Nuovo Cimento B84, 105 (1984)

[4] I. Lindgren : "The Rayleigh-Schrödinger perturbation and the linked diagrams theorem for a multiconfigurational model space" J.Phys. B7, 2441 (1974)

[5] H. Denk, M. Riederle : "A generalization of a theorem of Pringsheim" J. Approx. Th. 35, 355 (1982)

[6] M.A. Abdel-Raouf : "On the variational methods for bound states and scatteing problems" Phys. Rep. 84, n.3 (1982)

[7] V.Ya. Skorobogat'ko : "The theory of branched continued fractions and its application in computational mathematics" (in Russian) Nauka, Moscow 1983.

[8] H.M. Hoffman : "Probleme einer phaenomenologie-freien methode..." Habilitation Thesis, Erlangen University 1976.

The Generalized Schur Algorithm for the Superfast Solution of Toeplitz Systems

Gregory S. Ammar
Department of Mathematical Sciences
Northern Illinois University
DeKalb, Illinois 60115
U. S. A.

William B. Gragg *
Department of Mathematics
University of Kentucky
Lexington, Kentucky 40506
U. S. A.

Abstract

We review the connections between fast, $O(n^2)$, Toeplitz solvers and the classical theory of Szegö polynomials and Schur's algorithm. We then give a concise classically motivated presentation of the superfast, $O(n\log_2^2 n)$, Toeplitz solver that has recently been introduced independently by deHoog and Musicus. In particular, we describe this algorithm in terms of a generalization of Schur's classical algorithm.

1. Introduction

Let $M = [\mu_{j-k}] \in \mathbb{C}^{n \times n}$ be a *Toeplitz matrix*. The problem of solving the system of linear equations $Mx = b$ is important in many areas of pure and applied mathematics: orthogonal polynomials, Padé approximation, signal processing, linear filtering, linear prediction and time series analysis. See, for instance, [1, 3, 15, 20, 21, 23, 24]. There are several *fast*, $O(n^2)$, algorithms for solving such systems. This is in contrast with the $O(n^3)$ operations normally used to solve an arbitrary $n \times n$ system, for instance, by Gaussian factorization. Asymptotically *superfast* , $O(n\log_2^2 n)$, algorithms have been proposed for solving such systems [4, 5, 10, 22] but to our knowledge these methods have not yet been yet implemented.

In this paper we give a classically motivated presentation of the algorithm that has recently been independently presented by deHoog [10] and Musicus [22] in the case where the Toeplitz matrix M is (Hermitian) positive definite. Our treatment is based on the relations among positive definte Toeplitz matrices, Szegö polynomials and Schur's algorithm [25]. In particular, the deHoog-Musicus algorithm is naturally explained in terms of a

* Research supported in part by the National Science Foundation under grant DMS-8404980 and by the Seminar für Angewandte Mathematik of the ETH-Zürich.

generalization of Schur's algorithm. An analogous treatment of the positive definite *Hankel* case, $M = [\mu_{j+k}] = M^* > 0$, generalizing the algorithm of Chebyshev [7, 12], is given in [18].

In Section 2 we review the classical foundations of fast Toeplitz solvers. We present the generalized Schur algorithm in Section 3 and describe the use of the algorithm for the superfast solution of a positive definite Toeplitz system in section 4.

Before proceeding, we note that the restriction to positive definite Toeplitz systems is not as severe as it may seem. First, the positive definite case is of primary interest in most important applications, including discrete time Wiener filtering, autocorrelation problems, and Gaussian quadrature on the unit circle [16, 17]. Second, it is clear that most fast and superfast Toeplitz solvers are numerically unstable, and therefore unreliable, when applied to an arbitrary Toeplitz systems: [8, 9, 6]. In this connection we note that the algorithm of [5, 18], while potentially stable for positive definite *Hankel* systems, is *manifestly unstable* for positive definite Toeplitz systems. However, Cybenko has shown that the algorithms of Levinson, Durbin, and Trench are numerically stable for the class of positive definite Toeplitz matrices, and stability for this class can be expected in some superfast algorithms [6].

The implementation of the generalized Schur algorithm and the superfast (positive definite) Toeplitz solver of deHoog and Musicus is described in [2].

2. The Classical Foundations of Fast Toeplitz Solvers

2.1. Positive definite matrices and orthogonal polynomials.

Every complex (Hermitian) positive definite matrix $M = [\mu_{j,k}]$ can be factored uniquely as

$$M = L\, D\, L^*, \tag{2.1}$$

with $L = [\lambda_{j,k}]$ unit left triangular ($\lambda_{k,k} \equiv 1$) and $D = \mathrm{diag}[\delta_k]$ positive definite. The matrix $\hat{L} := L\, D^{1/2}$ is the *left Choleski factor* of M. Equivalently,

$$R^* M R = D \tag{2.2}$$

and

$$M^{-1} = R\, D^{-1} R^*, \tag{2.3}$$

with $R = [\rho_{j,k}] = L^{-*}$ unit right triangular. By abuse of terminology, we call (2.1) the *Choleski factorization* of M, and (2.2) or (2.3) the *inverse Choleski factorization* of M. (Actually, (2.3) is the *reverse* Choleski factorization of M^{-1}.) If either of these factorizations is known then the linear system $M x = b$ can be solved directly with at most n^2 multiplicative and additive operations (n^2 *flops*). Both factorizations are represented by the formula

$$\mathrm{T} := [\tau_{j,k}] := M\,R = L\,D. \tag{2.4}$$

In special cases, most notably when M is *Hankel,* $M = [\mu_{j+k}]$, or *Toeplitz,* $M = [\mu_{j-k}]$, it follows from *classical analysis* that these factorizations can be computed in $O(n^2)$ operations. We thus obtain *fast,* $O(n^2)$, algorithms for solving $M\,x = b$. In the Hankel case this point is moot: positive definite Hankel matrices are notoriously severely ill-conditioned (e.g., the *Hilbert matrix* with $\mu_k = \int_0^1 \lambda^k\, d\lambda = 1/(k+1)$). The situation can be quite different for positive definite Toeplitz matrices (e.g., $M = I$, the identity matrix).

We now describe the *orthogonal polynomials* associated with M. It is natural to number the indicies j and k from zero, and put

$$M := M_{n+1} := [\mu_{j,k}]_{j,k=0}^{n}.$$

More generally, for later use,

$$M_k := [\mu_{j,l}]_{j,l=0}^{k-1} \in \mathbb{C}^{k\times k} \quad (0 \le k-1 \le n)$$

is the kth *section* of M, and likewise for R_k and D_k. The positive definite matrix M determines an inner product $<\cdot,\cdot>$ for the complex vector space $\mathbb{C}_n[\lambda]$ of polynomials of degree at most n, on setting

$$<\lambda^j, \lambda^k> := \mu_{j,k} \quad (0 \le j \le n, 0 \le k \le n)$$

and extending $<\cdot,\cdot>$ to all of $(\mathbb{C}_n[\lambda])^2$ by requiring that it be linear in its second argument and conjugate linear in its first argument—like the Euclidean inner product y^*x for the complex vector space $\mathbb{C}^n$ of (column) n-vectors.

Now (2.2) states that the monic polynomials $\{\psi_k\}_0^n$ defined by

$$\psi_k(\lambda) := \textstyle\sum_j \rho_{j,k}\, \lambda^j \tag{2.5}$$

are orthogonal with respect to $<\cdot,\cdot>$:

$$<\psi_j, \psi_k> = \begin{cases} 0, & j \ne k, \\ \delta_k, & j = k \end{cases}.$$

Thus, $\{\psi_k\}_0^n$ is an orthogonal basis for $\mathbb{C}_n[\lambda]$, and the $(k+1)$th column of R contains the coefficients of the representation of ψ_k in terms of the standard basis $\{\lambda^j\}_0^n$ for $\mathbb{C}_n[\lambda]$. The columns of $\hat{R} := \hat{L}^{-*} = R\,D^{-1/2}$ likewise *generate* the ortho*normal* polynomials $\{\hat{\psi}_k\}_0^n$: $\hat{\psi}_k := \psi_k/\delta_k^{1/2}$. Finally, we see from (2.4) that

$$\begin{aligned} \tau_{j,k} &= \textstyle\sum_l \mu_{j,l}\rho_{l,k} \\ &= \textstyle\sum_l <\lambda^j,\lambda^l>\rho_{l,k} \\ &= <\lambda^j, \psi_k> = \lambda_{j,k}\delta_k \\ &= \begin{cases} 0, & j < k \\ \delta_k, & j = k \end{cases}. \end{aligned} \tag{2.6}$$

2.2. Inverse Choleski Factorization and Szegö Polynomials. We henceforth assume that $M = M^* = [\mu_{j-k}]_{j,k=0}^n > 0$ is a *Toeplitz matrix*. The orthogonal polynomials $\{\psi_k\}_0^n$ are then called the *Szegö polynomials* associated with M. They satisfy the *Szegö recurrence relation*

$$\psi_{k+1} = \lambda\, \psi_k + \gamma_{k+1}\tilde{\psi}_k, \tag{2.7a}$$

with $\tilde{\psi}_k(\lambda) = \lambda^k \overline{\psi}_k(1/\lambda)$ the polynomial obtained from ψ_k by conjugating and reversing the order of the coefficients,

$$\gamma_{k+1} = - < 1, \lambda\, \psi_k > / \delta_k, \tag{2.7b}$$

and

$$\delta_{k+1} = \delta_k\, (1 - |\gamma_{k+1}|^2). \tag{2.7c}$$

The numbers $\{\gamma_k\}_1^n$ are the *Schur parameters* associated with M. They determine the Szegö polynomials by (2.7a); note also that $\gamma_k \equiv \psi_k(0)$. From (2.7c) we see that $|\gamma_k| < 1\ (1 \le k \le n)$. The Schur parameters are referred to as *reflection coefficients* in the engineering literature, and as *partial correlation coefficients* in prediction theory.

Although we shall not use the result in this paper, it is known that there is a bounded nondecreasing function $m(\theta)$ with

$$< \beta, \alpha > = \frac{1}{2\pi} \int_0^{2\pi} \beta(\lambda)^* \alpha(\lambda)\, dm(\theta), \qquad \lambda = e^{i\theta}.$$

See, for instance, [1]. The Szegö polynomials are thus "orthogonal on the unit circle."

Formulas (2.7b) and (2.7c) follow rather directly from (2.7a), on using the orthogonality and the *isometry relation* $< \beta, \alpha > \equiv < \beta/\lambda, \alpha/\lambda >$, valid for $\alpha, \beta \in \mathbb{C}_n[\lambda]$ with $\alpha(0) = \beta(0) = 0$. However, a matrix theoretic proof of *all* of (2.7), based on the persymmetry of M, seems more efficient. The matrix $A \in \mathbb{C}^{n \times n}$ is *persymmetric* if it is invariant under reflection in its antidiagonal. This means that $A = A^P := J\, A^T J$, where $J := J_n$ is the $n \times n$ *reversal matrix* (obtained by reversing the columns of the $n \times n$ identity matrix).

Put

$$M_{k+1} =: \begin{bmatrix} M_k & \tilde{m}_k \\ \tilde{m}_k^* & \mu_0 \end{bmatrix} =: \begin{bmatrix} \mu_0 & m_k^* \\ m_k & M_k \end{bmatrix}$$

so that

$$m_k := [\, \mu_1, \mu_2, \cdots, \mu_k]^T$$

and

$$\tilde{m}_k := J_k\, \overline{m}_k.$$

Also put

$$R_{k+1} =: \begin{bmatrix} R_k & r_k \\ & 1 \end{bmatrix}, \quad r_{k+1} =: \begin{bmatrix} \gamma_{k+1} \\ s_k \end{bmatrix}$$

and

$$\tilde{r}_k := J_k \overline{r}_k .$$

Equating last columns in $M_{k+1} R_{k+1} = L_{k+1} D_{k+1}$ gives

$$M_k \; r_k + \tilde{m}_k = 0, \tag{2.8}$$
$$\tilde{m}_k^* \; r_k + \mu_0 = \delta_k . \tag{2.9}$$

Now

$$M_k = J_k \; M_k^T J_k = J_k \; \overline{M}_k \; J_k$$

is persymmetric and Hermitian. Hence (2.8) is equivalent with the *Yule-Walker equation*

$$m_k + M_k \; \tilde{r}_k = 0. \tag{2.10}$$

Now increase k by unity in (2.8) and use the second partitioning of M_{k+1} to get

$$\mu_0 \gamma_{k+1} + m_k^* s_k + \mu_{k+1}^* = 0, \tag{2.11}$$
$$m_k \gamma_{k+1} + M_k s_k + \tilde{m}_k = 0. \tag{2.12}$$

Subtracting (2.8) from (2.12), and using (2.10), we obtain

$$s_k = r_k + \tilde{r}_k \; \gamma_{k+1}, \tag{2.13}$$

that is,

$$r_{k+1} = \begin{bmatrix} 0 \\ r_k \end{bmatrix} + \begin{bmatrix} 1 \\ \tilde{r}_k \end{bmatrix} \gamma_{n+1}. \tag{2.14a}$$

This is the Szegö recurrence relation (2.7a). Using (2.13) in (2.11) we find, on account of (2.9), that

$$\begin{aligned} \delta_k \; \gamma_{k+1} &= - \; m_{k+1}^* \begin{bmatrix} r_k \\ 1 \end{bmatrix} \\ &= - \; \textstyle\sum_j \mu_{-j-1} \, \rho_{j,k} \\ &= - \; \textstyle\sum_j < 1, \lambda \, \psi_k >. \end{aligned} \tag{2.14b}$$

Finally, by (2.9), (2.14a) and (2.14b),

$$\begin{aligned} \delta_{k+1} &= \mu_0 + \tilde{m}_{k+1}^T \; \overline{r}_{k+1} \\ &= \mu_o + \tilde{m}_k^T \overline{r}_k + m_{k+1}^* \begin{bmatrix} r_k \\ 1 \end{bmatrix} \gamma_{k+1}^* \\ &= \delta_k \; (1 - \; |\gamma_{k+1}|^2). \end{aligned} \tag{2.14c}$$

Hence the Szegö recurrence relations (2.7) are equivalent with the matrix formulation (2.14) which, in the context of Toeplitz systems, is known as the *Levinson-Durbin algorithm* [21, 11, 14].

To solve the system $M\ x = b$ one puts

$$M_k\ x_k := b_k,\quad b_{k+1} := \begin{bmatrix} b_k \\ \beta_k \end{bmatrix},\quad b_{n+1} := b,$$

and finds from $x_{k+1} = M_{n+1}^{-1}\ b_{k+1}$ that

$$x_{k+1} = \begin{bmatrix} x_k \\ 0 \end{bmatrix} + \begin{bmatrix} r_k \\ 1 \end{bmatrix} \xi_{k+1}$$

with

$$\xi_{k+1} = [r_k^*, 1]\ b_{k+1} \,/\, \delta_k.$$

This algorithm applies *in general* to solve $M\ x = b$ when the inverse Choleski factorization of M is known; that is it makes no use of the Toeplitz structure of M.

The work for this two stage algorithm to solve $M_n\ x = b$ is about $2n^2$ flops, at most n^2 flops for each stage.

2.3. The Christoffel-Darboux-Szegö and Gohberg-Semencul Formulas. From (2.3) we see that the generating polynomial of $M^{-1} =: [\beta_{j,k}]$ is

$$\begin{aligned}\kappa_n(\lambda, \tau) &:= \textstyle\sum \beta_{j,k}\ \lambda^j\ \tau^{k*} \\ &= \textstyle\sum_0^n \hat{\psi}_k(\lambda)\ \hat{\psi}_k(\tau)^*.\end{aligned}$$

It is possile to express $\kappa_n(\lambda, \tau)$ solely in terms of the normalized polynomial $\hat{\psi}_n$. The unnormalized form of this result is

$$\delta_n\ (1 - \lambda\tau^*)\ \kappa_n(\lambda, \tau) = \tilde{\psi}_n(\lambda)\ \tilde{\psi}_n(\tau)^* - \lambda\ \tau^*\ \psi_n(\lambda)\ \psi_n(\tau)^*.$$

This is *Szegö's formula* [26]. It is the analog for Szegö polynomials of the *Christoffel-Darboux formula* for polynomials orthogonal on the real line. Its inductive verification reduces to

$$\tilde{\psi}_n(\lambda)\tilde{\psi}_n(\tau)^* - \lambda\tau^*\psi_n(\lambda)\psi_n(\tau)^* = (1 - |\gamma_n|^2)\ [\tilde{\psi}_{n-1}(\lambda)\tilde{\psi}_{n-1}(\tau)^* - \lambda\tau^*\psi_{n-1}(\lambda)\psi_{n-1}(\tau)^*],$$

which in turn is a direct consequence of (2.7a) and its equivalent:

$$\tilde{\psi}_{k+1} = \tilde{\psi}_k + \gamma_{k+1}^*\ \lambda\ \psi_k.$$

Setting now

$$\psi_n(\lambda) =: \textstyle\sum \rho_k\ \lambda^k,\quad \rho_k := \rho_{k,n},$$

we see that

$$\delta_n (\beta_{j,k} - \beta_{j-1,k-1}) = \rho^*_{n-j} \rho_{n-k} - \rho_{j-1} \rho^*_{k-1},$$

where elements with negative subscripts are zero. Thus

$$\delta_n \beta_{j,k} = \sum_l (\rho^*_{n+l-j} \rho_{n+l-k} - \rho_{j-l-1} \rho^*_{k-l-1})$$

and so

$$\delta_n M^{-1} = T_1^* T_1 - T_0 T_0^*$$

with *Toeplitz matrices*

$$T_0 := [\rho_{j-k-1}]^n_{j,k=0}$$

and

$$T_1 := [\rho_{n+j-k}]^n_{j,k=0}.$$

This is the Gohberg-Semencul formula [13]. Note that T_0 is *strictly* left triangular ($\rho_{-1} := 0$) and T_1 is unit right triangular ($\rho_n = \rho_{n,n} = 1$).

2.4. Choleski Factorization and Schur's Algorithm. We have seen that the Szegö recursions can be used to solve a positive definite Toepltiz system using the inverse Choleski factorization as well as the Gohberg-Semencul formula. We now describe an algorithm for finding the Choleski factors L_n and D_n of M_n (also see [22]). We will see that this algorithm is a manifestation of the classical algorithm of Schur. The algorithm presented below is in direct analogy with the derivation of Chebyshev's algorithm [7] for positive definite Hankel matrices, as presented, for instance, by Gautschi [12].

Extend the functional $<\cdot,\cdot>$ to *certain* pairs of Laurent polynomials by putting

$$< \lambda^j, \lambda^k > := \mu_{j-k}, \quad |j-k| \leq n.$$

Then

$$\tau_{j,k} := < \lambda^j, \psi_k >$$

and

$$\tilde{\tau}_{j,k} := < \lambda^j, \tilde{\psi}_k >$$

are defined for $0 \leq k \leq n$ and $-n+k < j < n$. In fact

$$\tau_{j,k} = \begin{cases} 0, & 0 \leq j < k \\ \delta_k, & j = k \\ \mu_j, & k = 0 \end{cases},$$

and moreover,

$$\begin{aligned} \tilde{\tau}_{j,k} &= \sum_{i=0}^k \rho^*_{i,k} < \lambda^j, \lambda^{k-i} > \\ &= (\sum_{i=0}^k \rho_{i,k} \mu_{k-i-j})^* \end{aligned}$$

$$=(\sum_{i=0}^{k} \rho_{i,k} < \lambda^{k-j},\lambda^{i}>)^*$$
$$=< \lambda^{k-j}, \psi_k >^* = \tau_{k-j,k}^*.$$

Now the Szegö recursion gives

$$\tau_{j,k} = <\lambda^j, \lambda\psi_{k-1}> + \gamma_k <\lambda^j, \tilde{\psi}_{k-1}>$$
$$= <\lambda^{j-1}, \psi_{k-1}> + \gamma_k \tilde{\tau}_{j,k-1}$$
$$= \tau_{j-1,k-1} + \gamma_k \tau_{k-j-1,k-1}^*,$$

and letting $j=0$ we obtain $\gamma_k = -\tau_{-1,k-1}/\delta_{k-1}$.

Since $L\ D = T := [\tau_{j,k}]_{j,k=0}^{n}$, the following algorithm can be used to obtain the Choleski factorization of a positive definite Toeplitz matrix.

Algorithm 2.2. (Fast Choleski Factorization).

input: $M = [\mu_{j-k}]_{j,k=0}^{n} > 0$,

$\tau_{0,0} = \mu_0$,

for $j = 1, 2, \cdots, n$

  $\tau_{j,0} = \mu_j$, $\tau_{-j,0} = \mu_j^*$,

  $\gamma_j = -\tau_{-1,j-1}/\tau_{j-1,j-1}$

  for $k = 1, 2, \cdots, j-1$

    $\tau_{j,k} = \tau_{j-1,k-1} + \gamma_k \tau_{k-j-1,k-1}^*$,

    $\tau_{k-j,k} = \tau_{k-j-1,k-1} + \gamma_k \tau_{j-1,k-1}^*$,

  $\tau_{j,j} = \tau_{j-1,j-1}(1 - |\gamma_j|^2)$

We now describe the classical algorithm of Schur [25]. The fast Choleski algorithm will then be shown to be equivalent with Schur's algorithm.

Let D $= \{\lambda : |\lambda| < 1\}$ be the open unit disk in the complex plane. A *Schur function* ϕ is holomorphic on D with $\phi(D)$ contained in the closure D^- of D. Schur's algorithm is a procedure that generates a (possibly finite) sequence $\{\phi_n\}$ of Schur functions by transforming an initial Schur function ϕ_0 by successive linear fractional transformations (LFT's). In this way a (formal) continued fraction representation of ϕ_0 is obtained.

Let $\phi = \phi_0$ be a Schur function, and put $\gamma = \gamma_1 := \phi_0(0)$. Note that $|\gamma| \leq 1$, and moreover, $|\gamma| = 1$ implies $\phi(\lambda) \equiv \gamma$ (by the maximum principle). In case $|\gamma| = 1$, Schur's algorithm terminates; if $|\gamma| < 1$ define

$$\phi_1 := \frac{1}{\lambda} \frac{\phi_0 - \gamma}{1 - \gamma^* \phi_0}.$$

To see that ϕ_1 is then a Schur function, note that the Möbius transformation

$(\gamma+\tau)/(1+\gamma^*\tau)$ maps D onto D with inverse function $(\tau-\gamma)/(1-\gamma^*\tau)$. The function $(\phi-\gamma)/(1-\gamma^*\phi)$ is therefore a Schur function that vanishes at $\lambda=0$, so by Schwarz' Lemma, ϕ_1 is also a Schur function. This construction represents one step of Schur's algorithm.

Schur's Algorithm.

input: an initial Schur function ϕ_0,

$\gamma_1 = \phi_0(0)$,

for $n=1,2,3,\cdots$ **while** $|\gamma_n|<1$

$$\phi_n(\lambda) = \frac{1}{\lambda}\,\frac{\phi_{n-1}(\lambda)-\gamma_n}{1-\gamma_n^*\phi_{n-1}(\lambda)},$$

$$\gamma_{n+1} = \phi_n(0).$$

Thus, $\phi_{n-1} = t_n(\phi_n)$, where

$$t_n(\tau) := \frac{\gamma_n+\lambda\tau}{1+\gamma_n^*\lambda\tau},$$

and in general, $\phi_0 = T_n(\phi_n)$ where $T_n = t_1 \circ t_2 \circ \cdots \circ t_n$. This composition of linear fractional transformations is n steps in the *Schur continued fraction* representation of ϕ_0. We refer to $T_n(0)$ and ϕ_n, respectively, as the *nth approximant* and *nth tail of* ϕ_0. The numbers γ_n are the *Schur parameters* associated with ϕ_0.

We can therefore view Schur's algorithm as producing a sequence of Schur functions ϕ_n, or equivalently, as producing a sequence of LFT's T_n. In Section 3 we will explicitly formulate Schur's algorithm in terms of the T_n.

In order to implement Schur's algorithm, we write the functions as quotients of power series,

$$\phi_n =: \frac{\alpha_n(\lambda)}{\beta_n(\lambda)} = \frac{\sum_k \alpha_{n,k}\lambda^k}{\sum_k \beta_{n,k}\lambda^k}.$$

Schur's algorithm as formulated above involves an infinite sequence of elementary operations on infinite power series α_n and β_n. However, it is not difficult to arrange the computations so that the coefficient pairs $(\alpha_{0,n},\beta_{0,n})$ are entered and processed in a *sequential* manner.

Algorithm 2.3 (progressive Schur algorithm).

input: the coefficients $\{\alpha_n\}_{n=0}^{\infty}$, $\{\beta_n\}_{n=0}^{\infty}$ of formal power series α and β such that $\phi_0 = \alpha/\beta$ is a Schur function,

for $n = 1, 2, 3, \ldots,$ **while** $|\gamma_n| < 1$

$\alpha_{0,n-1} = \alpha_{n-1}$, $\beta_{0,n-1} = \beta_{n-1}$,

for $k = 1, 2, 3, \cdots, n-1$

$$\begin{bmatrix} \alpha_{k,n-k-1} \\ \beta_{k,n-k} \end{bmatrix} = \begin{bmatrix} 1 & -\gamma_k \\ -\gamma_k^* & 1 \end{bmatrix} \begin{bmatrix} \alpha_{k-1,n-k} \\ \beta_{k-1,n-k} \end{bmatrix}$$

$\gamma_n = \alpha_{n-1,0}/\beta_{n-1,0}$,

$\sigma_n^2 = (1 - |\gamma_n|^2)$,

$\beta_{n,0} = \beta_{n-1,0}\sigma_n^2$.

Of course, in practice the first n coefficients of α_0 and β_0 will be input , and Schur's algorithm will be performed to obtain the first $n-k$ coefficients of the power series α_k and β_k (and the Schur parameter γ_k) for $k = 1, \cdots, n$.

The following proposition provides the connection between Schur's algorithm and the fast Choleski algorithm.

Proposition 2.1. Under the identifications $\alpha_{j,k} := -\tau_{-k-1,j}$, and $\beta_{j,k} := \tau_{j+k,j}^*$, the fast Choleski algorithm and the progressive Schur algorithm are equivalent. In other words, Algorithm 2.2 applied to the Hermitian matrix $M_n = [\mu_{j-k}]_{j,k=0}^{n}$ is equivalent with n steps of Algorithm 2.3 applied to a function $\phi_0 = \alpha_0/\beta_0$ with the first n coefficients of α_0 and β_0 given by $[-\tau_{-k-1}]_0^{n-1}$ and $[\tau_k^*]_0^{n-1}$, respectively.

The proof of this result follows immediately from the recursions. In particular,

$$\begin{bmatrix} -\tau_{k-j,k} \\ \tau_{j,k}^* \end{bmatrix} = \begin{bmatrix} 1 & -\gamma_k \\ -\gamma_k^* & 1 \end{bmatrix} \begin{bmatrix} -\tau_{k-j-1,k-1} \\ \tau_{j-1,k-1}^* \end{bmatrix}$$

for $j \geq k$. The equivalence of the two recursions implies the numerator and denominator of ϕ_0 are the first n terms in formal power series α_0 and β_0 such that α_0/β_0 is a Schur function. Thus, the computational procedure for finding the Choleski factorization of a positive definite Toeplitz matrix is equivalent with Schur's classical algorithm.

3. The Generalized Schur Algorithm.

3.1. Schur Polynomials. Schur's algorithm was described in the previous section as an iteration that generates the numerators and denominators of the Schur functions ϕ_n. We now reformulate Schur's algorithm in terms of the LFT's T_n defined by $\phi_0 = T_n(\phi_n)$.

Let ξ_n, η_n, $\tilde{\xi}_n$, $\tilde{\eta}_n$ be the polynomials such that

$$T_n(\tau) = \frac{\xi_n(\lambda) + \tilde{\eta}_n(\lambda)\,\tau}{\eta_n(\lambda) + \tilde{\xi}_n(\lambda)\,\tau} .$$

Since $T_0(\tau) = \tau$ and $T_n(\tau) = T_{n-1}(t_n(\tau))$ (where $t_n(\tau) = \dfrac{\gamma_n + \lambda\tau}{1 + \gamma_n^* \lambda\tau}$), we have the recurrence relations

$$\begin{bmatrix} \tilde{\eta}_n & \xi_n \\ \tilde{\xi}_n & \eta_n \end{bmatrix} = \begin{bmatrix} \tilde{\eta}_{n-1} & \xi_{n-1} \\ \tilde{\xi}_{n-1} & \eta_{n-1} \end{bmatrix} \begin{bmatrix} \lambda & \gamma_n \\ \gamma_n^*\lambda & 1 \end{bmatrix}, \quad \begin{bmatrix} \tilde{\eta}_0 & \xi_0 \\ \tilde{\xi}_0 & \eta_0 \end{bmatrix} = \begin{bmatrix} 1 & 0 \\ 0 & 1 \end{bmatrix}.$$

The results of the following proposition are easily obtained by induction.

Proposition 3.1. The polynomials $\tilde{\eta}_n$ and $\tilde{\xi}_n$ satisfy

$$\tilde{\xi}_n(\lambda) \equiv \lambda^n \overline{\xi}_n(1/\lambda), \quad \tilde{\eta}_n(\lambda) \equiv \lambda^n \overline{\eta}_n(1/\lambda).$$

Furthermore, for all $n \geq 1$,

$$deg(\xi_n) < n, \quad deg(\eta_n) < n ,$$
$$\xi_n(0) = \gamma_1 , \quad \eta_n(0) \equiv 1 .$$

We also have the *determinant formula*

$$\eta_n \tilde{\eta}_n - \xi_n \tilde{\xi}_n = \delta_n \lambda^n,$$

where $\delta_n = \sigma_1^2 \sigma_2^2 \cdots \sigma_n^2$.

We refer to ξ_n and η_n as the nth *Schur polynomials* associated with the Schur function ϕ_0.

Thus, we may view Schur's algorithm as generating a sequence of Schur functions, or equivalently, as generating a sequence of Schur polynomials which determine the LFT T_n. We now show that if the Schur function ϕ_0 is given as in Proposition 2.1 (where M is a positive definite Toeplitz matrix), then the Szegö polynomials are determined by the Schur polynomials. This will allow us to use Schur's algorithm to obtain the Gohberg-Semencul factorization of M^{-1}.

By transposing the recursions for the Schur polynomials and applying a diagonal scaling, we obtain

$$\begin{bmatrix} \tilde{\eta}_n & \tilde{\xi}_n/\lambda \\ \lambda\,\xi_n & \eta_n \end{bmatrix} = \begin{bmatrix} \lambda & \gamma_n^* \\ \lambda\,\gamma_n & 1 \end{bmatrix} \begin{bmatrix} \tilde{\eta}_{n-1} & \tilde{\xi}_{n-1}/\lambda \\ \lambda\,\xi_{n-1} & \eta_{n-1} \end{bmatrix}.$$

By the Szegö recurrence formula, we obtain

$$\begin{bmatrix} \psi_n \\ \tilde{\psi}_n \end{bmatrix} = \begin{bmatrix} \lambda & \gamma_n \\ \lambda\,\gamma_n^* & 1 \end{bmatrix} \begin{bmatrix} \psi_{n-1} \\ \tilde{\psi}_{n-1} \end{bmatrix}, \text{ or } \overline{\begin{bmatrix} \psi_n \\ \tilde{\psi}_n \end{bmatrix}} = \begin{bmatrix} \lambda & \gamma_n^* \\ \lambda\,\gamma_n & 1 \end{bmatrix} \overline{\begin{bmatrix} \psi_{n-1} \\ \tilde{\psi}_{n-1} \end{bmatrix}}.$$

From these recursions, together with the initial conditions

$$\begin{bmatrix} \tilde{\eta}_1 & \tilde{\xi}_1/\lambda \\ \lambda\xi_n & \eta_1 \end{bmatrix} = \begin{bmatrix} 1 & \gamma_1^* \\ \lambda\gamma_1 & 1 \end{bmatrix}, \quad \overline{\begin{bmatrix} \psi_1 \\ \tilde{\psi}_1 \end{bmatrix}} = \begin{bmatrix} 1 & \gamma_1^* \\ \lambda\gamma_1 & 1 \end{bmatrix} \begin{bmatrix} 1 \\ 1 \end{bmatrix},$$

we see that

$$\overline{\psi}_n = \tilde{\eta}_n + \tilde{\xi}_n/\lambda. \tag{3.1}$$

Thus, Schur's algorithm can be used to construct the Szegö polynomials.

3.2 The Generalized Schur Algorithm. As we have seen, Schur's classical algorithm generates a sequence of linear fractional transformations T_n. We now present a generalization of Schur's algorithm that allows us to generate T_{n+k} from T_n, where $k > 1$. The following simple lemma is needed in our derivation.

Lemma 3.1. Let $\phi_0 = \alpha_0/\beta_0$ be a Schur function, and let ξ_n and η_n be the nth Schur polynomials for ϕ_0. Then

$$\alpha_0\eta_n - \beta_0\xi_n = \delta_n\,\gamma_n\lambda^n + O(\lambda^{n+1})$$
$$\beta_0\tilde{\eta}_n - \alpha_0\tilde{\xi}_n = \delta_n\lambda^n + O(\lambda^{n+1}).$$

Proof: Since $T_n(\tau) = \dfrac{\xi_n + \tilde{\eta}_n\tau}{\eta_n + \tilde{\xi}_n\tau}$, we have by the determinant formula

$$\frac{T_n(\tau_0) - T_n(\tau)}{\tau_0 - \tau} = \frac{\delta_n\lambda^n}{(\eta_n + \tilde{\xi}_n\tau_0)(\eta_n + \tilde{\xi}_n\tau)}.$$

Setting $\tau_0 = \phi_n$ we get

$$\phi - T_n(\tau) = \frac{\delta_n(\phi_n - \tau)\lambda^n}{(\eta_n + \phi_n\tilde{\xi}_n)(\eta_n + \tilde{\xi}_n\tau)}.$$

Now let $\tau = 0$, and note that $\xi_n/\eta_n = T_n(0)$, $\eta_n(0) = 1$ and $\tilde{\xi}_n(0) = 0$ to obtain

$$\phi\eta_n - \xi_n = \frac{\delta_n \phi_n \lambda^n}{\eta_n + \phi_n \tilde{\xi}_n} = \gamma_n \delta_n \lambda^n + O(\lambda^{n+1})$$

as $\lambda \to 0$. Similarly, setting $\tau = \infty$ we obtain

$$\tilde{\eta}_n - \phi \tilde{\xi}_n = \frac{\delta_n \lambda^n}{\eta_n + \phi_n \tilde{\xi}_n} = \delta_n \lambda^n + O(\lambda^{n+1})$$

as $\lambda \to 0$. This completes the proof.

Let $\phi_0 = \phi$ be a Schur function, and let ϕ_n be the nth tail of ϕ (i.e., ϕ_n is the result of n steps of Schur's algorithm starting at ϕ_0). Also let $\xi_{0,n} = \xi_n$ and $\eta_{0,n} = \eta_n$ be the nth Schur polynomials of ϕ_0, so that $\phi_0 = \mathrm{T}_{0,n}(\phi_n)$, where

$$\mathrm{T}_{0,n}(\tau) = \frac{\xi_{0,n} + \tilde{\eta}_{0,n}\tau}{\eta_{0,n} + \tilde{\xi}_{0,n}\tau}$$

In order to construct $\mathrm{T}_{0,n+k} = \mathrm{T}_{n+k}$ we must first obtain ϕ_n from $\mathrm{T}_{0,n}$ and ϕ_0. We have

$$\phi_n = \frac{\alpha_n}{\beta_n} = \mathrm{T}_{0,n}^{-1}\left(\frac{\alpha_0}{\beta_0}\right) = \frac{\alpha_0 \eta_{0,n} - \beta_0 \xi_{0,n}}{\beta_0 \tilde{\eta}_{0,n} - \alpha_0 \tilde{\xi}_{0,n}} \quad . \tag{3.2}$$

By Lemma 3.1 both the numerator and denominator in (3.2) are divisible by λ^n. It is therefore natural to take

$$\begin{aligned} \alpha_n &= (\alpha_0 \eta_{0,n} - \beta_0 \xi_{0,n}) / \lambda^n \ , \\ \beta_n &= (\beta_0 \tilde{\eta}_{0,n} - \alpha_0 \tilde{\xi}_{0,n}) / \lambda^n \ . \end{aligned} \tag{3.3}$$

Thus, formula (3.3) enables us to obtain the n-th tail ϕ_n of ϕ_0 from ϕ_0 and $\mathrm{T}_{0,n}$.

Since ϕ_n is a Schur function, we can obtain $\mathrm{T}_{n,k}$, the LFT that results from k steps of Schur's algorithm applied to ϕ_n. We then have $\phi_n = \mathrm{T}_{n,k}(\phi_{n+k})$, (i.e., the kth tail of ϕ_n is equal to the $(n+k)$th tail of ϕ_0). Once we have $\mathrm{T}_{n,k}$, we can construct $\mathrm{T}_{0,n+k}$ by simply composing thc LFT's. In particular,

$$\begin{aligned} \xi_{0,n+k} &= \tilde{\eta}_{0,n}\xi_{n,k} + \xi_{0,n}\eta_{n,k} \\ \eta_{0,n+k} &= \tilde{\xi}_{0,n}\xi_{n,k} + \eta_{0,n}\eta_{n,k} \end{aligned} \tag{3.4}$$

The generalized Schur algorithm is a doubling procedure based on the recursions (3.3) and (3.4) that generates T_n for $n = 1, 2, 4, \cdots, 2^p, \cdots$. As in the case of the classical Schur algorithm, the computations are to be organized so that the coefficients of the formal power series α_0 and β_0 enter in a sequential fashion. However, instead of entering one at a time, the coefficients enter in groups, each group being twice as large as the previous one.

For the formal power series $\alpha_n = \sum_{j=0}^{\infty} \alpha_{n,j} \lambda^j$, let $\alpha_n^{(k)}$ denote the polynomial $\sum_{j=0}^{k-1} \alpha_{n,j} \lambda^j$ of degree less than k, and define $\beta_n^{(k)}$ similarly. We can describe 'the algorithm as follows.

Generalized Schur Algorithm.

input: $\alpha_0^{(2^p)}$ and $\beta_0^{(2^p)}$, where $\phi_0 = \alpha_0 / \beta_0$ is a Schur function,

$\xi_{0,1} = \gamma_1 = \alpha_0^{(1)} / \beta_0^{(1)}$, $\eta_{0,1} = 1$,

for $n = 1, 2, 4, \cdots, 2^{p-1}$

1: compute $\alpha_n^{(n)}$, $\beta_n^{(n)}$, which are respectively given by the first n coefficients of the polynomials

$$(\alpha_0^{(2n)} \eta_{0,n} - \beta_0^{(2n)} \xi_{0,n}) / \lambda^n ,$$
$$(\beta_0^{(2n)} \tilde{\eta}_{0,n} - \alpha_0^{(2n)} \tilde{\xi}_{0,n}) / \lambda^n .$$

2: compute $\xi_{n,n}$ and $\eta_{n,n}$ from $\alpha_n^{(n)}$ and $\beta_n^{(n)}$ as $\xi_{0,n}$ and $\eta_{0,n}$ were obtained from $\alpha_0^{(n)}$ and $\beta_0^{(n)}$ (this is the doubling step).

3: compute

$$\xi_{0,2n} = \tilde{\eta}_{0,n} \xi_{n,n} + \xi_{0,n} \eta_{n,n} ,$$
$$\eta_{0,2n} = \tilde{\xi}_{0,n} \xi_{n,n} + \eta_{0,n} \eta_{n,n} .$$

Recall that in the progressive Schur algorithm, the input polynomials $\alpha_0^{(n)}$ and $\beta_0^{(n)}$ determine $\alpha_k^{(n-k)}$ and $\beta_k^{(n-k)}$ (as well as $\gamma_k = \alpha_{k,0}/\beta_{k,0}$) for $k = 1, \cdots, n$. By considering the doubling process of the generalized Schur algorithm, we see that, given $\alpha_0^{(n)}$ and $\beta_0^{(n)}$ where n is a power of two, the number of coefficients of α_k and β_k that are computed depends on the bianary representation of the integer k. For example, the first $n/2$ coefficients of $\alpha_{n/2}$ and $\beta_{n/2}$ are calculated, while only the constant terms of α_k and β_k are calculated if k is odd. Nevertheless, *all* n *Schur parameters* $\gamma_k = \xi_{k,1} = \alpha_k^{(1)}/\beta_k^{(1)}$ *are computed in the generalized Schur algorithm.* This is important because the Schur parameters are often of significance for physical and mathematical reasons.

4. The Superfast Solution of a Positive Definite Toeplitz System

The efficient implementation of the generalized Schur algorithm is achieved by using fast Fourier transform (FFT) techniques to perform the polynomial recursions (3.3) and (3.4). A detailed description of this procedure is given in [2], where it is shown that the Schur polynomials ξ_n and η_n can be calculated using $2n \lg^2 n + O(n \lg n)$ complex multplications and $4n \lg^2 n + O(n \lg n)$ complex additions (where $\lg n \equiv \log_2 n$).

The following algorithm describes the use of the generalized Schur algorithm for the superfast solution of a positive definite Toeplitz system of equations. This algorithm is equivalent with the superfast Toeplitz solver that is presented by deHoog [10] and by Musicus [22] when applied to a positive definite matrix.

Algorithm 4.1. Let $M = [\mu_{j-k}]_{j,k=0}^{n} = M^* > 0$ where $n = 2^\nu$. The following procedure will calculate the solution of the system of equations $Mx = b$.

Set $\alpha_{0,j} := -\mu_{j+1}^*$; $\beta_{0,j} := \mu_j^*$, $(j = 0, 1, \cdots, n-1)$.

Phase 1: Use the generalized Schur algorithm to calculate ξ_n and η_n. Then obtain ψ_n from equation (3.1).

Phase 2: Solve $Mx = b$ using the Gohberg-Semencul decomposition of M^{-1} and fast Fourier transform techniques.

Phase 2 can be performed using $O(n \lg n)$ operations as described in, for example, Jain [19]. Moreover, Phase 2 can be repeated to solve $Mx = b$ for another right-hand side b. The technique of iterative improvement can therefore be efficiently implemented in this algorithm. (Of course, this is true of any algorithm that uses the Gohberg-Semencul formula in its solution phase, as in [19] and [5].)

Thus, the algorithm of deHoog and Musicus applied to a positive definite Toeplitz matrix is naturally described in terms of the generalized Schur algorithm. This algorithm therefore shares the classical roots of many of the fast and superfast algorithms.

References

[1] N. I. Akhiezer, The Classical Moment Problem, Oliver and Boyd, Edinburgh, 1965.

[2] G. S. Ammar and W. B. Gragg, *Implementation and Use of the Generalized Schur Algorithm,* in Computational and Combinatorial Methods in Systems Theory, C. I. Byrnes and A. Lindquist, eds., North-Holland, Amsterdam, 1986, pp. 265-280.

[3] B. D. O. Anderson and J. B. Moore, Optimal Filtering, Prentice-Hall, Englewood Cliffs, NJ, 1979.

[4] R. R. Bitmead and B. D. O. Anderson, *Asymptotically Fast Solution of Toeplitz and Related Systems of Linear Equations,* Linear Algebra Appl., 34 (1980), 103-116.

[5] R. P. Brent, F. G. Gustavson and D. Y. Y. Yun, *Fast Solution of Toeplitz Systems of Equations and Computation of Pade'Approximants,* J. Algorithms, 1 (1980) 259-295.

[6] J. R. Bunch, *Stability of Methods for Solving Toeplitz Systems of Equations,* SIAM J. Sci. Statist. Comput., 6 (1985), 349-364.

[7] P. L. Chebyshev, *Sur l'Interpolation par la Methode des Moindres Carrés,* Mém. Acad. Impér. Sci. St. Pétersbourg, 1 (1859), 1-24.

[8] G. Cybenko, Error Analysis of some Signal Processing Algorithms, Ph. D. Thesis, Princeton Univ., Princeton, NJ, 1978.

[9] G. Cybenko, *The Numerical Stability of the Levinson-Durbin Algoritm for Toeplitz Systems of Equations,* SIAM J. Sci. Statist. Comput., 1 (1980), 303-319.

[10] F. de Hoog, *On the Solution of Toeplitz Systems of Equations,* Lin. Algebra Appl., to appear.

[11] J. Durbin, *The Fitting of Time-Series Models,* Rev. Inst. Internat. Statist., 28 (1959), 229-249.

[12] W. Gautschi, *On Generating Orthogonal Polynomials,* SIAM J. Sci. Statist. Comput., 3 (1982), 289-317.

[13] I. C. Gohberg and I. A. Fel'dman, Convolution Equations and Projection Methods for their Solution, American Mathematical Society, Providence, RI, 1974.

[14] G. H. Golub and C. F. Van Loan, Matrix Computations, John Hopkins University Press, Baltimore, MD, 1984.

[15] W. B. Gragg, *The Pade' Table and its Relation to Certain Algorithms of Numerical Analysis,* SIAM Rev., 14 (1972), 1-62.

[16] W. B. Gragg, *Positive definite Toeplitz Matrices, the Arnoldi Process for Isometric Operators, and Gaussian Quadrature on the Unit Circle (in Russian),* in Numerical Methods in Linear Algebra (E.S. Nikolaev editor), Moscow University Press, 1982, 16-32.

[17] W. B. Gragg, *The QR Algorithm for Unitary Hessenberg Matrices,* J. Comput. Appl. Math., to appear.

[18] W. B. Gragg, F. G. Gustavson, D. D. Warner and D. Y. Y. Yun, *On Fast Computation of Superdiagonal Pade' Fractions,* Math. Programming Stud., 18 (1982), 39-42.

[19] J. R. Jain, *An Efficient Algorithm for a Large Toeplitz Set of Linear Equations,* IEEE Trans. Acoust. Speech Signal Process., 27 (1979), 612-615.

[20] T. Kailath, *A View of Three Decades of Linear Filtering Theory,* IEEE Trans. Inform. Theory, 20 (1974), 146-181.

[21] N. Levinson, *The Wiener RMS (Root-Mean-Square) Error Criterion in Filter Design and Prediction,* J. Math. Phys., 25 (1947), 261-278.

[22] B. R. Musicus, *Levinson and Fast Choleski Algorithms for Toeplitz and Almost Toeplitz Matrices,* Report, Res. Lab. of Electronics, M.I.T., 1984.

[23] A. V. Oppenheim, Applications of Digital Signal Processing, Prentice-Hall, Englewood Cliffs, NJ, 1978.

[24] E. Parzen, *Autoregressive Spectral Estimation,* in Time Series in the Frequency Domain, D.R. Brillinger and P.R. Krishnaiah, eds., North-Holland, Amsterdam, 1983.

[25] I. Schur, *Uber Potenzreihen, die in Innern des Einheitskrises Beschränkt Sind,* J. Reine Angew. Math., 147 (1917), 205-232.

[26] G. Szegö, Orthogonal Polynomials, American Mathematical Society, Providence, RI, 1939.

STRONG UNICITY IN NONLINEAR APPROXIMATION

Ryszard Smarzewski
Department of Mathematics
M.Curie-Sklodowska University
20-031 Lublin, Poland

1. INTRODUCTION

In 1963 Newman and Shapiro [14] introduced the concept of strong unicity in approximation theory by demonstrating that a best polynomial approximation to a function in C[a,b] is a strongly unique best approximation. Since then, this subject has been studied extensively and various generalizations and applications of the Newman and Shapiro's result are known (see, e.g., [1,3,6,12-13,15-17,24]). In particular, in the last years there has been a great deal of interest in strong unicity constants for the spaces of polynomials and splines (see [16] and references therein). The main reason for this is the fact that strong unicity constants play an important role in estimating numerically the accuracy of a given approximation with respect to the best approximation [6,17]. It should be noticed that almost all papers on the subject are concerned with strongly unique best approximations in the space of continuous functions on a compact Hausdorff space equipped with the uniform norm and the Lebesgue space L_1, because of Wulbert's observation [24] that a best approximation in a proper linear subspace of dimension greater than 1 of a smooth space cannot be strongly unique.

This paper is a continuation of [19] - [23], where the systematic study of strong unicity in arbitrary Banach spaces was initiated. We recall that the study was based on a new definition of strongly unique best approximations. This definition extends the classical definition of Newman and Shapiro [14] and is especially convenient to investigate strong unicity in the classical smooth function spaces such as the Lebesgue, Hardy and Sobolev spaces. The main purpose of this paper is to give strong unicity inequalities for nonlinear best approximations in the Lebesgue, Hardy and Sobolev spaces with possibly the best strong unicity constants.

2. NOTATION AND PRELIMINARIES

Let M be a nonempty proper subset of a linear normed space X. An element $m \in M$ is said to be a best approximation in M to an element $x \in X$ if

(2.1) $$\|x - m\| \leqslant \|x - y\|$$

for all y in M. If the set $P_M(x)$ of all such elements m is nonempty, then one can define the mapping $P_M : x \longrightarrow P_M(x)$ of X into 2^M which is called a metric projection. Denote the domain of P_M by D_M. Clearly, we have $D_M \supset M$. Throughout this paper we shall assume that g is an increasing convex function defined on the interval $[0,\infty)$ and such that $g(0)=0$. Following [19] , an element $m \in M$ is said to be a strongly unique best approximation in M to an element $x \in X$ if there exists a constant $c=c(x)>0$ such that the inequality

(2.2) $$g(\|x - m\|) \leqslant g(\|x - y\|) - cg(\|m - y\|)$$

holds for all y in M. Denote the set of all elements x in X having the strongly unique best approximation in M by D_M^o. Clearly, we have $D_M \supset D_M^o \supset M$. A positive constant c_g, such that $c(x) \geqslant c_g$ for all x in D_M^o, is called a strong unicity constant. It is clear that the strongly unique best approximation m to x is the unique best approximation to x, i.e., $P_M(x)=\{m\}$. However, the converse statement may be false (cf., e.g., [1,6,12,24]). In order to study the converse statement we have to restrict our attention to subsets M of X which are suns [8]. We recall that M is called a sun if $m \in P_M(x)$ implies $m \in P_M(m+s(x-m))$ for every $s>0$. Clearly, M is a sun if and only if the inequalities

(2.3) $$\|x - m\| \leqslant \|x - [(1-t)m + ty]\| , \quad y \in M,$$

hold for all $t=1/s>0$, $x \in D_M$ and $m \in P_M(x)$.

Now, let us denote

(2.4) $$\tau_g(x,y) = \tau_{g,X}(x,y) := \lim_{t \longrightarrow 0+} [g(\|x + ty\|) - g(\|x\|)]/t .$$

In the particular case when $g(t)=t^p$, we shall write τ_p instead of τ_g.

LEMMA 2.1. The right derivative $\tau_g(x,y)$ exists and

$$(2.5) \qquad \tau_g(x,y) \leqslant \left[g(\|x + ty\|) - g(\|x\|)\right] / t$$

for all $t > 0$ and $x, y \in X$.

Proof. By convexity of the norm we have

$$\|x + sy\| \leqslant (1 - \tfrac{s}{t})\|x\| + \tfrac{s}{t}\|x + ty\|$$

for any $0 < s \leqslant t$. Since g is increasing and convex, it directly follows that

$$(2.6) \qquad \left[g(\|x + sy\|) - g(\|x\|)\right]/s \leqslant \left[g(\|x + ty\|) - g(\|x\|)\right]/t$$

for any $0 < s \leqslant t$. Hence it remains to show that the left-hand side, say w_s, of inequality (2.6) is bounded from below by a constant independent of s. Without loss of generality, we may suppose that $\|x+sy\| < \|x\|$. Then we get

$$w_s \geqslant -\|y\| \frac{g(\|x + sy\|)-g(\|x\|)}{\|x + sy\| - \|x\|} \geqslant -\|y\| \left[g(2\|x\|)-g(\|x\|)\right]/\|x\|,$$

where we used the well known inequality for convex functions [10, Exercise 3, p.125] in the last inequality.

The lemma can be readily applied to prove the following characterization of best approximations which will be used in the next sections.

THEOREM 2.1. Let M be a sun in X. Then an element $m \in M$ is a best approximation in M to an element $x \in X$ if and only if

$$(2.7) \qquad \tau_g(x - m, m - y) \geqslant 0$$

for all y in M.

Proof. If $m \in P_M(x)$, then it follows from (2.3) that

$$(2.8) \qquad \left[g(\|x - m + t(m - y)\|) - g(\|x - m\|)\right]/t \geqslant 0$$

for all $y \in M$ and $t > 0$. Hence by (2.4) we get (2.7). Conversely, if (2.7) holds then we can use (2.5) to get inequality (2.8), which is equivalent to (2.1) in the case when $t=1$.

The hypothesis of Theorem 2.1 can not be weakened in general. Indeed, if M

is not a sun then, in view of (2.3), there exist a real $s > 0$, $x \in D_M$ and $m \in P_M(x)$ such that

$$[g(\|x - m + t(m - y)\|) - g(\|x - m\|)]/t < 0 \tag{2.9}$$

for $y \in M \setminus \{m\}$ and $t=s$. Hence by (2.6) we conclude that (2.9) is valid also for all $t \in (0,s)$. Therefore, letting $t \longrightarrow 0+$ in (2.9) we get $\tau_g(x-m, m-y) < 0$. This in conjunction with Theorem 2.1 gives the following characterization of suns.

THEOREM 2.2. A subset M of X is a sun if and only if

$$\inf_{y \in M} \tau_g(x - m, m - y) \geqslant 0$$

for all $x \in D_M$ and $m \in P_M(x)$.

It should be noticed that Theorems 2.1 and 2.2 are well-known (cf., e.g., [4, 5]) in the particular case when $g(t)=t$.

3. STRONGLY UNIQUE BEST APPROXIMATION IN L_p SPACES

Let $L_p=L_p(S,\Sigma,\mu)$, $1 < p < \infty$, be the space of all μ- measurable extended scalar valued functions (equivalence classes) x on S such that

$$\|x\| = \|x\|_p := \left(\int_S |x(s)|^p \mu(ds)\right)^{1/p} ,$$

where (S,Σ,μ) is a positive measure space. It is well known [24] that the classical definition of Newman and Shapiro [14] of strongly unique best approximations, obtained by setting $g(t)=t$ into (2.2), is useless when $X=L_p$, $1 < p < \infty$. On the other hand, we have recently proved in [20] that

$$\tau_p(y,x-y) \leqslant \|x\|^p - \|y\|^p - c_p \|x-y\|^p \; ; \; x,y \in L_p \; , \; p \geqslant 2,$$

where L_p is the real Lebesgue space. The best positive constant $c_p \leqslant 1$ in this inequality is equal to

$$c_p = (p-1)(1+s)^{2-p} , \tag{3.1}$$

where $s=s(p)$ is the unique solution of the equation

$$- t^{p-1} + (p-1)t + p - 2 = 0$$

in the interval $(1, \infty)$ for $p > 2$, and $s(2)=1$. The inequality is also true in the complex L_p spaces, which can be proved as follows.

LEMMA 3.1. If $p > 2$, then

$$(3.2)\qquad p|v|^{p-2}\,[v_1(u_1-v_1) + v_2(u_2-v_2)] \leqslant |u|^p - |v|^p - c_p|u-v|^p$$

for every complex numbers $u=u_1+iu_2$ and $v=v_1+iv_2$, where c_p is as in (3.1). The constant c_p is the best possible.

Proof. If $v=0$, then the best constant c_p in (3.2) is equal to 1. Therefore, we may suppose that $v \neq 0$. Denote

$$s = \begin{cases} 0\ , \text{ if } u = 0, \\ (u_1v_1 + u_2v_2)/(|u|\,|v|), \text{ otherwise.} \end{cases}$$

Then, dividing both sides of inequality (3.2) by $|v|^p$, we can rewrite the inequality in the equivalent form

$$(3.3)\qquad h(t,s) := t^p - c_p(t^2-2ts+1)^{p/2} - pts + p - 1 \geqslant 0\ ,$$

where $t = |u|/|v| \geqslant 0$ and $-1 \leqslant s \leqslant 1$. Since

$$(3.4)\qquad \frac{\partial^2 h}{\partial s^2} = -\,p(p-2)c_p t^2(t^2-2ts+1)^{(p-4)/2} \leqslant 0\ ,$$

it follows that inequality (3.3) should be verified only for $s = 1$ and $s = -1$. Thus inequality (3.3) is equivalent to the inequality

$$f(t) := |t|^p - c_p|t-1|^p - pt + p - 1 \geqslant 0,\ t \in \mathbb{R},$$

because $f(t)=h(t,1)$ and $f(-t)=h(t,-1)$ for $t \geqslant 0$. But this inequality coincides with the inequality (3.5) from [20]. Hence we can apply Lemma 3.1 from [20] to finish the proof.

Since $\tau_p(x,y) = p\,\|x\|^{p-1}\,\tau_1(x,y)$, it follows from [11, Formula 3, p.351] that the right derivative τ_p can be written in the form

$$(3.5)\qquad \tau_p(x,y) = p \int_S |x(s)|^{p-2}[x_1(s)y_1(s) + x_2(s)y_2(s)\,]\,\mu(ds)$$

for all $x = x_1 + ix_2$ and $y = y_1 + iy_2$ in L_p, $p > 1$.

THEOREM 3.1. For every $x,y \in L_p$, $p \geqslant 2$, we have

(3.6) $$\tau_p(y,x-y) \leqslant \|x\|^p - \|y\|^p - c_p\|x-y\|^p ,$$

where c_p is as in (3.1).

Proof. If p=2, then L_p is a Hilbert space, and so one can readily verify [20] that we have equality in (3.6). Otherwise, we apply Lemma 3.1 replacing u by x(s) and v by y(s). Then we obtain

$$p|y(s)|^{p-2}\{y_1(s)[x_1(s)-y_1(s)] + y_2(s)[x_2(s)-y_2(s)]\} \leqslant |x(s)|^p - |y(s)|^p - c_p|x(s)-y(s)|^p$$

for every $s \in S \setminus B$, where B is the set of measure zero of all points s in S such that values $x(s)=x_1(s)+ix_2(s)$ or $y(s)=y_1(s)+iy_2(s)$ are not finite. Finally, we integrate both sides of this inequality, and then apply (3.5) to the left-hand side in order to get (3.6).

Now, we can put x := x-y and y := x-m into (3.6) and apply Theorem 2.1 to get the theorem

THEOREM 3.2. Let M be a sun in L_p, $p \geqslant 2$. If m is a best approximation in M to an element $x \in L_p$, then the inequality

$$\|x - m\|^p \leqslant \|x - y\|^p - c_p\|m - y\|^p$$

holds for every y in M.

The theorem was first proved in [20] for a closed convex subset M of the real space L_p, $p \geqslant 2$. Note that Theorem 3.2 shows that if M is a sun in L_p, $p \geqslant 2$, then the best approximation $m=P_M(x)$ is the strongly unique best approximation (with respect to $g(t)=t^p$) in M to each $x \in D_M$, and the strong unicity functional $c = c(x)$, $x \in D_M$, occurring in (2.2) is bounded from below by the constant c_p, i.e., c_p is a strong unicity constant and $D_M^o = D_M$.

A counterpart of Theorem 3.2 for L_p spaces, $1 < p < 2$, has been recently presented in [22], and then improved in [23]. The constants $c_p = (p-1)/64$ and $c_p = p(p-1)/4$, $1 < p < 2$, given in these papers are not optimal. Moreover, it follows from the inequality

$$2^{2-p} \leqslant c_p \leqslant (p-1)2^{2-p}, \ p \geqslant 2,$$

given in [20] that these constants c_p do not converge to the proper value $c_2=1$ as $p \longrightarrow 2-$. Now we are going to establish the counterpart of Theorem 3.2 for a sun in L_p, $1<p<2$, which has not this disadventage. For this purpose, we need two following auxiliary lemmas.

LEMMA 3.2. If $1<p<2$ and $t \geqslant s \geqslant 0$, then

$$(t^p - s^p)(\tfrac{t+s}{2})^{2-p} \leqslant t^2 - s^2 .$$

Proof. If $s=0$, then the inequality is trivial. Otherwise, we can rewrite it in the equivalent form

$$f(p) := z^2 - 1 - (z^p - 1)(\tfrac{z+1}{2})^{2-p} \geqslant 0, \quad 1<p<2 ,$$

where $z = t/s \geqslant 1$. Since $f(2)=0$, the proof will be completed if we show that $f(p)$ does not increase as p increases. For this purpose, let us denote

$$h(p)=(z^p - 1)\ln \frac{z+1}{2} - z^p \ln z \quad \text{and} \quad r(z)=h(1).$$

Then we have

$$h'(p)= z^p(\ln \frac{z+1}{2} - \ln z)\ln z \leqslant 0$$

and

$$r'(z)= \ln \frac{z+1}{2} - \ln z - \frac{2}{z+1} < 0$$

for every $z \geqslant 1$. Hence we obtain

$$h(p) \leqslant h(1) = r(z) \leqslant r(1)=0$$

and

$$f'(p) =(\tfrac{z+1}{2})^{2-p} h(p) \leqslant 0 \tag{3.7}$$

for every $p>1$, which completes the proof.

LEMMA 3.3. If $1<p<2$, then the best positive constant c_p, such that the inequality

$$[\tfrac{1}{2}(|u|+|v|)]^{2-p}\{|u|^p - |v|^p - p|v|^{p-2}[v_1(u_1-v_1)+v_2(u_2-v_2)]\} \geqslant c_p|u-v|^2 \tag{3.8}$$

holds for all complex numbers $u=u_1+iu_2$ and $v=v_1+iv_2$, satisfies the estimates

$$2^{p-3}p(p-1) < c_p < p(p-1)/2 . \tag{3.9}$$

Proof. If $v=0$, then inequality (3.8) holds with the best constant $c_p=2^{p-2}$

which satisfies estimates (3.9). Otherwise, we can divide (3.8) by $|v|^2$ to obtain the equivalent inequality

$$h(t,s) := t^p - d(t+1)^{p-2}(t^2-2ts+1) - pts + p - 1 \geqslant 0,$$

where t,s ($t \geqslant 0$ and $-1 \leqslant s \leqslant 1$) are as in (3.3) and

$$d = 2^{2-p} c_p .$$

Since h is a linear function of variable s, it follows that the last inequality ought to be verified only for s=1 and s= -1. Hence inequality (3.8) is equivalent to the inequality

(3.10) $$f(t) := |t|^p - d(|t|+1)^{p-2}(t-1)^2 - p(t-1) - 1 \geqslant 0, \quad t \in \mathbb{R},$$

because $f(t)=h(t,1)$ and $f(-t)=h(t,-1)$ for $t \geqslant 0$. Since $f(1)=0$, it follows that inequality (3.10) is true in a neigborhood of the point t=1 if and only if the function f has a minimum at t=1. Note that

$$f'(1)=0 , \quad f''(1)=p(p-1) - 2^{p-1}d \quad \text{and} \quad f'''(1)=(p-2)(f''(1)-2^{p-2}d) .$$

Hence f has a minimum at t=1 if and only if

$$d < 2^{1-p}p(p-1),$$

which gives the upper estimate for c_p in (3.9). Therefore, by (3.10), we conclude that the proof will be completed if we show that the inequality

(3.11) $$d(t) := \left[|t|^p - p(t-1) - 1\right] (|t|+1)^{2-p}/(t-1)^2 > p(p-1)/2$$

holds for every real $t \neq 1$. If $t \geqslant 0$, then according to Taylor's theorem we have

$$t^p = 1 + p(t-1) + \frac{p(p-1)}{2} z^{p-2}(t-1)^2,$$

where z lies strictly between t and 1. Hence

$$t^p - p(t-1) - 1 > \frac{p(p-1)}{2} t^{p-2}(t-1)^2$$

for every $t > 1$, and

(3.12) $$t^p - p(t-1) - 1 > \frac{p(p-1)}{2} (t-1)^2$$

for every $t \in [0,1)$. By applying these inequalities to d(t), we obtain

$$d(t) > \frac{p(p-1)}{2}\left(\frac{t+1}{t}\right)^{2-p} > \frac{p(p-1)}{2}$$

and

$$d(t) > \frac{p(p-1)}{2}(t+1)^{2-p} \geqslant \frac{p(p-1)}{2}$$

for every $t > 1$ and $t \in [0,1)$, respectively. Finally, if $t < 0$ then

$$r(t) := (1-t)^{p+1}d'(t)/p = -(-t)^{p-1} - (p-1)(t-1) - 1.$$

Since the point $t = -1$ is the unique minimum of r and $r(-1)=2(p-2) < 0$, it follows that $d'(t) < 0$ for all $t < 0$. Hence we have $d(t) \geqslant d(0)$ for all $t < 0$, which in view of (3.12) completes the proof of (3.11).

The lemma can be applied to obtain the following counterpart of inequality (3.6) for L_p spaces, $1 < p < 2$.

<u>THEOREM 3.3.</u> Let $1 < p < 2$ and let the positive constant c_p be as in Lemma 3.3. Then for every $x, y \in L_p$, we have

$$\|\tfrac{1}{2}(|x|+|y|)\|^{2-p}\,[\,\|x\|^p - \|y\|^p - \tau_p(y,x-y)] \geqslant c_p\|x-y\|^2. \tag{3.13}$$

<u>Proof.</u> If f, h, r are nonnegative functions in L_p such that

$$f^{1-p/2}h^{p/2} \geqslant r$$

almost everywhere on S, then it follows from Hölder's inequality applied to $f^{p(1-p/2)}h^{p\cdot p/2}$ that

$$\|f\|^{1-p/2}\|h\|^{p/2} \geqslant \|r\|. \tag{3.14}$$

Now apply Lemma 3.3 replacing u by $x(s)=x_1(s)+ix_2(s)$ and v by $y(s)=y_1(s)+iy_2(s)$. Then we obtain

$$[\tfrac{1}{2}(|x(s)|+|y(s)|)]^{2-p}\,\{|x(s)|^p - |y(s)|^p - p|y(s)|^{p-2}\,[y_1(s)(x_1(s)-y_1(s)) + y_2(s)(x_2(s)-y_2(s))]\} \geqslant c_p|x(s)-y(s)|^2$$

for every $s \in S \setminus B$, where B is the set of measure zero consisting of all points s in S such that values $x(s)$ or $y(s)$ are not finite. Finally, taking square root on both sides of the last inequality, applying (3.14) to the left-hand product, using (3.5), and then squaring both sides, we obtain (3.13).

Note that the constant c_p in Theorem 3.3 converges to 1 as $p \longrightarrow 2-$. Thus inequality (3.6) coincides in the case when $p=2$ with the inequality which can be obtained from (3.13) when $p \longrightarrow 2-$.

THEOREM 3.4. Let M be a sun in L_p, $1 < p < 2$, and let the positive constant c_p be as in Lemma 3.3. Then a best approximation m in M to any element $x \in L_p$ satisfies the inequality

$$\|x - m\|^2 \leq \|x - y\|^2 - c_p \|m - y\|^2 \tag{3.15}$$

for all y in M.

Proof. Replacing x by x-y and y by x-m in (3.13), and then using Theorem 2.1 and the triangle inequality for the norm, we obtain

$$\left[\tfrac{1}{2}(\|x-y\| + \|x-m\|)\right]^{2-p}(\|x-y\|^p - \|x-m\|^p) \geq c_p \|x-y\|^2$$

for all y in M. Hence by Lemma 3.2 we derive (3.15).

Let us note that Theorem 3.4 shows that if M is a sun in L_p, $1 < p < 2$, then the best approximation $m = P_M(x)$ is the strongly unique best approximation (with respect to $g(t)=t^2$) in M to any $x \in D_M$, and c_p is a strong unicity constant. Moreover, by Lemma 3.3 it is clear that the constant c_p occurring in Theorems 3.3 and 3.4 can be replaced by $2^{p-3}p(p-1)$.

4. STRONGLY UNIQUE BEST APPROXIMATION IN SOBOLEV AND HARDY SPACES

Let T be an open subset of R^n and let $H^{k,p} = H^{k,p}(T)$, $k \geq 0$ and $1 < p < \infty$, denotes the Sobolev space [2, p.149] of distributions x such that $D^j x \in L_p(T)$ for all multiindexes $j = (j_1, \dots, j_n)$ ($|j| \leq k$), equipped with the norm

$$\|x\| = \Big(\sum_{|j| \leq k} \|D^j x\|_p^p\Big)^{1/p}, \tag{4.1}$$

where $\|\cdot\|_p$ denotes the norm in the Lebesgue space $L_p(T)$. If $g(t)=t^p$, then the right derivative τ_g in the space $H^{k,p}$ will be denoted by $\tau_{p,k}$. By (2.4) and (4.1), we directly obtain

$$\tau_{p,k}(x,y) = \sum_{|j| \leq k} \tau_p(D^j x, D^j y) \;;\; x,y \in H^{k,p}, \tag{4.2}$$

where τ_p is the right derivative of p-th power of the norm in $L_p(T)$. We shall assume in this section that the positive constant c_p is as in (3.1) and Lemma 3.3 when $p \geq 2$ and $1 < p < 2$, respectively. When $p \geq 2$, then by applying inequality

(3.6) to $D^j x$ and $D^j y$, we obtain

$$\tau_p(D^j y, D^j(x-y)) \leqslant \|D^j x\|_p^p - \|D^j y\|_p^p - c_p \|D^j(x-y)\|_p^p$$

for all x,y in $H^{k,p}$. If we sum up these inequalities over $|j| \leqslant k$ and use (4.1)-(4.2), then we obtain

$$\tau_{p,k}(y,x-y) \leqslant \|x\|^p - \|y\|^p - c_p \|x-y\|^p \tag{4.3}$$

for all x,y in $H^{k,p}$, $p \geqslant 2$.

THEOREM 4.1. For every $x,y \in H^{k,p}$, $1 < p < 2$, we have

$$\left[\tfrac{1}{2}(\|x\| + \|y\|)\right]^{2-p} \left[\|x\|^p - \|y\|^p - \tau_{p,k}(y,x-y)\right] \geqslant c_p \|x-y\|^2 . \tag{4.4}$$

Proof. If $\|x\| + \|y\| = 0$, then $\|x-y\| = 0$. Hence inequality (4.4) is trivial in this case. Otherwise, by Theorem 3.3 we have

$$\|D^j x\|_p^p - \|D^j y\|_p^p - \tau_p(D^j y, D^j(x-y)) \tag{4.5}$$
$$\geqslant c_p \|D^j(x-y)\|_p^2 \left[\tfrac{1}{2}(\|D^j x\|_p + \|D^j y\|_p)\right]^{p-2}$$

for all $|j| \leqslant k$ and x,y in $H^{k,p}$ such that $\|D^j x\|_p + \|D^j y\|_p \neq 0$. Now summing these inequalities over j's, applying the Radon inquality [9, Theorem 51, p.61],

$$\sum t_j^{2/p} s_j^{1-2/p} \geqslant \left(\sum t_j\right)^{2/p}\left(\sum s_j\right)^{1-2/p}; \; t_j \geqslant 0, \; s_j > 0,$$

to the right-hand side, and then using (4.1)-(4.2), we obtain

$$\|x\|^p - \|y\|^p - \tau_{p,k}(y,x-y) \geqslant c_p \|x-y\|^2 \left\{\sum \left[\tfrac{1}{2}(\|D^j x\|_p + \|D^j y\|_p)\right]^p\right\}^{1-2/p} \geqslant c_p \|x-y\|^2 \left(\frac{\|x\| + \|y\|}{2}\right)^{p-2},$$

where we used Minkowski's inequality to derive the last inequality. This completes the proof.

The inequalities (4.3)-(4.4) can be used to show strong unicity of nonlinear best approximations in the Sobolev spaces. More precisely, we have

THEOREM 4.2. Let M be a sun in $H^{k,p}$ ($k \geqslant 0$ and $1 < p < \infty$), and let m be a best approximation in M to an element $x \in H^{k,p}$. Then the inequality

$$\|x - m\|^q \leqslant \|x - y\|^q - c_p \|m - y\|^q \tag{4.6}$$

holds for all y in M, where $q = \max(2,p)$.

Proof. Substitution of x-y for x and x-m for y into inequality (4.3) and application of Theorem 2.1 yields (4.6) in the case when $p \geq 2$. The same substitution into inequality (4.4) and application of Theorem 2.1 and Lemma 3.2 implies (4.6) for $p \in (1,2)$.

Another proof of Theorem 4.2 has been presented in [23] . But the constant

$$c_p = \begin{cases} p(p-1)/4 \ , & \text{if } 1 < p \leq 2, \\ 2^{1-p} \ , & \text{if } p \geq 2, \end{cases}$$

given in [23] , is smaller than the constant c_p obtained in this paper.

REMARK 4.1 The techniques used in the proofs of inequalities (4.3)-(4.4) and Theorem 4.2 can be also applied to study strong unicity in the space X constructed as follows. Let $X_1, X_2, \dots, X_n$ $(n \geq 2)$ be a given sequence of linear normed spaces such that either inequality (4.3) or inequality (4.4) is satisfied for some p and c_p independent of the index k in X_k, $k=1,2,\dots,n$. Then we define

$$X = X_p := (X_1 \oplus X_2 \oplus \dots \oplus X_n)_{l_p} .$$

Hence we have

$$\|x\| = \left(\sum_{k=1}^{n} \|x_k\|_{X_k}^p \right)^{1/p}$$

and

$$\tau_{p,X}(x,y) = \sum_{k=1}^{n} \tau_{p,X_k}(x_k,y_k)$$

for any $x=(x_1,\dots,x_n)$ and $y=(y_1,\dots,y_n)$ in X. Now, using these identities instead of (4.1)-(4.2), we can repeat mutatis mutandis the proofs on inequalities (4.3)-(4.4) and (4.6) in order to show that these inequalities hold also in the space X.

Now, let H^p, $1 < p < \infty$, be the Hardy space of all functions x analytic in the unit disc $|z| < 1$ of the complex plane and such that

$$\|x\| := \lim_{r \to 1-} \left(\frac{1}{2\pi} \int_0^{2\pi} |x(re^{it})|^p \, dt \right)^{1/p} < \infty .$$

The space H^p is isometrically isomorphic with a subspace of the normalized Lebesgue space $L_p(0,2\pi)$ (see [7]). Hence by Theorems 3.2 and 3.4 we have

<u>THEOREM 4.3.</u> Let M be a sun in the Hardy space H^p, $1<p<\infty$, and let m be a best approximation in M to an element $x \in H^p$. Then the inequality

$$\|x - m\|^q \leqslant \|x - y\|^q - c_p \|m - y\|^q$$

holds for all y in M, where $q = \max(2,p)$.

Finally, let M be a sun in the space X_p, where $X_p = L_p$, $H^{k,p}$ or H^p ($k \geqslant 0$ and $1<p<\infty$). Then Theorems 3.2, 3.4, 4.2 and 4.3 in conjunction with Theorem 5.1 from [23] yield

<u>THEOREM 4.4.</u> Let M be a sun in the space X_p, $1<p<\infty$, such that $0 \in M$. Then the metric projection P_M is locally Hölder continuous, i.e.,

$$\|P_M(x_1)-P_M(x_2)\| \leqslant d\, r^{1-1/q} \|x_1 - x_2\|^{1/q}$$

for all x_1, x_2 in the ball $B(r)=\{x \in D_M : \|x\| \leqslant r\}$, where $q = \max(2,p)$ and

$$d = (q/c_p)^{1/q}(1 + c_p^{-q})^{1-1/q} \leqslant 2 + (c_p)^{-q}.$$

It should be noticed that strong unicity can be also applied to establish the rate of convergence of numerical algorithms for computing best approximations in the space X_p, $1<p<\infty$. Indeed, let m be a best approximation in a sun $M \subset X_p$ to an element $x \in D_M$. Moreover, let $\{m_k\}$ be a minimizing sequence for the functional

$$f(y) = \|x - y\|^q, \quad y \in M,$$

produced by a numerical algorithm, i.e., let

$$e_k := f(m_k) \longrightarrow e := \inf\{f(y) : y \in M\}.$$

Then, replacing y by m_k in the strong unicity inequalities for the space X_p given in Theorems 3.2, 3.4, 4.2 or 4.3, we obtain

$$\|m - m_k\|^q \leqslant (e_k - e)/c_p,$$

where $q = \max(2,p)$.

5. REVERSED INEQUALITIES TO STRONG UNICITY INEQUALITIES

We observed in [20] that an inequality of type (3.6) is false when $1 < p < 2$. Next, we proved in [21] that if $1 < p \leqslant 2$ then the reversed inequality

$$\tau_p(y,x-y) \geqslant \|x\|^p - \|y\|^p - c_p\|x-y\|^p$$

holds for all x,y in the real L_p space, where the positive constant c_p is defined by (3.1). We recall that the estimates for c_p presented in [21] imply that

$$1 < c_p < 2 \text{ for every } 1 < p < 2,$$

$$\lim_{p \to 1+} c_p = 2 \quad \text{and} \quad \lim_{p \to 2-} c_p = 1.$$

In order to show that the inequality remains valid for the complex L_p space, we need the lemma.

LEMMA 5.1. If $1 < p < 2$, then

$$(5.1) \qquad p|v|^{p-2}\left[v_1(u_1-v_1) + v_2(u_2-v_2)\right] \geqslant |u|^p - |v|^p - c_p|u-v|^p$$

for every complex numbers $u = u_1+iu_2$ and $v = v_1+iv_2$, where c_p is as in (3.1).

Proof. Without loss of generality we may suppose that $v \neq 0$. Define the functions $h(t,s)$ and $f(t)$ as in the proof of Lemma 3.1, and note that $\partial^2 h/\partial s^2 \geqslant 0$ by (3.4). Therefore, we can repeat mutatis mutandis the proof of Lemma 3.1 in order to show that inequality (5.1) is valid if and only if $f(t) \leqslant 0$ for all real t. Hence we can apply Lemma 2.1 from [21] to complete the proof of (5.1).

By using Lemma 5.1 instead of Lemma 3.1 in the proof of Theorem 3.1, we get

THEOREM 5.1. For every $x,y \in L_p$, $1 < p \leqslant 2$, we have

$$(5.2) \qquad \tau_p(y,x-y) \geqslant \|x\|^p - \|y\|^p - c_p\|x-y\|^p$$

where c_p is as in (3.1).

From (5.2) we readily obtain

$$(5.3) \qquad \tau_{p,k}(y,x-y) \geqslant \|x\|^p - \|y\|^p - c_p\|x-y\|^p$$

for all x,y in $H^{k,p}$, $1 < p \leqslant 2$. Since this inequality can be established in the same way as (4.3), we omit the proof. Now we consider the case $p > 2$. We begin with the following two lemmas.

LEMMA 5.2. If $p > 2$ and $t \geq s \geq 0$, then

$$(t^p - s^p)\left(\frac{t+s}{2}\right)^{2-p} \geq t^2 - s^2 .$$

Proof. If $s=0$, then the proof is trivial. Otherwise, define $f(p)$, $p \geq 2$, as in the proof of Lemma 3.2. Then we have $f'(p) \leq 0$ by (3.7). Since $f(2)=0$, it follows that $f(p) \leq 0$ for $p > 2$, which completes the proof.

LEMMA 5.3. If $p > 2$, then the best positive constant c_p, such that the inequality

$$|u|^p - |v|^p - p|v|^{p-2}\left[v_1(u_1-v_1)+v_2(u_2-v_2)\right] \leq c_p|u-v|^2\left[\tfrac{1}{2}(|u|+|v|)\right]^{p-2} \tag{5.4}$$

is valid for all complex numbers $u = u_1+iu_2$ and $v = v_1+iv_2$, satisfies

$$p(p-1)/2 < c_p < 2^{p-3}p(p-1). \tag{5.5}$$

Proof. If $v=0$, then the best constant c_p in (5.4) is equal to 2^{p-2}. Clearly, this constant satisfies estimates (5.5). Now, suppose that $v \neq 0$ and observe that the similar reasoning as in the proof of Lemma 3.3 shows that inequality (5.4) is equivalent to the inequality $f(t) \leq 0$; $t \in \mathbb{R}$, where the function f is as in (3.10). Since $f(1)=f'(1)=0$, it follows that inequality (5.4) can be true only if the function f has a maximum at the point $t=1$. As in the proof of Lemma 3.3, we conclude that this is equivalent to $f''(1) < 0$, which gives the lower estimate for c_p. In order to prove the upper estimate, one can modify the part of the proof of Lemma 3.3 below (3.11). Since this requires only to replace all established inequalities in the text by the reversed inequalities, we omit the details.

THEOREM 5.2. Let $p > 2$, and let the positive constant c_p be as in Lemma 5.3. Then for every $x, y \in L_p$, we have

$$\tau_p(y,x-y) \geq \|x\|^p - \|y\|^p - c_p\|x-y\|^2\left\|\tfrac{1}{2}(|x|+|y|)\right\|^{p-2}. \tag{5.6}$$

Proof. Let us replace u and v in inequality (5.4) with $x(s)$ and $y(s)$, respectively. Then, integrating both sides of the inequality and applying (3.5), we obtain

$$\|x\|^p - \|y\|^p - \tau_p(y,x-y) \leqslant c_p \int_S |x(s) - y(s)|^{p\cdot 2/p} \left[\tfrac{1}{2}(|x(s)| + |y(s)|)\right]^{p(1-2/p)} \mu(ds) \leqslant c_p \|x-y\|^2 \|\tfrac{1}{2}(|x| + |y|)\|^{p-2},$$

where we used Hölder's inequality to derive the last inequality.

If $x,y \in H^{k,p}$ $(p > 2)$, then inserting $D^j x$ and $D^j y$ $(|j| \leqslant k)$ into inequality (5.6), we get the inequalities reversed to inequalities (4.5). Next, we sum up these inequalities, apply Minkowski's inequality on the right-hand side, and then use (4.1)-(4.2). As a result, we obtain

THEOREM 5.3. Let $p > 2$ and $k \geqslant 0$, and let the positive constant c_p be as in Lemma 5.3. Then for every $x,y \in H^{k,p}$, we have

$$\tau_{p,k}(y,x-y) \geqslant \|x\|^p - \|y\|^p - c_p \|x-y\|^2 \left[\tfrac{1}{2}(\|x\| + \|y\|)\right]^{p-2}.$$

We assume in the next theorem that the positive constant c_p is as in (3.1) and Lemma 5.3 if $1 < p \leqslant 2$ and $p > 2$, respectively.

THEOREM 5.4. Let m be a best approximation in a linear subspace M of X to an element $x \in X$, where $X = L_p$ or $X = H^{k,p}$ $(k \geqslant 0$ and $1 < p < \infty)$. Then we have

(5.7) $$\|x - m\|^q \geqslant \|x - y\|^q - c_p \|m - y\|^q$$

for all y in M, where $q = \min(2,p)$.

Proof. Let m be a best approximation in a linear subspace M of L_p to an element $x \in L_p$, $1 < p < \infty$. Since the norm in the L_p spaces is Gateaux differentiable [11, p.351], it follows from Kolmogorov's criterion [18, p.90] that $\tau_1(x-m,m-y) = 0$ for all y in M. Hence

(5.8) $$\tau_p(x-m,m-y) = p\|x-m\|_p^{p-1} \tau_1(x-m,m-y) = 0, \quad y \in M.$$

Consequently, substitution of $x-y$ for x and $x-m$ for y into (5.2) yields (5.7) in the case when $1 < p \leqslant 2$. The same substitution into inequality (5.6) and application of (5.8) and Lemma 5.2 implies (5.7) for $p > 2$. Similarly, if M is a linear subspace of $H^{k,p}$ and $x \in H^{k,p}$, then we deduce (5.7) from inequality (5.3),

Theorem 5.3, Lemma 5.2 and the fact that

$$\tau_{p,k}(x-m,m-y) = 0 \tag{5.9}$$

for all $y \in M$ and $m \in P_M(x)$. Thus it remains to prove identity (5.9). From (4.2), (5.8) and Theorem 1.1 we have

$$0 \leqslant \tau_{p,k}(x-m,m-y) = p \sum_{|j| \leqslant k} \|D^j(x-m)\|_p^{p-1} \tau_1(D^j(x-m),D^j(m-y)) \tag{5.10}$$

for all y in M. On the other hand, the Gateaux derivative τ_1 in L_p, $1 < p < \infty$, is a homogeneous functional of the second variable. Therefore, replacing y by 2m-y in (5.10), we obtain

$$0 \leqslant \tau_{p,k}(x-m,y-m) = -p \sum_{|j| \leqslant k} \|D^j(x-m)\|_p^{p-1} \tau_1(D^j(x-m),D^j(m-y)) = -\tau_{p,k}(x-m,m-y)$$

for all y in M. This in conjunction with (5.10) completes the proof of (5.9).

The theorem has been recently proved in [21] for real L_p spaces, $1 < p < 2$. Finally, we note that Theorem 5.4 is valid also in Hardy spaces H^p, $1 < p < \infty$.

6. STRONGLY UNIQUE BEST APPROXIMATION IN LINEAR NORMED SPACES

In order to study strongly unique best approximations in an arbitrary linear normed space X, it is sometimes convenient to establish first a characterization theorem for them (see, e.g., [3]). In this section we restrict our attention to characterization of <u>strongly unique best approximations of order p</u> $\geqslant 1$, i.e., such that $g(t)=t^p$ in the definition (2.2). This seems to be justified, because of results presented in [23] and Sections 3 and 4 of this paper. Now, we introduce the definition of p-stars, which parallels the definition of suns and extends the definition of strong suns given by Mah in [13].

Namely, a subset M of X is said to be a <u>p-star</u> if, whenever $m \in M$ is a strongly unique best approximation of order p in M to some $x \in X$, then m is also a best strongly unique best approximation of order p in M to any element x_s of the ray $\{m + s(x-m) : s > 0\}$, with a strong uniqueness constant such that

$$c_s := c(x_s) = s^{p-1}(c + o(1)), \quad s \longrightarrow +\infty, \tag{6.1}$$

where c is an absolute positive constant.

By (2.2) it can be readily seen that M is a p-star if and only if, for every $x \in X$ having a strongly unique best approximation of order p in M, say m, there exist positive constants c_s satisfying (6.1) and such that the inequalities

$$(6.2) \qquad \|x-m\|^p \leq \|x-[(1-t)m+ty]\|^p - (c_s/s^p)\|m-y\|^p, \; y \in M,$$

hold for all $t = 1/s > 0$.

THEOREM 6.1. Let $p \geq 1$ and let M be a p-star in a linear normed space X. Then m is a strongly unique best approximation of order p in M to an element $x \in X$ if and only if there exists a positive constant $c > 0$ such that

$$(6.3) \qquad \tau_p(x-m, m-y) \geq c\|m-y\|^p$$

for all y in M.

Proof. If (6.3) holds, then (2.5) implies

$$\|x-m+t(m-y)\|^p - \|x-m\|^p \geq tc\|m-y\|^p$$

for any $t > 0$ and $y \in M$. Hence we can put $t=1$ into this inequality to get (2.2). Conversely, let M be a p-star, and let m be a strongly unique best approximation of order p in M to an element $x \in X$. Then, dividing (6.2) by $t > 0$ and passing with t to zero, we obtain (6.3) by (2.4) and (6.1).

The theorem has been proved recently by Mah [13] in the particular case when $p=1$. If M is not a p-star, then in view of (6.2) and (2.6) there exists an element $x \in X$ having a strongly unique best approximation m of order p in M such that the expression

$$(\|x-m+t(m-y)\|^p - \|x-m\|^p)/t$$

has an arbitrary small positive upper bound if $t \longrightarrow 0+$. Hence by (2.5) and Theorem 6.1 one can obtain

THEOREM 6.2. A subset M of X is a p-star, $p \geq 1$, if and only if

$$\inf_{y \in M} \tau_p(x-m, m-y)/\|m-y\|^p > 0$$

for all $x \in X \setminus M$ and $m \in M$ such that m is a strongly unique best approximation of order p in M to x .

REFERENCES

1. J.R. Angelos and D. Schmidt, The prevalence of strong uniqueness in L^1, preprint.
2. J. Barros-Neto, An Introduction to the Theory of Distributions, Marcel Dekker, Inc., New York 1973.
3. M.C. Bartelt and H.W. Mclaughlin, Characterization of strong unicity in approximation theory, J. Approx. Theory 9 (1973), 255-266.
4. B. Brosowski and R. Wegmann, Charakterisierung bester Approximationen in normierten Räumen, J. Approx. Theory 3 (1970), 369-397.
5. B. Brosowski and F. Deutsch, On some geometrical properties of suns, J. Approx. Theory 10 (1974), 245-267.
6. E.W. Cheney, Introduction to Approximation Theory, McGraw-Hill, New York 1966.
7. W.L. Duren, Theory of H^p Spaces, Academic Press, New York 1970.
8. N. Efimov and S. Steckin, Some properties of Chebyshev sets, Dokl. Akad. Nauk SSSR 118 (1958), 17-19.
9. G.H. Hardy, J.E. Littlewood and G. Pólya, Inequalities, Cambridge 1934.
10. G. Klambauer, Real Analysis, Elsievier, New York 1973.
11. G. Köthe, Topological Vector Spaces I, Springer-Verlag, Berlin 1969.
12. P.F. Mah, Strong uniqueness in nonlinear approximation, J. Approx. Theory 41 (1984), 91-99.
13. P.F. Mah, Characterization of the strongly unique best approximations, preprint.
14. D.J. Newman and H.S. Shapiro, Some theorems on Chebyshev approximation, Duke Math. J. 30 (1963), 673-684.
15. G. Nürnberger and I. Singer, Uniqueness and strong uniqueness of best approximations by spline subspaces and other subspaces, J. Math. Anal. Appl. 90 (1982), 171-184.
16. G. Nürnberger, Strong unicity constants for spline functions, Numer. Funct. Anal. Optimiz. 5 (1982-83), 319-347.
17. G. Nürnberger, Strong unicity of best approximations: A numerical aspect, Numer. Funct. Anal. Optimiz. 6 (1983), 399-421.
18. I. Singer, Best Approximation in Normed Linear Spaces by Elements of Linear Subspaces, Springer-Verlag, Berlin 1970 .
19. R. Smarzewski, Strongly unique best approximation in Banach spaces, J. Approx. Theory, in press.
20. R. Smarzewski, Strongly unique minimization of functionals in Banach spaces with applications to theory of approximation and fixed points, J. Math. Anal. Appl., in press.
21. R. Smarzewski, On the best approximation in L_p spaces, J. Approx. Theory, in press.

22. R. Smarzewski and B. Prus, Strongly unique best approximations and centers in uniformly convex spaces, J. Math. Anal. Appl., in press.

23. R. Smarzewski, Strongly unique best approximation in Banach spaces II, to appear.

24. D.E. Wulbert, Uniqueness and differential characterization of approximation from manifolds of functions, Amer. J. Math. 18 (1971), 350-366.

Vol. 1090: Differential Geometry of Submanifolds. Proceedings, 1984. Edited by K. Kenmotsu. VI, 132 pages. 1984.

Vol. 1091: Multifunctions and Integrands. Proceedings, 1983. Edited by G. Salinetti. V, 234 pages. 1984.

Vol. 1092: Complete Intersections. Seminar, 1983. Edited by S. Greco and R. Strano. VII, 299 pages. 1984.

Vol. 1093: A. Prestel, Lectures on Formally Real Fields. XI, 125 pages. 1984.

Vol. 1094: Analyse Complexe. Proceedings, 1983. Edité par E. Amar, R. Gay et Nguyen Thanh Van. IX, 184 pages. 1984.

Vol. 1095: Stochastic Analysis and Applications. Proceedings, 1983. Edited by A. Truman and D. Williams. V, 199 pages. 1984.

Vol. 1096: Théorie du Potentiel. Proceedings, 1983. Edité par G. Mokobodzki et D. Pinchon. IX, 601 pages. 1984.

Vol. 1097: R. M. Dudley, H. Kunita, F. Ledrappier, École d'Éte de Probabilités de Saint-Flour XII – 1982. Edité par P. L. Hennequin. X, 396 pages. 1984.

Vol. 1098: Groups – Korea 1983. Proceedings. Edited by A. C. Kim and B. H. Neumann. VII, 183 pages. 1984.

Vol. 1099: C. M. Ringel, Tame Algebras and Integral Quadratic Forms. XIII, 376 pages. 1984.

Vol. 1100: V. Ivrii, Precise Spectral Asymptotics for Elliptic Operators Acting in Fiberings over Manifolds with Boundary. V, 237 pages. 1984.

Vol. 1101: V. Cossart, J. Giraud, U. Orbanz, Resolution of Surface Singularities. Seminar. VII, 132 pages. 1984.

Vol. 1102: A. Verona, Stratified Mappings – Structure and Triangulability. IX, 160 pages. 1984.

Vol. 1103: Models and Sets. Proceedings, Logic Colloquium, 1983, Part I. Edited by G. H. Müller and M. M. Richter. VIII, 484 pages. 1984.

Vol. 1104: Computation and Proof Theory. Proceedings, Logic Colloquium, 1983, Part II. Edited by M. M. Richter, E. Börger, W. Oberschelp, B. Schinzel and W. Thomas. VIII, 475 pages. 1984.

Vol. 1105: Rational Approximation and Interpolation. Proceedings, 1983. Edited by P. R. Graves-Morris, E. B. Saff and R. S. Varga. XII, 528 pages. 1984.

Vol. 1106: C. T. Chong, Techniques of Admissible Recursion Theory. IX, 214 pages. 1984.

Vol. 1107: Nonlinear Analysis and Optimization. Proceedings, 1982. Edited by C. Vinti. V, 224 pages. 1984.

Vol. 1108: Global Analysis – Studies and Applications I. Edited by Yu. G. Borisovich and Yu. E. Gliklikh. V, 301 pages. 1984.

Vol. 1109: Stochastic Aspects of Classical and Quantum Systems. Proceedings, 1983. Edited by S. Albeverio, P. Combe and M. Sirugue-Collin. IX, 227 pages. 1985.

Vol. 1110: R. Jajte, Strong Limit Theorems in Non-Commutative Probability. VI, 152 pages. 1985.

Vol. 1111: Arbeitstagung Bonn 1984. Proceedings. Edited by F. Hirzebruch, J. Schwermer and S. Suter. V, 481 pages. 1985.

Vol. 1112: Products of Conjugacy Classes in Groups. Edited by Z. Arad and M. Herzog. V, 244 pages. 1985.

Vol. 1113: P. Antosik, C. Swartz, Matrix Methods in Analysis. IV, 114 pages. 1985.

Vol. 1114: Zahlentheoretische Analysis. Seminar. Herausgegeben von E. Hlawka. V, 157 Seiten. 1985.

Vol. 1115: J. Moulin Ollagnier, Ergodic Theory and Statistical Mechanics. VI, 147 pages. 1985.

Vol. 1116: S. Stolz, Hochzusammenhängende Mannigfaltigkeiten und ihre Ränder. XXIII, 134 Seiten. 1985.

Vol. 1117: D. J. Aldous, J. A. Ibragimov, J. Jacod, Ecole d'Été de Probabilités de Saint-Flour XIII – 1983. Édité par P. L. Hennequin. IX, 409 pages. 1985.

Vol. 1118: Grossissements de filtrations: exemples et applications. Seminaire, 1982/83. Edité par Th. Jeulin et M. Yor. V, 315 pages. 1985.

Vol. 1119: Recent Mathematical Methods in Dynamic Programming. Proceedings, 1984. Edited by I. Capuzzo Dolcetta, W. H. Fleming and T. Zolezzi. VI, 202 pages. 1985.

Vol. 1120: K. Jarosz, Perturbations of Banach Algebras. V, 118 pages. 1985.

Vol. 1121: Singularities and Constructive Methods for Their Treatment. Proceedings, 1983. Edited by P. Grisvard, W. Wendland and J. R. Whiteman. IX, 346 pages. 1985.

Vol. 1122: Number Theory. Proceedings, 1984. Edited by K. Alladi. VII, 217 pages. 1985.

Vol. 1123: Séminaire de Probabilités XIX 1983/84. Proceedings. Edité par J. Azéma et M. Yor. IV, 504 pages. 1985.

Vol. 1124: Algebraic Geometry, Sitges (Barcelona) 1983. Proceedings. Edited by E. Casas-Alvero, G. E. Welters and S. Xambó-Descamps. XI, 416 pages. 1985.

Vol. 1125: Dynamical Systems and Bifurcations. Proceedings, 1984. Edited by B. L. J. Braaksma, H. W. Broer and F. Takens. V, 129 pages. 1985.

Vol. 1126: Algebraic and Geometric Topology. Proceedings, 1983. Edited by A. Ranicki, N. Levitt and F. Quinn. V, 423 pages. 1985.

Vol. 1127: Numerical Methods in Fluid Dynamics. Seminar. Edited by F. Brezzi, VII, 333 pages. 1985.

Vol. 1128: J. Elschner, Singular Ordinary Differential Operators and Pseudodifferential Equations. 200 pages. 1985.

Vol. 1129: Numerical Analysis, Lancaster 1984. Proceedings. Edited by P. R. Turner. XIV, 179 pages. 1985.

Vol. 1130: Methods in Mathematical Logic. Proceedings, 1983. Edited by C. A. Di Prisco. VII, 407 pages. 1985.

Vol. 1131: K. Sundaresan, S. Swaminathan, Geometry and Nonlinear Analysis in Banach Spaces. III, 116 pages. 1985.

Vol. 1132: Operator Algebras and their Connections with Topology and Ergodic Theory. Proceedings, 1983. Edited by H. Araki, C. C. Moore, Ş. Strătilă and C. Voiculescu. VI, 594 pages. 1985.

Vol. 1133: K. C. Kiwiel, Methods of Descent for Nondifferentiable Optimization. VI, 362 pages. 1985.

Vol. 1134: G. P. Galdi, S. Rionero, Weighted Energy Methods in Fluid Dynamics and Elasticity. VII, 126 pages. 1985.

Vol. 1135: Number Theory, New York 1983–84. Seminar. Edited by D. V. Chudnovsky, G. V. Chudnovsky, H. Cohn and M. B. Nathanson. V, 283 pages. 1985.

Vol. 1136: Quantum Probability and Applications II. Proceedings, 1984. Edited by L. Accardi and W. von Waldenfels. VI, 534 pages. 1985.

Vol. 1137: Xiao G., Surfaces fibrées en courbes de genre deux. IX, 103 pages. 1985.

Vol. 1138: A. Ocneanu, Actions of Discrete Amenable Groups on von Neumann Algebras. V, 115 pages. 1985.

Vol. 1139: Differential Geometric Methods in Mathematical Physics. Proceedings, 1983. Edited by H. D. Doebner and J. D. Hennig. VI, 337 pages. 1985.

Vol. 1140: S. Donkin, Rational Representations of Algebraic Groups. VII, 254 pages. 1985.

Vol. 1141: Recursion Theory Week. Proceedings, 1984. Edited by H.-D. Ebbinghaus, G. H. Müller and G. E. Sacks. IX, 418 pages. 1985.

Vol. 1142: Orders and their Applications. Proceedings, 1984. Edited by I. Reiner and K. W. Roggenkamp. X, 306 pages. 1985.

Vol. 1143: A. Krieg, Modular Forms on Half-Spaces of Quaternions. XIII, 203 pages. 1985.

Vol. 1144: Knot Theory and Manifolds. Proceedings, 1983. Edited by D. Rolfsen. V, 163 pages. 1985.

Vol. 1145: G. Winkler, Choquet Order and Simplices. VI, 143 pages. 1985.

Vol. 1146: Séminaire d'Algèbre Paul Dubreil et Marie-Paule Malliavin. Proceedings, 1983–1984. Edité par M.-P. Malliavin. IV, 420 pages. 1985.

Vol. 1147: M. Wschebor, Surfaces Aléatoires. VII, 111 pages. 1985.

Vol. 1148: Mark A. Kon, Probability Distributions in Quantum Statistical Mechanics. V, 121 pages. 1985.

Vol. 1149: Universal Algebra and Lattice Theory. Proceedings, 1984. Edited by S. D. Comer. VI, 282 pages. 1985.

Vol. 1150: B. Kawohl, Rearrangements and Convexity of Level Sets in PDE. V, 136 pages. 1985.

Vol 1151: Ordinary and Partial Differential Equations. Proceedings, 1984. Edited by B.D. Sleeman and R.J. Jarvis. XIV, 357 pages. 1985.

Vol. 1152: H. Widom, Asymptotic Expansions for Pseudodifferential Operators on Bounded Domains. V, 150 pages. 1985.

Vol. 1153: Probability in Banach Spaces V. Proceedings, 1984. Edited by A. Beck, R. Dudley, M. Hahn, J. Kuelbs and M. Marcus. VI, 457 pages. 1985.

Vol. 1154: D.S. Naidu, A.K. Rao, Singular Pertubation Analysis of Discrete Control Systems. IX, 195 pages. 1985.

Vol. 1155: Stability Problems for Stochastic Models. Proceedings, 1984. Edited by V.V. Kalashnikov and V.M. Zolotarev. VI, 447 pages. 1985.

Vol. 1156: Global Differential Geometry and Global Analysis 1984. Proceedings, 1984. Edited by D. Ferus, R.B. Gardner, S. Helgason and U. Simon. V, 339 pages. 1985.

Vol. 1157: H. Levine, Classifying Immersions into $\mathbb{R}^4$ over Stable Maps of 3-Manifolds into $\mathbb{R}^2$. V, 163 pages. 1985.

Vol. 1158: Stochastic Processes – Mathematics and Physics. Proceedings, 1984. Edited by S. Albeverio, Ph. Blanchard and L. Streit. VI, 230 pages. 1986.

Vol. 1159: Schrödinger Operators, Como 1984. Seminar. Edited by S. Graffi. VIII, 272 pages. 1986.

Vol. 1160: J.-C. van der Meer, The Hamiltonian Hopf Bifurcation. VI, 115 pages. 1985.

Vol. 1161: Harmonic Mappings and Minimal Immersions, Montecatini 1984. Seminar. Edited by E. Giusti. VII, 285 pages. 1985.

Vol. 1162: S.J.L. van Eijndhoven, J. de Graaf, Trajectory Spaces, Generalized Functions and Unbounded Operators. IV, 272 pages. 1985.

Vol. 1163: Iteration Theory and its Functional Equations. Proceedings, 1984. Edited by R. Liedl, L. Reich and Gy. Targonski. VIII, 231 pages. 1985.

Vol. 1164: M. Meschiari, J.H. Rawnsley, S. Salamon, Geometry Seminar "Luigi Bianchi" II – 1984. Edited by E. Vesentini. VI, 224 pages. 1985.

Vol. 1165: Seminar on Deformations. Proceedings, 1982/84. Edited by J. Ławrynowicz. IX, 331 pages. 1985.

Vol. 1166: Banach Spaces. Proceedings, 1984. Edited by N. Kalton and E. Saab. VI, 199 pages. 1985.

Vol. 1167: Geometry and Topology. Proceedings, 1983–84. Edited by J. Alexander and J. Harer. VI, 292 pages. 1985.

Vol. 1168: S.S. Agaian, Hadamard Matrices and their Applications. III, 227 pages. 1985.

Vol. 1169: W.A. Light, E.W. Cheney, Approximation Theory in Tensor Product Spaces. VII, 157 pages. 1985.

Vol. 1170: B.S. Thomson, Real Functions. VII, 229 pages. 1985.

Vol. 1171: Polynômes Orthogonaux et Applications. Proceedings, 1984. Edité par C. Brezinski, A. Draux, A.P. Magnus, P. Maroni et A. Ronveaux. XXXVII, 584 pages. 1985.

Vol. 1172: Algebraic Topology, Göttingen 1984. Proceedings. Edited by L. Smith. VI, 209 pages. 1985.

Vol. 1173: H. Delfs, M. Knebusch, Locally Semialgebraic Spaces. XVI, 329 pages. 1985.

Vol. 1174: Categories in Continuum Physics, Buffalo 1982. Seminar. Edited by F.W. Lawvere and S.H. Schanuel. V, 126 pages. 1986.

Vol. 1175: K. Mathiak, Valuations of Skew Fields and Projective Hjelmslev Spaces. VII, 116pages. 1986.

Vol. 1176: R.R. Bruner, J.P. May, J.E. McClure, M. Steinberger, H_∞ Ring Spectra and their Applications. VII, 388 pages. 1986.

Vol. 1177: Representation Theory I. Finite Dimensional Algebras. Proceedings, 1984. Edited by V. Dlab, P. Gabriel and G. Michler. XV, 340 pages. 1986.

Vol. 1178: Representation Theory II. Groups and Orders. Proceedings, 1984. Edited by V. Dlab, P. Gabriel and G. Michler. XV, 370 pages. 1986.

Vol. 1179: Shi J.-Y. The Kazhdan-Lusztig Cells in Certain Affine Weyl Groups. X, 307 pages. 1986.

Vol. 1180: R. Carmona, H. Kesten, J.B. Walsh, École d'Été de Probabilités de Saint-Flour XIV – 1984. Édité par P.L. Hennequin. X, 438 pages. 1986.

Vol. 1181: Buildings and the Geometry of Diagrams, Como 1984. Seminar. Edited by L. Rosati. VII, 277 pages. 1986.

Vol. 1182: S. Shelah, Around Classification Theory of Models. VII, 279 pages. 1986.

Vol. 1183: Algebra, Algebraic Topology and their Interactions. Proceedings, 1983. Edited by J.-E. Roos. XI, 396 pages. 1986.

Vol. 1184: W. Arendt, A. Grabosch, G. Greiner, U. Groh, H.P. Lotz, U. Moustakas, R. Nagel, F. Neubrander, U. Schlotterbeck, One-parameter Semigroups of Positive Operators. Edited by R. Nagel. X, 460 pages. 1986.

Vol. 1185: Group Theory, Beijing 1984. Proceedings. Edited by Tuan H.F. V, 403 pages. 1986.

Vol. 1186: Lyapunov Exponents. Proceedings, 1984. Edited by L. Arnold and V. Wihstutz. VI, 374 pages. 1986.

Vol. 1187: Y. Diers, Categories of Boolean Sheaves of Simple Algebras. VI, 168 pages. 1986.

Vol. 1188: Fonctions de Plusieurs Variables Complexes V. Séminaire, 1979–85. Edité par François Norguet. VI, 306 pages. 1986.

Vol. 1189: J. Lukeš, J. Malý, L. Zajíček, Fine Topology Methods in Real Analysis and Potential Theory. X, 472 pages. 1986.

Vol. 1190: Optimization and Related Fields. Proceedings, 1984. Edited by R. Conti, E. De Giorgi and F. Giannessi. VIII, 419 pages. 1986.

Vol. 1191: A.R. Its, V.Yu. Novokshenov, The Isomonodromic Deformation Method in the Theory of Painlevé Equations. IV, 313 pages. 1986.

Vol. 1192: Equadiff 6. Proceedings, 1985. Edited by J. Vosmansky and M. Zlámal. XXIII, 404 pages. 1986.

Vol. 1193: Geometrical and Statistical Aspects of Probability in Banach Spaces. Proceedings, 1985. Edited by X. Femique, B. Heinkel, M.B. Marcus and P.A. Meyer. IV, 128 pages. 1986.

Vol. 1194: Complex Analysis and Algebraic Geometry. Proceedings, 1985. Edited by H. Grauert. VI, 235 pages. 1986.

Vol.1195: J.M. Barbosa, A.G. Colares, Minimal Surfaces in $\mathbb{R}^3$. X, 124 pages. 1986.

Vol. 1196: E. Casas-Alvero, S. Xambó-Descamps, The Enumerative Theory of Conics after Halphen. IX, 130 pages. 1986.

Vol. 1197: Ring Theory. Proceedings, 1985. Edited by F.M.J. van Oystaeyen. V, 231 pages. 1986.

Vol. 1198: Séminaire d'Analyse, P. Lelong – P. Dolbeault – H. Skoda. Seminar 1983/84. X, 260 pages. 1986.

Vol. 1199: Analytic Theory of Continued Fractions II. Proceedings, 1985. Edited by W.J. Thron. VI, 299 pages. 1986.

Vol. 1200: V.D. Milman, G. Schechtman, Asymptotic Theory of Finite Dimensional Normed Spaces. With an Appendix by M. Gromov. VIII, 156 pages. 1986.